ELEMENTARY MECHANICS OF FLUIDS

BY

HUNTER ROUSE

Carver Professor Emeritus
Institute of Hydraulic Research
The University of Iowa
Iowa City

DOVER PUBLICATIONS, INC.
NEW YORK

Published in Canada by General Publishing Company, Ltd., 30 Lesmill Road, Don Mills, Toronto, Ontario.
Published in the United Kingdom by Constable and Company, Ltd., 10 Orange Street, London, WC2H 7EG.

This Dover edition, first published in 1978, is an unabridged and slightly corrected republication of the work originally published by John Wiley & Sons, Inc., in 1946.

International Standard Book Number: 0-486-63699-2
Library of Congress Catalog Card Number: 78-57159

Manufactured in the United States of America
Dover Publications, Inc.
180 Varick Street
New York, N.Y. 10014

PREFACE

Mathematics, physics, and applied mechanics have always been regarded as essential courses in the education of an engineer, but there has been a general tendency in the past to focus primary attention upon various specialized phases of professional endeavor. The engineer of today, however, is confronted with a variety of new and ever-changing problems that are seldom restricted to one particular field. Since the background of fundamentals necessary to the solution of such problems is exceedingly difficult to acquire after graduation, whereas the specialized skills of engineering practice are best obtainable through postgraduate experience, technical education is now evidencing a marked trend toward more thorough grounding in the basic engineering sciences.

Closely associated with this trend has been a growing realization of the importance of flow phenomena in almost every field of application. Once considered a special interest of but a single professional group, the science of fluid motion is gradually attaining recognition as one of the basic subjects necessary to all engineers. Since the principles of fluid motion stem from the same physical laws as the principles of motion of rigid and elastic solids, the study of such motion is properly becoming an essential branch of engineering mechanics.

The mechanics of fluids is evidently not merely traditional hydraulics under another name, or even a combination of hydraulics with certain aspects of aerodynamics, but rather as fundamental a treatment of fluid behavior as the mechanics of solids is of the behavior of rigid and elastic bodies. In the effort to provide a textbook for an elementary mechanics course in fluid motion, the author assembled in 1940 a preliminary series of notes for undergraduate study at the University of Iowa. After two years of use the notes were considerably revised, and during the following years further revisions and additions were made in preparation for publication in book form. The present volume therefore represents the "third edition" of the original notes, the major portion of which reflects experience gained through five years of teaching by various members of the mechanics staff.

The word "elementary" appearing in the title has been used advisedly, since the subject matter presumes no previous training in flow principles—and, in fact, no prior knowledge of mechanics beyond that taught in engineering physics. The book does not, however, represent an over-simplification of what is at best a complex subject. For instance, in the belief that mathematics is taught to engineers for a definite purpose, no effort has been made to avoid use of the calculus when such use is advantageous. Needless to say, however, a mathematical background beyond that normally given to college sophomores is not required, nor is any greater mental capacity necessary than that demanded by other subjects of the fundamental engineering curriculum.

The author trusts that a twofold endeavor will consistently be apparent to the reader: first, the logical and systematic development of a coherent chain of flow principles, from the simplest aspects of velocity and acceleration to the final link with thermodynamics; second, the practical application of every principle to problems encountered in various phases of engineering endeavor. The text contains many illustrative examples of such application. The problems, moreover, are so designed as to further the student's own power of analysis, since in few instances will the routine duplication of type solutions be found possible. In fact, despite the relative simplicity of the exercises, only through a firm grasp of the principles involved can the correct solutions be obtained.

Although the time devoted to the original notes was but two semester hours, from three to five hours will permit more satisfactory coverage of the present material. Moreover, for really effective grounding in the subject, one or two semester hours of laboratory experiments specifically illustrating the principal topics covered by the book will be found desirable. In a limited course, needless to say, a number of the sections (such as 8, 23, 25, 37, 45, 46, and 47), as well as the last two chapters of the book, could be omitted without destroying the continuity of the remainder. As a matter of fact, many of these passages have been included primarily to call the reader's attention to related material which is equally essential but generally beyond the scope of an elementary course.

Were proper credit given to all who have aided in the preparation of this volume, the authorship would indeed be a multiple one. Dean F. M. Dawson, Professor J. W. Howe, Professor A. A. Kalinske, Professor C. J. Posey, and Dr. W. S. Hamilton each taught one or more classes from the earlier editions of the notes and offered much pertinent criticism which greatly affected the form of the final manuscript. Many of the original problems were composed by Professors Howe and Kalinske, and Professor J. S. McNown, as well as Professor Howe, performed individually all problem solutions before the material reached foundry proof; the last two, moreover, assisted greatly in eliminating errors from successive stages of manuscript and proof. By far the majority of the drawings represent the able workmanship of Mr. Jack T. Enburg, formerly on the Iowa staff in engineering drawing. For a vast amount of painstaking effort in preparing the manuscript for publication, the author is indebted to Miss Leona Amelon, secretary of the Iowa Institute of Hydraulic Research.

Except where otherwise noted, the photographs appearing in the book were made by the author in the laboratories of the Iowa Institute, largely through use of equipment devoted to student instruction in fluid mechanics. The task of enlarging and matching prints from the original negatives was competently handled by Mr. F. W. Kent, photographer of the University. The author gratefully acknowledges the kindness of officials of the U. S. Engineer Office, the Aberdeen Proving Ground, the National Advisory Committee for Aeronautics, and the David Taylor Model Basin in permitting official photographs of their research projects to be reproduced; particular gratitude is due the Model Basin for preparation of the frontispiece photograph especially for this volume.

HUNTER ROUSE

IOWA CITY, IOWA
July, 1945

CONTENTS

CHAPTER I

INTRODUCTION TO THE STUDY OF FLUID MOTION

1. MECHANICS OF FLUIDS AS AN ENGINEERING SCIENCE

Historical development. Some two hundred years ago, mankind's centuries of experience with the flow of water began to crystallize in scientific form. Despite this common origin, however, two distinct schools of thought gradually evolved. On the one hand, through the convenient creation of an "ideal fluid," mathematical physicists developed the theoretical science known as *classical hydrodynamics*. On the other hand, claiming that idealized theories were of no practical use without empirical correction factors, engineers developed from experimental findings the applied science known as *hydraulics*, for the specific fields of water supply, irrigation, river control, and water power.

As other engineers became confronted with problems involving the flow of oil and gas in pipes or of air in ventilating systems, hydraulics formulas were gradually extended to fill these needs. The design of modern aircraft, on the contrary, permitted from the outset much smaller "factors of ignorance" than most engineering fields. Not only were existing principles of hydraulics often too inexact for use in aeronautics, but these principles were not sufficiently descriptive of fluid motion in general to be adaptable to the motion of bodies through air. Fortunately, rather than develop purely empirical methods to supply this need, physicists and engineers began instead to expand the basic concepts of fluid motion into a science which now includes as specialized phases both aerodynamics and hydraulics, as well as a host of other applied fields.

This science has come to be known as the *mechanics of fluids*, a subject parallel to the mechanics of solids and engineering materials and built upon the same fundamental laws of motion. Unlike empirical hydraulics, therefore, it stems primarily from basic physical principles; unlike the purely mathematical treatment of classical hydrodynamics, on the other hand, the science is closely correlated with experimental studies which both complement and substantiate the fundamental analysis. In a word, the mechanics of fluids avoids the weaknesses of its forebears, but draws heavily upon the strength of each for much of its material.

Present-day scope. So broad in scope is this relatively new science that it forms a basis for such varied fields as meteorology, oceanography, ballistics, lubrication, marine engineering, and even certain phases of geology—not to mention hydraulic engineering and aeronautics. In fact, a single example, the phenomenon known as turbulence, will suffice to show the almost universal importance of fluid motion to present-day civilization. Turbulence, or the presence of eddies in a moving fluid, gives rise to the twofold effect of a pronounced mixing of the fluid and a subsequent dissipation of energy. Like solid friction, fluid turbulence is sometimes a blessing to mankind, and sometimes a curse. Without the mixing which eddies produce, both the water in boiler tubes and the air surrounding the earth would be very poor distributors of heat; steam engines would prove too costly to run, and the atmosphere would be incapable of supporting life. On the other hand, dust storms would not then occur, nor would rivers transport their tremendous loads of silt from the foothills to the sea. Without means of producing eddies in the process of propulsion, moreover, a swimmer—or an ocean liner—could make little headway, just as a car would remain at rest if friction provided no tractive force. Yet, paradoxically, were turbulence not produced by the motion of a body through a fluid, the process of streamlining would be quite unnecessary. Even the act of breathing depends upon turbulence, for without the violent mixing which accompanies exhalation no fresh air could be inhaled thereafter unless one moved to a new location. If but one aspect of fluid motion can be of such general importance, it should be evident that the principles of fluid motion as a whole must govern a vast realm of human endeavor. A clear understanding of these principles is therefore essential to the modern scientist and engineer.

2. FUNDAMENTAL CHARACTERISTICS OF FLOW

Units of measurement. Fluid motion, like other phases of mechanics, may fully be described in units of *length*, *time*, and *force*. For instance, the form of an observation or barrage balloon and its elevation above the earth can be specified in detail solely in terms of the length dimension; the length and time dimensions together provide a basis for expressing its rate of ascent and the wind velocity encountered at any elevation; and the lift of the balloon and the drag of balloon and cable, in units of force, complete the description of the practical features of the motion. Likewise, the shape of a lubricated shaft and bearing can be stated in terms of geometrical measurements, the rotational speed in terms of time, and the shear and pressure of the fluid lubricant in terms of length and force. Once such flow characteristics are known, any problem of fluid motion is, for all practical purposes, completely solved.

There are fairly simple means of measuring length, time, and force—whether individually or in combination—in laboratory, shop, and field, so that the characteristics of practically any kind of fluid motion may be determined either directly or indirectly. The measurement of boundary form, as the reader is well aware, requires in effect only a protractor and a linear scale, whether one deals with the minute fractions of an inch involved in bearing clearances or with the thousands of feet over which atmospheric disturbances may extend. The basic unit of time is generally considered to be the second, and such short-period phenomena as the induced vibration of a submarine periscope may be of this order of magnitude; on the other hand, flood movements in rivers may be followed more conveniently with a calendar than with a stop watch, while tides have daily, monthly, and even seasonal cycles. Force, normally measured with a calibrated spring, likewise varies greatly in magnitude from one flow phenomenon to the next, the resistance to the settling of dust particles in air being a minute fraction of an ounce as compared to the thrust of many tons exerted by the screws of an ocean liner.

The measurement of fluid *velocity*, a combination of length and time, is somewhat more involved. For example, one may time the movement of a float over a known distance, or count the revolutions per minute of a sensitive propeller which has been calibrated in a flow of known speed. Through simultaneous use of many such floats or current meters, moreover, one can obtain a picture of the rate and direction of movement over a large region, as in maps of prevailing winds or ocean currents. Another combination of length and time units is illustrated by *volume rate of flow:* the number of drops of liquid emerging per second from a pipette, or the number of million gallons of water passing per day through the supply mains of a large city.

Quite as often as one measures the force exerted upon a body by a fluid, one is interested in determining the *intensity* of such force at a given locality. The local intensity of the *pressure* exerted upon a body by the atmosphere, for instance, is usually measured in pounds per square inch, although far below the ocean surface, or in high-pressure equipment, the intensity of pressure is so great that it is often evaluated in "atmospheres," or multiples of the normal atmospheric intensity. Dimensionally similar to pressure intensity is the intensity of *shear*, such as the unit tangential stress exerted by the moving fluid on a lubricated bearing or on the wall of a pipe.

Problems of flow prediction. It should be evident to the reader that the correct measurement of such flow characteristics is very essential in the study of existing states of fluid motion—for instance, in checking the efficiency of a hydraulic turbine or testing the performance of a new method of fluid transmission. Nevertheless, flow measurement is not the ultimate goal of the modern mechanics of fluids, but merely a necessary tool, for science and engineering today require not so much the measurement as the accurate prediction of one or another characteristic of motion from known or assumed conditions. For example, in designing a plumbing system for a large building it is not sufficient to note after installation if, and at what rate, water is available at the top floor; the system must be so designed prior to construction that the required flow can be guaranteed, yet without waste of material or power. Similarly, the drag and lift of a model airplane may easily be measured in a wind tunnel, but only if the results are properly converted to the actual scale of the prototype will the tests have any practical significance. Atmospheric pressures and wind velocities, likewise, can readily be recorded during a storm, but a meteorologist is not worth his salt if he cannot foretell such conditions of motion days in advance. In the large-scale control of rivers, in the streamlining of high-velocity craft, in submarine signaling and detection—in fact, in any problem which involves the motion of a fluid medium—it is obviously necessary to understand the essential principles of such motion in order to predict with fair certainty what the characteristics of flow will be under any given conditions.

3. FLUID PROPERTIES AS A GUIDE TO STUDY

Mechanical properties of fluids. Were the conditions of motion of different fluids completely unrelated, the task of flow analysis would indeed be hopeless. All fluids, however, possess essentially the same mechanical properties, differing from one another only in degree. Water, for example, is well known to be less dense than mercury but considerably more dense than air; it follows that the inertial characteristics of water will be less pronounced than those of mercury and more pronounced than those of air—in other words, in comparison with water it should be relatively simple to set air in motion but relatively difficult to produce the same rate of motion in mercury. The basic role of *mass* in resisting such acceleration is nevertheless identical in all three instances, and is embodied in a single principle of fluid mechanics.

Water, furthermore, is known to flow down a slope as a result of *weight*, or gravitational attraction. Few realize, however, that a relatively slight difference in fluid weight due to local temperature changes in the atmosphere or in the ocean will likewise result in the "downhill" flow of the colder within the warmer medium. The principle of gravitational action in river hydraulics thus has its counterpart in meteorology and oceanography, to mention only two of many fields.

Likewise, although the *viscosity* of lubricating oil is obviously far higher than that of alcohol, tests on the flow of oil through a pipe may, through present-day knowledge, be used to predict with great accuracy the corresponding flow characteristics of alcohol—or of steam, or natural gas—because the principle of viscous resistance is the same for any true fluid. Indeed, the wind resistance of skyscrapers or suspension bridges could just as well be investigated on models towed through water as on models held in a stream of air.

Gases, finally, are well known to be readily compressible, but liquids are normally considered of fairly constant density even under considerable pressure. Yet the fact that the *elastic modulus* of water, for instance, is well below that of steel indicates that it is also compressible, even though to a much lower degree than air. As a result, the propagation of elastic waves in the sea follows quite the same basic laws as the propagation of sound in the atmosphere, however much the actual velocity of propagation may differ. Indeed, the analogy may even be extended to include other fluid properties, for the *capillary* waves in front of a model bridge pier, the *gravity* waves at the bow of a ship, and the *sound* waves at the nose of a projectile may be analyzed through essentially the same principles of fluid mechanics.

Influence of fluid properties upon flow characteristics. Before proceeding to the study of each fluid property, it is necessary for the reader to become familiar with methods of visualizing a *flow pattern*, such as that produced, let us say, by the speeding of an automobile along a highway or by the flow of water over a spillway. As a first approximation, such patterns of fluid motion may be obtained directly from the geometrical form of the flow boundaries, and in the following chapter will be found a detailed treatment of the corresponding velocities and accelerations without consideration of the fluid forces which are involved. Fluid density is introduced in the chapter thereafter, but the sole force discussed is that due to the pressure variation which accompanies the acceleration of a fluid according to the various patterns of motion already studied.

Were fluid properties other than density non-existent, as is conveniently assumed in classical hydrodynamics, further study would be unnecessary. As a matter of fact, the principles developed in these earlier chapters will be found sufficient for the solution of a number of practical flow problems. But in order to recognize these problems, and as well to deal effectively with the many which do not fall in this category, one must study in detail the effect of each additional fluid property upon the basic patterns of motion already discussed.

Therefore, the influence of fluid weight, or gravitational attraction, upon the distribution of pressure and velocity is then described for the same elementary boundary forms, whereafter use is made of a simplified method of analyzing flow in conduits and open channels. Fluid viscosity is next introduced, leading directly to the principles of fluid turbulence, boundary resistance, lift, and propulsion, which play extremely important roles in present-day analysis. The influence of surface tension, perhaps the least important of the fluid properties, is then briefly discussed. Finally, the effect of compressibility upon the flow pattern is described, with particular attention to motion at velocities greater than that of sound; since this phase of fluid mechanics verges upon the realm of *thermodynamics*, only those phenomena are treated which bear a close relationship to the earlier portions of the text.

As is true in all phases of mechanics, the basic principles of fluid motion are subject to mathematical formulation as well as physical description. The principles considered essential in the present book are therefore derived through use of the elementary calculus required of all undergraduate engineers, and every derivation is accompanied by a detailed physical interpretation. Despite the resulting precision of each individual principle, however, it is generally difficult and often impossible to combine these principles in the rigorous solution of flow problems involving the influence of more than one fluid property. As the reader advances, he will accordingly become aware of the extent to which engineers and scientists must still use simplifying approximations in the final analysis of fluid motion. For this reason, if for no other, satisfactory results must invariably depend upon a sound grasp of the fundamental principles herein derived.

QUESTIONS FOR CLASS DISCUSSION

1. Under what circumstances did man probably first realize a need for knowledge of the basic principles of fluid motion?

2. Cite examples of flow phenomena encountered in (*a*) everyday life and (*b*) engineering practice.

3. Suggest how an understanding of fluid behavior might advantageously be applied in the following fields: bridge design; oil-well drilling; sanitary engineering; mining and metallurgy; soil conservation; grading of abrasives; chemical engineering; geology.

4. Is the motion of the atmosphere always turbulent? By what visual means readily at hand may such motion be observed?

5. Why are modern vehicles streamlined? Cite cases in which streamlining is (*a*) scientifically sound and (*b*) merely aesthetic.

6. Is man's blood stream probably turbulent like the flow through a moderately large boiler tube, or is it more reasonable to conclude that it exemplifies another means of transferring heat?

7. High winds tend to lift roofs from buildings rather than to force them inward. Explain this effect in terms of pressure distribution.

8. Suggest a likely cause of the "singing" of telephone wires in the wind.

9. Enumerate the fluid properties which can influence the behavior of a fluid in motion. Suggest flow phenomena illustrating the influence of each property.

10. It is a well-known fact that one can float more easily in salt water than in fresh. Should one also be able to swim faster in salt water?

11. Should mercury or water be expected to flow more rapidly down a channel of given slope?

12. Why is it necessary in winter to use a "lighter" oil for automobiles than in summer? To what property does the term "light" refer?

13. Would you imagine the viscous resistance to the flow of air to be greater or less than the viscous resistance to the flow of water?

14. Is the pressure intensity (*a*) within a bubble of gas, and (*b*) within a drop of liquid, probably greater than, equal to, or less than that of the surrounding medium?

15. Projectiles are known to whistle or scream during flight. Should the sound wave follow a projectile or precede it?

16. *Gulliver's Travels* is sometimes criticized on the basis of conversion principles now used in model experiments. Should gravitational attraction seem relatively greater to a Lilliputian or to a Brobdingnagian? To which would the day seem shorter? To which would water appear more viscous? Which would have less difficulty with surface tension? Assuming a height ratio of 1 : 100, estimate their relative weights, velocities of walking, and frequencies of breathing.

SELECTED REFERENCES

GIACOMELLI, R., and PISTOLESI, E. "Historical Sketch." *Aerodynamic Theory*, Vol. I, Springer, 1934.

BARDSLEY, C. E. "Historical Résumé of the Development of the Science of Hydraulics." *Publication 39*, Engineering Experiment Station, Oklahoma A & M College, 1939.

KÁRMÁN, TH. VON. "The Role of Fluid Mechanics in Modern Warfare." *Proceedings of the Second Hydraulics Conference*, Bulletin 27, University of Iowa Studies in Engineering, 1943.

CHAPTER II

FLUID VELOCITY AND ACCELERATION

4. VELOCITY AND THE STREAM LINE

Visualization of the flow pattern. Smoke emerging from a chimney on a windy day permits a fascinating visual study of the motion of the surrounding air. The movement of the atmosphere on a larger scale is likewise made visible by the behavior of clouds, and currents in a

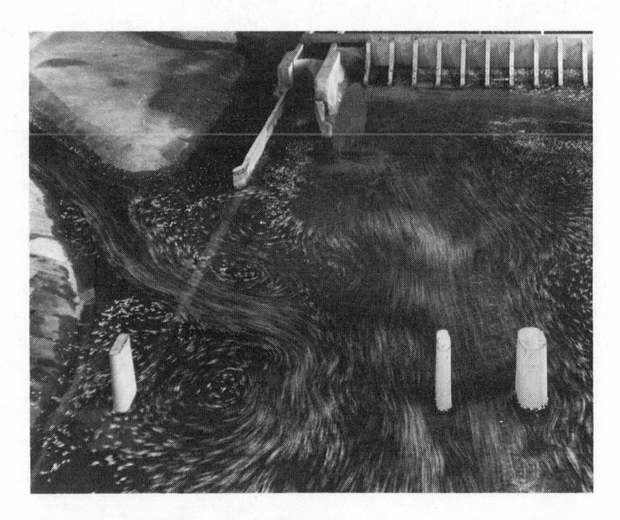

PLATE I. Currents in a river model shown by the motion of confetti scattered on the water surface; from a navigation study by the U. S. Engineer Sub-Office in the Hydraulics Laboratory at Iowa City.

river or canal by the silt and detritus carried in suspension. In each of these examples, one can either watch the trend of the movement in a given zone, or else choose at random a small portion of the suspended matter and follow it along its path, thereby obtaining a mental record of particular phases of the motion.

In the majority of flow phenomena, however, such visible agents as soot and silt are not so conveniently suspended in the moving fluid.

In the laboratory, to be sure, it is common practice to introduce droplets of oil or shiny particles of aluminum for visual or photographic observation, but for general purposes one's only recourse is to construct on paper or in the imagination a system of flow lines showing the nature of the motion in any desired region. Such a system of lines may seem at first quite a haphazard affair; the motion which it indicates, nevertheless, must be in complete accord with principles of mechanics, the investigation of which is the purpose of this book. While the ultimate goal of the present chapter is the determination of the flow pattern around or between boundaries of any given form, it is evidently first necessary to devise means of interpreting such a pattern once it is at hand.

Velocity vectors and components. The motion of a fluid, like that of a solid, is described quantitatively in terms of the characteristic known as *velocity*. In dealing with a solid, however, it is generally sufficient to measure the velocity of the body as a whole, whereas the motion of a fluid may be quite different at different points of observation. At any such point, nevertheless, the velocity completely defines the rate of motion at a given instant, in that it is a measure not only of the *speed* at which the fluid is passing that point but also of the *direction* in which the fluid is moving. Velocity is thus a *vector* quantity, for it possesses both magnitude and direction. The vector is usually represented by an arrow, as shown in Fig. 1, the length of the arrow being proportional to the magnitude of the velocity, and the orientation of the arrow indicating the direction of the flow.

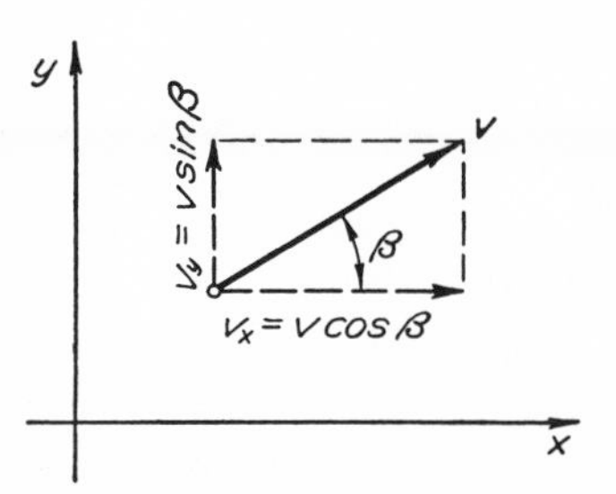

FIG. 1. Velocity vector and rectilinear components.

Owing to its vector properties, a velocity may be resolved into components in any desired directions (such as the rectilinear directions x and y of Fig. 1). Likewise, velocity components at a given point may be combined vectorially to yield their resultant, a procedure which is of particular importance in problems involving *relative motion*. For instance, to an observer in an airplane the air at a point which is fixed in relation to the plane will appear to have a definite direction and speed; the actual air velocity, however, is the vector sum of the velocity of the air relative to the plane and the velocity of the plane itself, as indicated in Fig. 2. Such vector addition is perhaps most readily accomplished by resolving each velocity into rectilinear components, which may then be added algebraically to obtain the components of the resultant.

Stream lines. If one had a series of photographs of smoke leaving a chimney, one might roughly indicate thereon by means of a series of

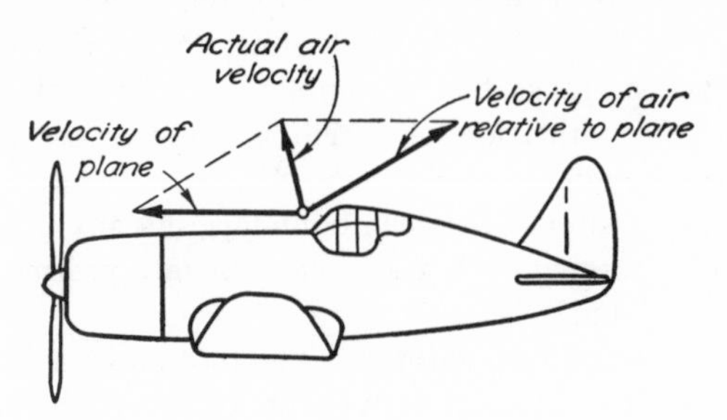

FIG. 2. Vector diagram for relative motion.

arrows the speed and direction of the air currents at a number of typical points. Velocity diagrams of this nature would be complete, of course, only if arrows were drawn at each and every possible point, but then the result would be hopelessly confused. On the other hand, a very satisfactory representation of the flow as a whole at any instant would be obtained by sketching in a series of curves in such manner that the velocity vectors for all points lying upon the curves would meet them tangentially, as shown in Fig. 3. Such curves are known as *stream lines*.

A stream line may thus be defined as a line which shows, through tangency to the velocity vector, the instantaneous direction of flow at every point over its entire length. It follows as a corollary that there can be no flow across a stream line at any point; in other words, the velocity vector necessarily has a zero component at right angles to the stream line to which it is tangent. In general, instantaneous stream lines will converge or diverge as they curve through space, for the velocity usually varies in magnitude and direction from point to point throughout a moving fluid. Once the stream-line pattern is at hand, of course, it is no longer necessary to include the individual velocity vectors, because the *direction* of flow in every region may be

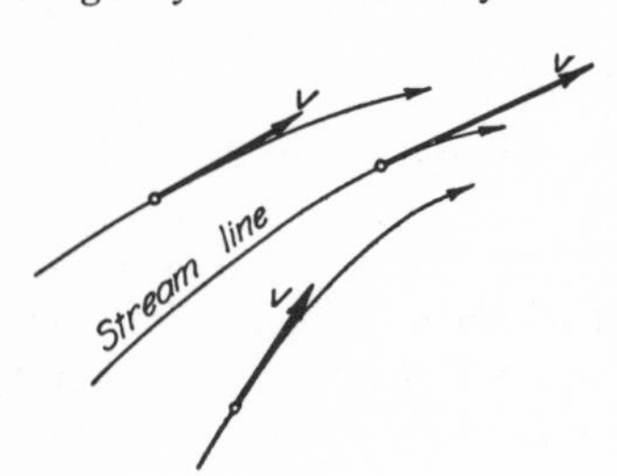

FIG. 3. Stream lines and velocity vectors.

seen from this pattern at a glance; that the stream-line configuration also indicates the *magnitude* of the velocity at all points will be shown in the following section.

Example 1. Tests upon a bomb held stationary in the air stream of a wind tunnel yielded (by means of a series of smoke filaments) the stream lines shown. At point a, where the inclination of the stream line is 30°, the velocity v_R relative

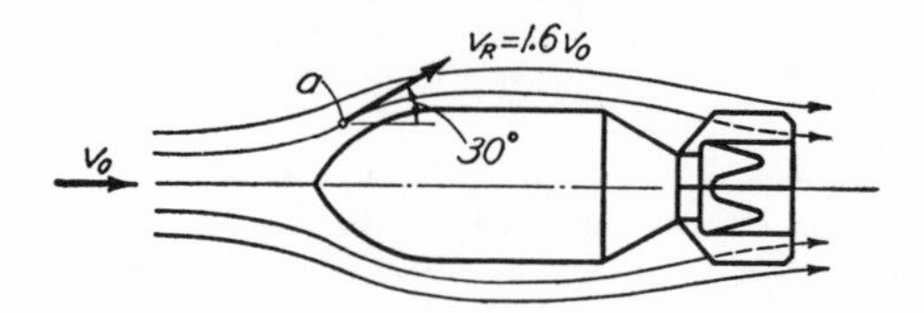

to the bomb was found to be 1.6 times as great as the velocity v_0 of the approaching air. What resultant velocity would this indicate when the bomb was falling at the rate of 500 feet per second? Sketch the corresponding instantaneous stream line.

Since the rate of fall corresponds to the velocity of the air stream in the wind-tunnel tests (i.e., $v_B = v_0$) the relative velocity at point a will be

$$v_R = 1.6\, v_B = 1.6 \times 500 = 800 \text{ fps}$$

The resultant velocity is then the vector sum of the relative velocity and the bomb velocity:

$$v = v_R \leftrightarrow v_B$$

In terms of vertical and horizontal components,

$$v_y = v_R \cos 30° - v_B$$
$$= 800 \times 0.866 - 500 = 193 \text{ fps}$$

$$v_x = v_R \sin 30° = 800 \times 0.5 = 400 \text{ fps}$$

$$v = \sqrt{v_x^2 + v_y^2} = \sqrt{400^2 + 193^2} = 444 \text{ fps}$$

$$\tan \beta = \frac{v_y}{v_x} = \frac{193}{400} = 0.482$$

$$\beta = 25° \, 45'$$

The instantaneous stream line would therefore be inclined at the angle β to the horizontal at this point, as indicated in the sketch.

PROBLEMS

1. When a jet of water is deflected by a plate held at right angles to the jet axis, the speed of the water relative to the plate is the same before and after deflection. If the plate is moved in the direction of the jet at one-half the absolute jet velocity, through what absolute angle will the jet be deflected?

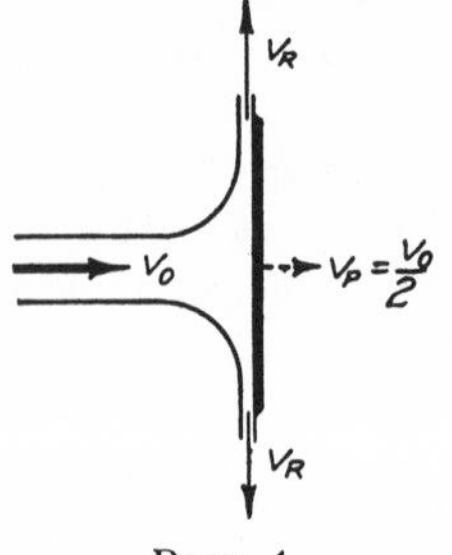

PROB. 1.

2. An airplane is observed to travel due north at a speed of 150 miles per hour in a 50-mile-per-hour wind from the northwest. What is the apparent wind velocity observed by the pilot?

3. Jets at the two ends of the 18-inch rotating arm of a lawn sprinkler direct the water in the plane of rotation at an angle of 45° relative to the arm. If the relative jet velocity is 30 feet per second and the arm makes 150 revolutions per minute, determine the actual velocity of the water as it leaves the sprinkler.

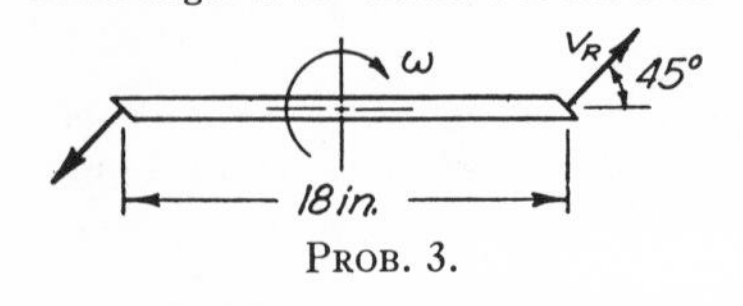

PROB. 3.

4. The air around the front of a baseball necessarily has the same velocity as that part of the ball with which it is in contact. If a ball is thrown horizontally with a speed of 75 feet per second and a back spin of 15 revolutions per second, what will be the air velocity at the foremost point of contact? (Assume a ball diameter of 3 inches.)

5. Water passes through a series of moving vanes inclined at an angle of 60° to their direction of motion. If the water velocity relative to the vanes is 15 feet per second and the vanes move with a speed of 10 feet per second, determine the actual velocity of the water and show in a sketch the inclination of the stream lines.

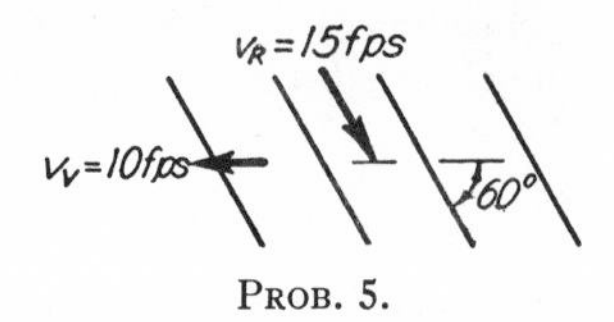

PROB. 5.

5. EQUATIONS OF CONTINUITY

Continuity of flow through a stream tube. Just as stream lines may be passed through a moving fluid in such manner as to indicate the direction of motion at every point, a tubelike surface bounded by such stream lines may be considered to enclose an elementary portion of the flow. Such an imaginary surface, known as a *stream tube*, is illustrated in Fig. 4. If the cross-sectional area of the tube is sufficiently small, the velocity at its midpoint will closely represent in magnitude and direction the average velocity for the section as a whole. In differential terms, therefore,

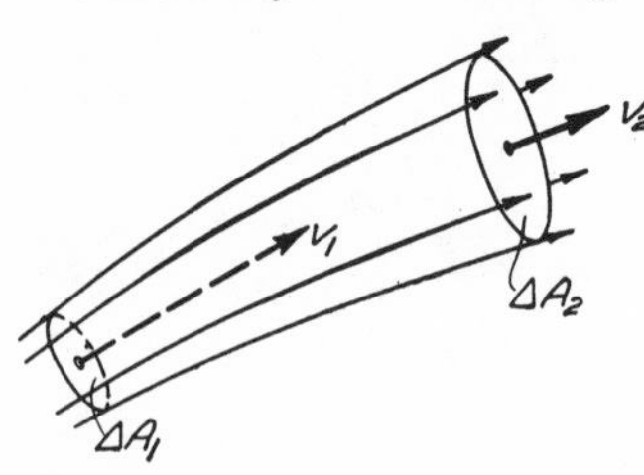

FIG. 4. Velocity variation through a stream tube.

the *volume rate of flow* past any normal cross section of an elementary stream tube will be

$$dQ = v\,dA \tag{1}$$

in which dQ has the dimension of volume per unit time or [length^3/time], v that of distance per unit time or [length/time], and dA that of area or [length^2].

Since the stream tube is bounded by stream lines, it is evident from the definition of the stream line that no fluid whatever can pass through the tube wall. From the law of conservation of mass, moreover, it is clear that fluid matter can be neither created nor destroyed. If, for the present, it is further specified that fluid matter is not expanded or compressed during motion, then at any instant flow must take place past all successive cross sections of a stream tube at the same mass rate. In other words, the mass of fluid passing one cross section per unit of time must equal, simultaneously, the mass per unit time passing every other cross section. But, if the mass per unit volume of the fluid is assumed for the present to remain constant, the volume of fluid passing every section of a given tube per unit time must then also be the same. Therefore, if the mass density of the fluid does not change, the volume rates of flow past all successive cross sections of a stream tube must be equal at any instant. This elementary principle is expressed in the following basic *equation of continuity*:

$$v_1\,dA_1 = v_2\,dA_2 = v_3\,dA_3 \cdots \tag{2}$$

Evidently, the instantaneous velocity of flow must vary *inversely* with the cross-sectional area of the tube.

Before proceeding farther, two circumstances warrant clarification: Equations such as (1) and (2) frequently involve areas and lengths of differential magnitude, whereas the corresponding figures invariably show areas and lengths of small but finite magnitude; this is due primarily to the fact that quantities which are sufficiently minute to make the differential equations exact cannot well be illustrated to scale without losing all semblance of stream-line curvature; on the other hand, the equations are often applied to stream tubes of finite cross-sectional area as a first approximation, the actual error remaining small so long as the curvature of the tube and the velocity variation across the section are not excessive. It may also have been noted that the stream lines and stream tubes appearing in the illustrations show no suggestion of the *turbulence* so often present in fluid motion. Although the concept of stream lines is quite as applicable to flow with turbulence as to flow without, in the former instance the pattern as a whole can become extremely complicated; when turbulence exists, therefore, it is customary to represent by the stream lines only the *average* pattern of motion, a procedure which will be followed throughout this book.

Continuity relationships for two-dimensional flow. In some cases of flow the velocity vector will have components in only two coordinate directions, the component in the third direction being everywhere equal to zero. Under such circumstances the flow is said to be *two-dimensional*, since all the stream lines must then lie in parallel planes. In two-dimensional flow, therefore, all such parallel planes will display the same flow pattern. The intersection of a stream tube with one such plane will be simply two stream lines, as indicated in Fig. 5, and the product of the velocity and the normal distance dn between these lines will correspond to a volume rate of flow *per unit width of section*,

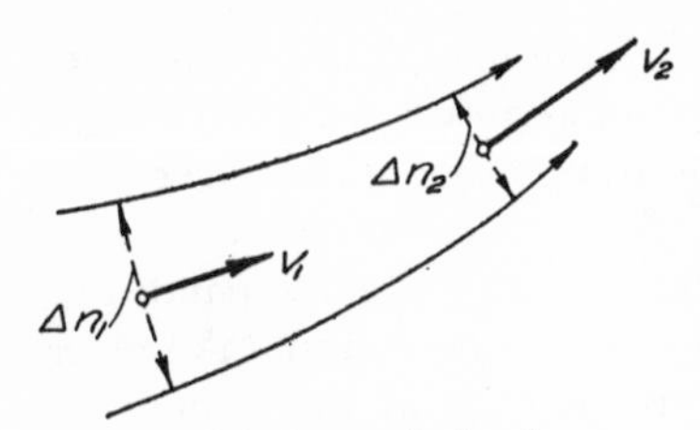

FIG. 5. Velocity variation in two-dimensional flow.

$$dq = v\,dn \tag{3}$$

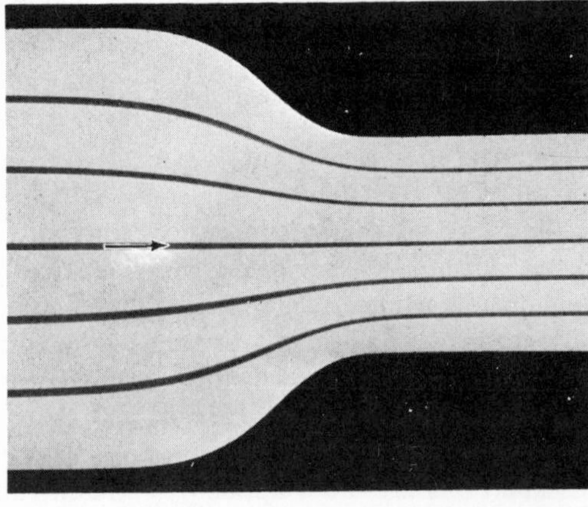

PLATE II. Stream lines at a two-dimensional boundary contraction, made visible by the injection of dye into water flowing between closely spaced glass plates.

dq necessarily having the dimension [volume/time/length] or simply [length^2/time]. The differential equation of continuity for two-dimensional motion thus becomes

$$v_1\,dn_1 = v_2\,dn_2 = v_3\,dn_3 \cdots \tag{4}$$

Assuming that a pattern of stream lines, such as that for the two-dimensional boundary contraction shown in Fig. 6, has been determined experimentally, means are now at hand of evaluating the velocity at every point in terms of that in some reference zone. For instance, if the variation in the velocity v_0 across the parallel section at the left of Fig. 6 is known, the variation in the velocity across any other section may be calculated from the stream-line spacing by means of the

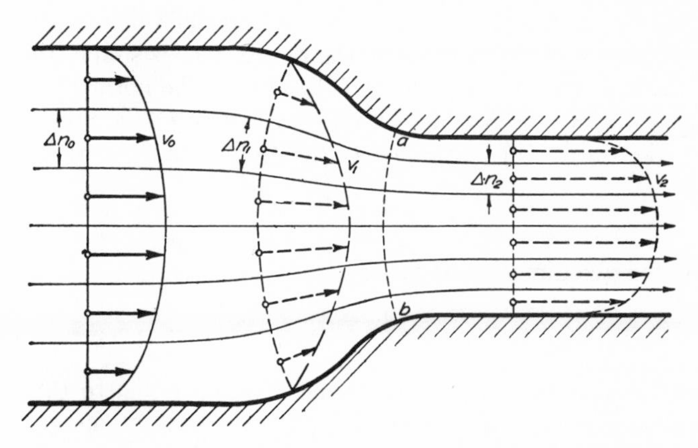

FIG. 6. Determination of the velocity distribution from the stream-line spacing.

following approximate form of Eq. (4), in which Δn_0 and Δn represent distances between the same two stream lines at successive sections:

$$v \approx v_0 \frac{\Delta n_0}{\Delta n} \tag{5}$$

Typical velocity-distribution curves so obtained are shown in the figure; in each instance the relative spacing of the stream lines was determined from the diagram by direct measurement, and the corresponding velocity was assumed to occur midway between each pair of stream lines. Noteworthy is the fact that divergence of the stream lines increases the velocity variation across a normal section, whereas convergence of the stream lines tends to equalize the velocity distribution.

Evaluation of the total rate of flow. Once the velocity distribution across a normal section is known, the *total* flow passing the section per unit time may be evaluated through integration of Eq. (1):

$$Q = \int v\, dA \tag{6}$$

The counterpart of this equation for two-dimentional flow is evidently

$$q = \int v \, dn \tag{7}$$

If the velocity distribution can be expressed in algebraic form, the corresponding integral may be evaluated analytically. Otherwise, the integration must be performed graphically (for instance, by adding together the increments $v \, \Delta n$ across any section of Fig. 6)—unless (as is nearly true at the right of Fig. 6) the velocity is constant across the section, under which circumstance the rate of flow is simply the product of the velocity at any point and the area of the normal section.

If the total rate of flow is divided by the area of the normal cross section, the result will be the *mean velocity* for the section; that is,

$$V = \frac{Q}{A} \tag{8}$$

whence, for two-dimensional flow,

$$V = \frac{q}{n} \tag{9}$$

Evidently, the mean velocity is inversely proportional to the size of the flow section. Equations (2) and (4) may now be integrated by inspection to yield the following useful forms of the continuity principle:

$$V_1A_1 = V_2A_2 = V_3A_3 \cdots \tag{10}$$

and, for two-dimensional flow,

$$V_1n_1 = V_2n_2 = V_3n_3 \cdots \tag{11}$$

In applying Eqs. (10) and (11), the fact cannot be too strongly borne in mind that these relationships are generally valid only if the areas of successive cross sections are bounded by the same stream lines.

Example 2. If the velocity of oil flowing between convergent plates varies across any normal section according to the parabolic equation

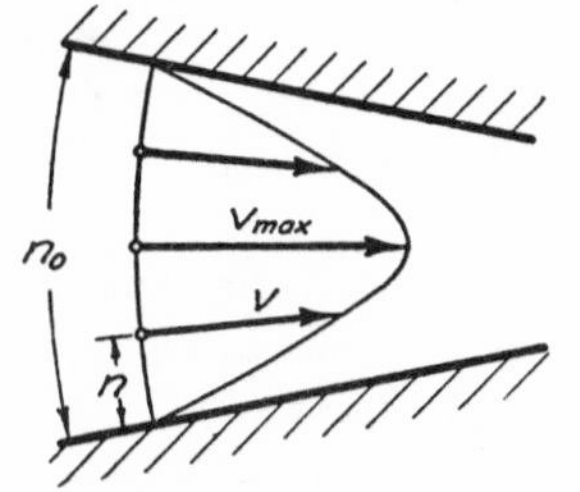

$$v = v_{\max} \frac{4n}{n_0^2} (n_0 - n)$$

and if $v_{\max} = 0.5$ foot per second where $n_0 = 1$ inch, determine: (*a*) the total rate of flow, assuming the boundaries to have the constant breadth $b = 9$ inches; (*b*) the mean velocity for this section; and (*c*) the mean velocity for the section at which $n_0 = 0.25$ inch.

(*a*) According to Eq. (7), the rate of flow per unit breadth of boundary must be found from the integral of $dq = v\,dn$ over a section normal to the stream lines; that is,

$$q = \int_0^{n_0} v\,dn = 4\frac{v_{max}}{n_0^2}\int_0^{n_0} n(n_0 - n)dn = \tfrac{2}{3}\,v_{max}\,n_0$$

Introducing the given magnitudes of n_0 and v_{max}, reduced to consistent units,

$$q = \tfrac{2}{3} \times 0.5 \times \tfrac{1}{12} = 0.0278 \text{ cfs/ft}$$

Finally, since the required rate of flow is equal to the flow per unit breadth times the breadth of section,

$$Q = qb = 0.0278 \times \tfrac{9}{12} = 0.0209 \text{ cfs}$$

(*b*) From Eq. (9) the mean velocity for the normal section will be

$$V = \frac{q}{n_0} = \frac{0.0278}{1/12} = 0.334 \text{ fps}$$

(*c*) Since the rate of flow past all cross sections must be the same, the mean velocity for the second section will necessarily be increased in proportion to the reduction in boundary spacing; that is, from Eq. (11),

$$(Vn_0)_1 = (Vn_0)_2 \quad \text{and} \quad V_2 = 0.334 \times \frac{1}{0.25} = 1.33 \text{ fps}$$

It follows that v_{max} must also vary inversely with n_0.

Example 3. In making velocity measurements at the crest of a spillway, it was necessary to traverse a vertical section rather than a section at right angles

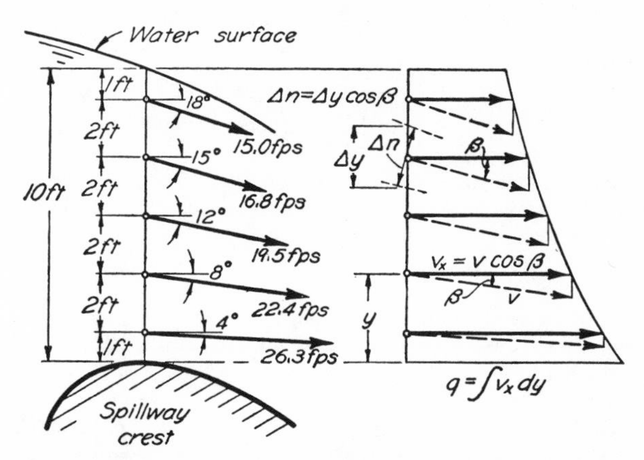

to the stream lines, the direction, magnitude, and location of the measured velocity vectors being as shown in the diagram. What rate of flow per unit length of crest is indicated by these measurements?

The rate of flow may be evaluated as a first approximation by multiplying each velocity magnitude by the normal distance $\Delta n \approx \Delta y \cos \beta$ and adding the resulting values of Δq. Thus

$$q = \Sigma \Delta q = \Sigma(v\ \Delta n) = 15 \times 2 \cos 18° + 16.8 \times 2 \cos 15°$$
$$+ 19.5 \times 2 \cos 12° + 22.4 \times 2 \cos 8° + 26.3 \times 2 \cos 4° = 196 \text{ cfs/ft}$$

But $v(\Delta y \cos \beta) = (v \cos \beta)\Delta y = v_x\ \Delta y$; hence, the foregoing process amounts in reality to approximating the integral $q = \int v_x\, dy$. Therefore, if the horizontal components of the vectors are plotted, as shown, the area enclosed by the curve will represent a more exact value of q. As determined by planimeter, this value is 195 cfs/ft.

PROBLEMS

6. Natural gas is pumped through a pipe at the rate of 200 cubic feet per minute. If the pipe diameter at successive sections is 12 inches, 9 inches, and 6 inches, what are the corresponding mean velocities of flow?

7. Water flows down the 60° face of a spillway having a horizontal crest length of 40 feet. If the vertical depth of flow is found to be 1.5 feet at a section for which the mean velocity is 25 feet per second, estimate the total rate of flow over the spillway.

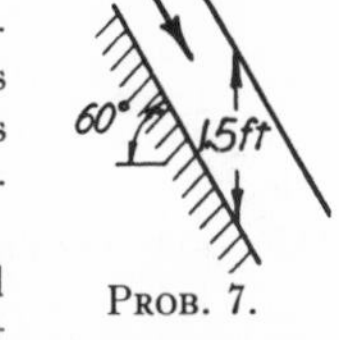

PROB. 7.

8. The mean velocity of flow in a wide river is 5 feet per second at a section of 8-foot depth. What is the rate of flow per unit width? What mean velocity should prevail at a section where dredging had increased the depth to 12 feet?

PROB. 9.

9. Liquid flows through an orifice in the bottom of a cylindrical tank at a velocity of 12 feet per second. If the jet diameter is 1 inch, what is the rate of flow? If the tank is 10 inches in diameter, at what rate is the liquid surface falling?

10. Water enters the lock of a canal through 250 rectangular ports each having an outlet area of 3 square feet, the lock itself being 800 feet long and 80 feet wide. If 6 feet per minute is the maximum rate at which the water level rises in the lock during filling, determine the corresponding mean velocity of efflux from the ports.

11. Determine and plot to scale the velocity distribution over section a–b of Fig. 6.

12. Measurements of velocity in a wide river yielded the vertical distribution curve shown in the accompanying sketch. Evaluate by graphical or numerical integration the rate of flow per unit width.

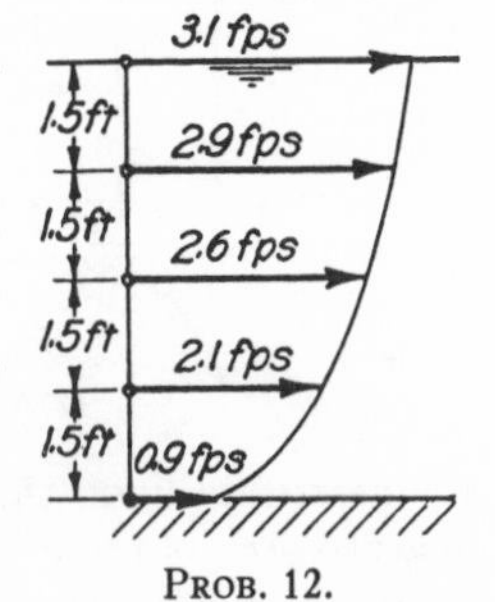

PROB. 12.

13. What rate of flow per unit width through the boundary transition of Fig. 6 would be indicated by the velocity-distribution curve at the left if $n_0 = 5$ feet and $(v_0)_{max} = 8$ feet per second?

14. The velocity of heavy oil flowing through a tube of radius r_0 is known to vary with distance r from the centerline according to the expression $v = 2V(r_0^2 - r^2)/r_0^2$. Determine the maximum (i.e., centerline) velocity for the condition that $Q = 0.02$ cubic foot per second and $r_0 = 1$ inch.

15. Air is exhausted through a 1-inch tube into a diffusor consisting of two parallel disks 6 inches in diameter and ¼ inch apart. Assuming radial stream lines within the diffusor, determine the mean velocity within the tube which will produce a mean exit velocity of 2 feet per second.

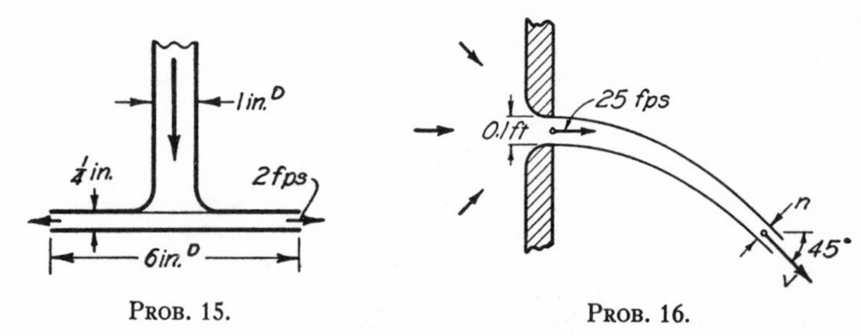

PROB. 15. PROB. 16.

16. A sheet of water emerging horizontally from a long slot has a thickness of 0.1 foot and a mean velocity of 25 feet per second. If the horizontal component of the velocity remains constant as the jet is deflected by gravity, determine the velocity v and thickness n of the sheet at a section having a slope of 45°.

17. As a first approximation, liquid may be assumed to approach an opening in the wall of a reservoir at the same rate from all directions. If such an opening corresponds to the intake of a pump drawing 3 cubic feet per second, estimate the velocity at radial distances of 1 foot and 3 feet from the midpoint of the intake.

18. Assuming that the water approaches the slot of Problem 16 radially in vertical planes, estimate the velocity at distances of 1 foot and 3 feet from the slot.

19. The accompanying sketch represents a vertical section through a draft tube discharging water from a turbine, the boundaries being surfaces of revolution about a vertical axis. Determine the mean velocities at sections A and B when the rate of flow is 150 cubic feet per second.

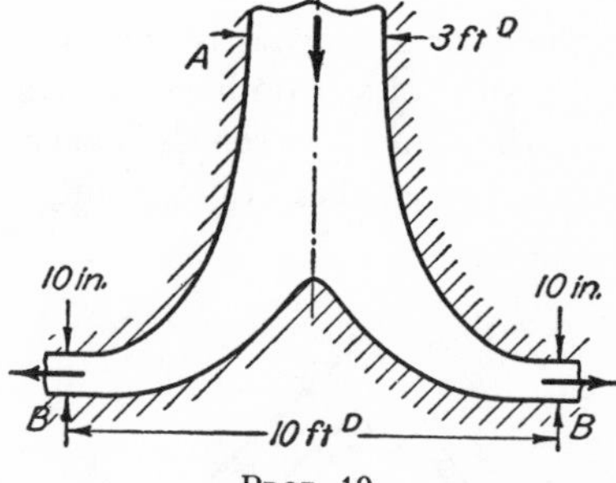

PROB. 19.

6. SIGNIFICANCE OF THE FLOW NET

Elementary flow nets. With means now at hand of obtaining the velocity distribution from a given stream-line pattern, the next step is to seek a method of determining the form of the stream lines for any boundary geometry. The only absolute method of determination, to be sure, entails observation of the flow itself, by means of smoke, dye, or other visible agents. In many instances, however, effective use may be made of a simple graphical process based upon mathematical principles of classical hydrodynamics. Since these principles embody, as a matter of fact, the only general means of even approximating a flow pattern without recourse to actual field or laboratory measurement, their graphical representation warrants careful attention at this point.

For two-dimensional motion, such graphical representation of the

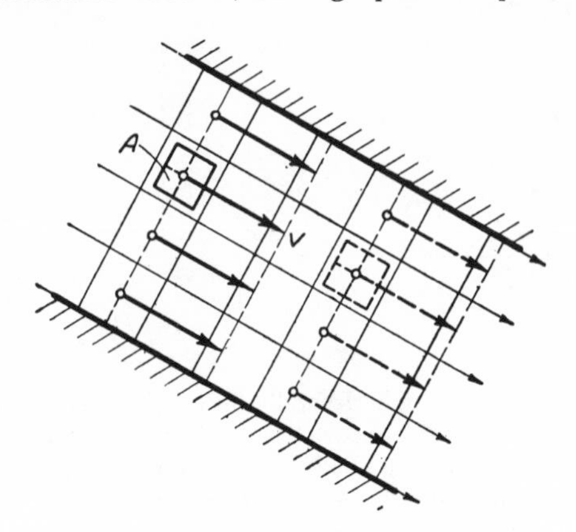

FIG. 7. Flow net for parallel boundaries.

mathematical analysis is known as a *flow net*. This consists, in brief, of a system of stream lines so spaced that the incremental rate of flow Δq is the same between each successive pair, and a system of normal lines so spaced that at any point the distance Δs between normal lines

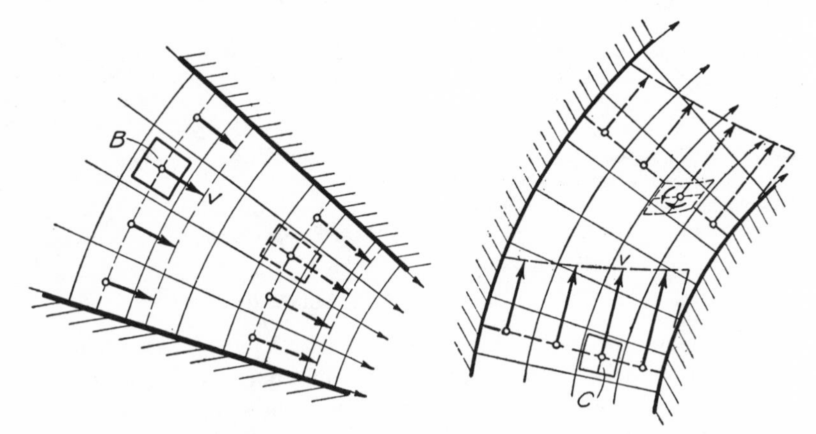

FIG. 8. Flow net for convergent boundaries.

FIG. 9. Flow net for coaxial boundaries.

equals the distance Δn between stream lines. The velocity, under such circumstances, would then be inversely proportional to the distance between either the stream lines or the normal lines throughout the flow, and Eq. (5) would assume the more general form, applicable between any two points in the entire field of motion,

$$\frac{v}{v_0} \approx \frac{\Delta n_0}{\Delta n} \approx \frac{\Delta s_0}{\Delta s} \tag{12}$$

For example, the net for flow between parallel boundaries would consist, as shown in Fig. 7, of a series of square meshes of constant size, indicating the same velocity at every point. Were the same boundaries to converge, as in Fig. 8, the meshes would no longer be exactly square, but the median lines of any mesh would still have essentially the same length; the velocity would evidently be constant along any normal line but would vary as the stream lines converged. If, on the other hand, the boundaries were coaxial cylinders, the stream lines and normal lines of Fig. 8 would simply be interchanged, as indicated in Fig. 9; the velocity would then be constant along the circular stream lines but would vary inversely with the radius in the normal direction.

Irrotational motion. The primary characteristic of flow patterns represented by such nets is that they correspond, hydrodynamically

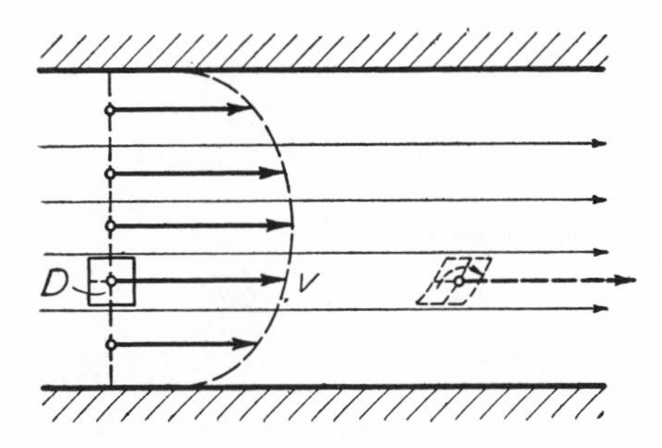

FIG. 10. Rotational flow between parallel boundaries.

speaking, to *irrotational motion*—that is, to flow in which every fluid element has a zero angular velocity about its own mass center. It is not necessary to go into the mathematical nature of rotation to see that it does not occur in any of the three elementary cases cited. The parallel flow of Fig. 7 is obviously irrotational, since the median lines of element A do not change their orientation as the element moves through space. Element B in Fig. 8 is evidently *deformed* during motion, but the median lines do not vary in orientation. Although element C in Fig. 9 may at first glance seem to rotate, careful inspection will show that it also deforms in such a way that the *average* orientation of the median lines still remains unchanged.

If, on the contrary, one considers an example of parallel flow in which the velocity is lower near the boundaries than in the central region (Fig. 10), not only will an element display a distinct rotation of one median line, but if the stream lines are properly spaced (i.e., so that Δq = constant) it will be found impossible to construct a square-meshed net. Likewise, if flow between coaxial boundaries (Fig. 11) is more rapid at the outer than at the inner boundary, the same general situation will be found to prevail. Such cases, therefore, represent motion which is definitely not irrotational.

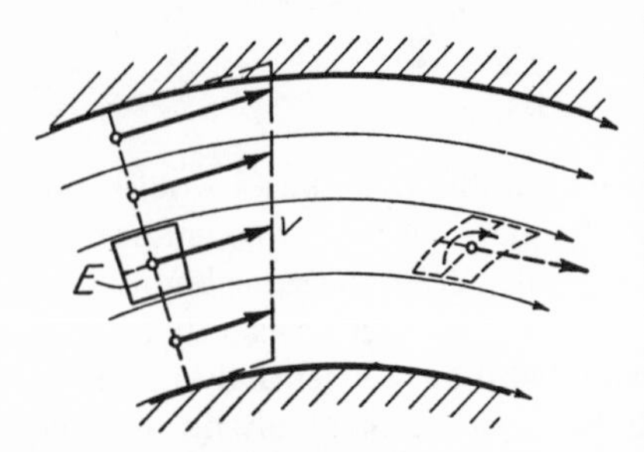

FIG. 11. Rotational flow between coaxial boundaries.

Flow-net construction and interpretation. Although these elementary examples of irrotational flow are easily described mathematically, the primary usefulness of the flow net is in connection with boundary forms not readily subject to mathematical analysis. The basis for such use, however, lies in the following mathematical fact: If the limiting stream lines coincide with fixed boundaries of given form, one and only one pattern of irrotational flow can exist. The flow net which defines this pattern may therefore be obtained by constructing a system of

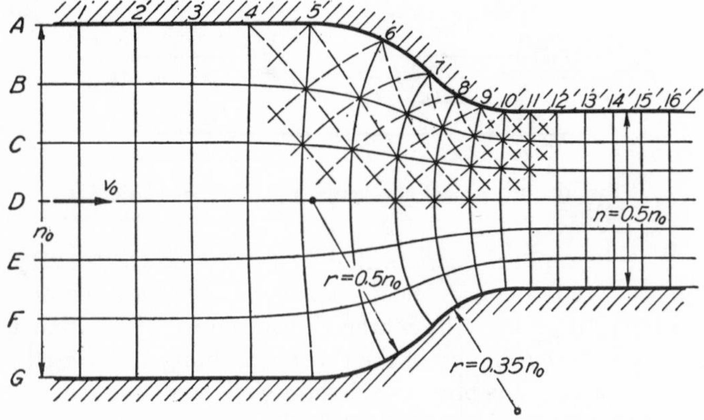

FIG. 12. Flow net for the boundary transition of Fig. 6.

stream lines and a system of normal lines in such manner as to approximate a square-meshed net. Stream lines are first sketched in by eye, followed by normal lines, the test of accuracy in sketching being the criterion that the two systems of lines not only must meet at right angles at every intersection but also must be equally spaced in every zone. The spacing could be exactly equal, of course, only if the meshes were infinitesimal in size, and only then would these meshes form perfect squares. However, such a degree of mesh perfection is beyond all practical interest, since in zones where the meshes depart considerably from squareness one need only sketch in the diagonals (which must also form a rectilinear network) to check the accuracy of construction. The beginner may have to make considerable use of the eraser before

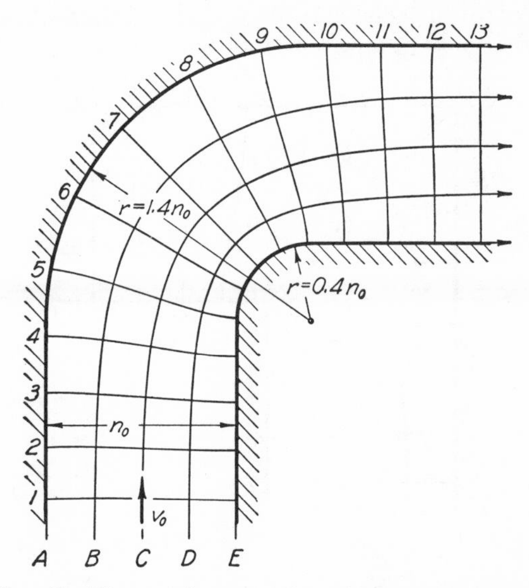

FIG. 13. Flow net for a two-dimensional conduit bend.

producing a satisfactory net, but the number of successive approximations will be reduced to a minimum as the eye becomes experienced.

A typical net is shown in Fig. 12, for the same boundaries as in Fig. 6, from which the influence of the boundary form upon the local spacing of the stream lines will at once be evident. For instance, inspection will indicate that both the maximum spacing and the minimum spacing of the normal lines (and hence of the stream lines) are found at the boundaries in the zones of maximum curvature. Such conditions are, in fact, quite general. Moreover, high velocities, or small meshes of the net, are invariably associated with boundaries which curve away from the flow, and low velocities, or large meshes, with boundaries which curve toward the flow. This is clearly seen by comparison of the meshes along the inner and outer boundaries of the flow net for the 90° bend shown in Fig. 13. The more rapid the curvature, of course, the more pronounced will be the velocity change. If, for instance, the radius of boundary curvature in Fig. 13 were reduced to zero, as in Fig. 14, the spacing of the lines at the outer corner would become, relatively speaking, infinitely great, while at the inner corner the spacing would be infinitesimal. The infinite velocity corresponding to the infinitesimal spacing at the inner corner is obviously of no physical significance. On the other hand, the zero velocity indicated by the infinite spacing at the outer corner is by no means physically impossible; indeed, a boundary angle of this nature invariably tends to produce what is known as a zone of *stagnation*.

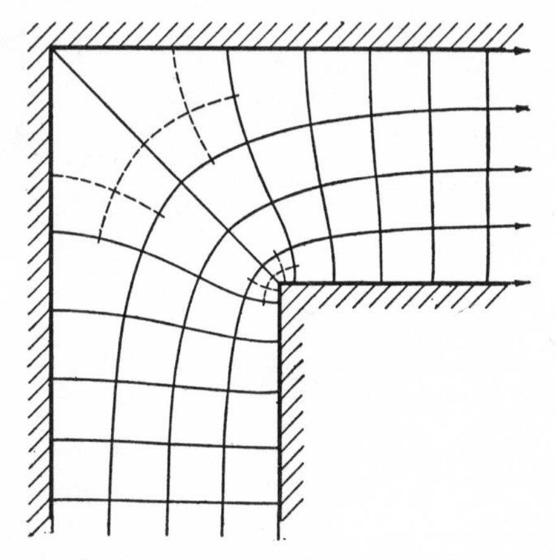

FIG. 14. Flow-net configuration at boundary corners.

Separation in divergent flow. Since the flow net was stated to apply only to irrotational flow, a question naturally arises as to its usefulness when a given state of motion actually is not irrotational. So long as the boundary form is such that the stream lines rapidly converge, the velocity distribution will become more and more nearly like that of irrotational flow; the net will then be found to yield a very close approximation to the actual flow pattern, as may be seen by comparing Figs. 6 and 12. If, on the other hand, the flow net indicates appreciable divergence of the stream lines (as in flow from right to left between the boundaries of Fig. 12), in the case of rotational motion the velocity near the boundary is generally so low that further velocity reduction in this region is physically impossible—despite the indications of the net. In other words, although a primary assumption in flow-net construction is that the outermost stream lines coincide with the fixed boundaries, a primary characteristic of any divergent boundary is that it does not effectively guide a moving fluid.

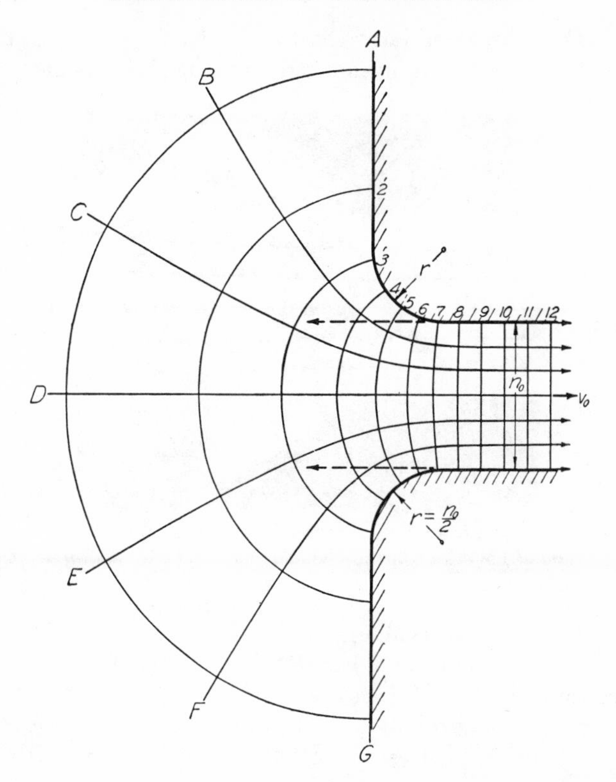

FIG. 15. Flow net for a two-dimensional rounded inlet.

Thus, as a result of the very rotation which the flow net ignores and which invariably exists at a fixed boundary, the flow tends to leave the boundary wherever the net shows marked divergence of the stream lines in the boundary region. This phenomenon is known as *separation*.

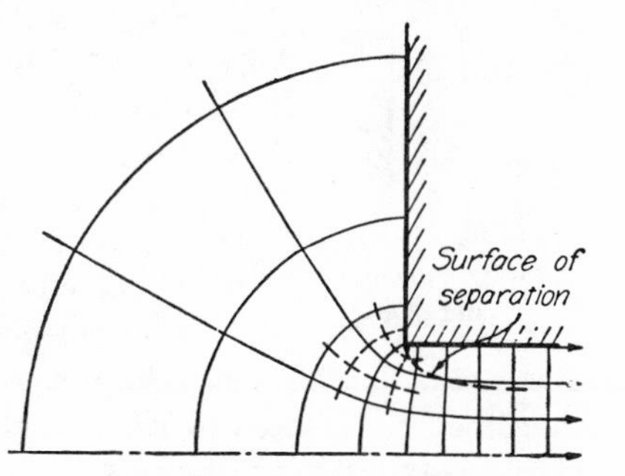

FIG. 16. Separation at an abrupt inlet as indicated by flow net.

Consider, for example, the juncture between a conduit and a large reservoir, for which the two-dimensional flow net is shown in Fig. 15. So long as the flow is from left to right, i.e., from the reservoir into the conduit, the actual pattern will be very nearly the same as that indicated by the converging stream lines of the net. Were the flow from right to left, on the contrary, the flow net itself would remain unchanged but the fluid would actually

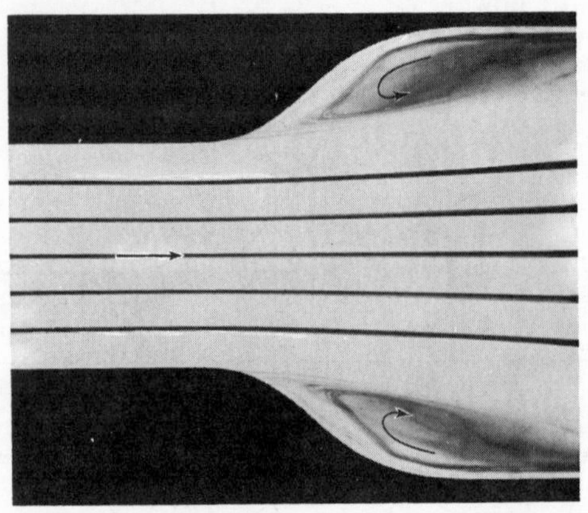

Plate III. Separation at a rapid enlargement of cross section, produced by reversing the direction of flow through the boundaries of Plate II; note zone of backflow colored by dye.

pass into the reservoir as an essentially parallel jet, failing completely to follow the divergent boundaries. If, moreover, the juncture were not properly rounded, as indicated in Fig. 16, the stream lines near the corner would diverge in both directions, with the result that flow even from left to right would separate from the boundary, as shown by the heavy broken line, rather than diverge so rapidly. As for the abrupt 90° bend of Fig. 14, the stream lines are seen to diverge as they approach the outer corner and after they pass the inner corner, the flow net thus showing conclusively that the two zones of separation indicated in Fig. 17 will exist.

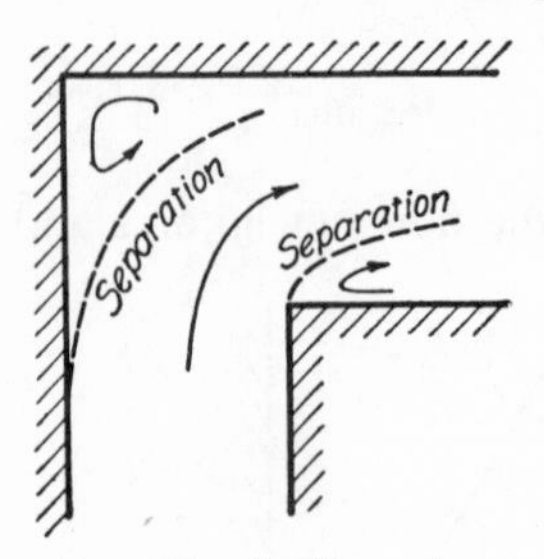

Fig. 17. Regions of separation indicated by flow net of Fig. 13.

Summary of flow-net principles. As will be discussed at a later point in this book, the phenomenon of separation generally involves a decrease in flow efficiency, for not only is the actual flow section smaller than that represented by the boundaries themselves, but also the zone of reverse flow beyond the surface of separation may produce intermittent disturbances in the form of rotating masses of fluid which work their way into the central zone as they are carried downstream. Were it possible to define the average form of this surface of separation, the flow net might still be used to approximate the resulting velocity distribution; indeed, this method is employed in connection with jets in the following chapter and with immersed bodies later in the text. For the present, however, only the following basic characteristics of the flow net need be kept in mind:

1. The flow net will yield a close approximation to the actual flow pattern as long as the stream lines do not diverge appreciably in the boundary zone.
2. The more uniform the distribution of velocity in zones of parallel flow, and the more rapidly the boundaries thereafter cause the stream lines to converge, the more accurately will the flow net indicate the actual velocity distribution in all zones.
3. The flow net cannot be used to determine the velocity distribution near divergent boundaries if the form of the surface of separation is not known.
4. Since separation represents a reduction in flow efficiency, the most desirable boundary design is that for which the flow net most closely approximates the actual pattern of motion.

Example 4. Determine and plot to scale the ratio of the boundary velocity v to the velocity of approach v_0 for the flow net shown in Fig. 12.

According to Eq. (12) the ratio v/v_0 varies inversely with either the ratio $\Delta n/\Delta n_0$ or the ratio $\Delta s/\Delta s_0$. Owing to the finite size of the flow-net meshes, the

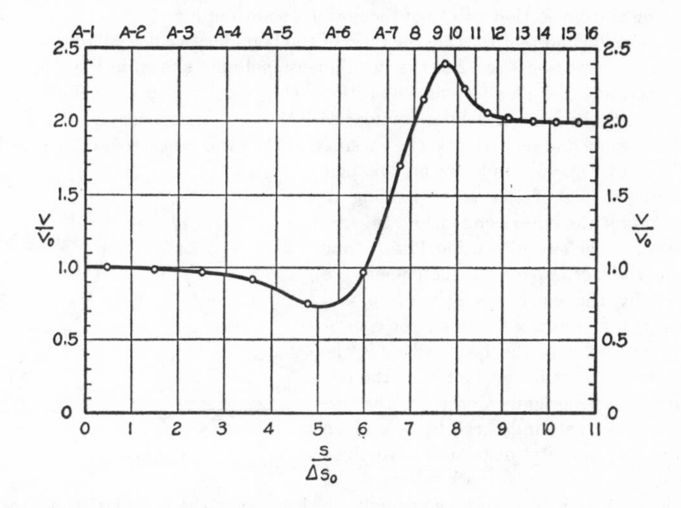

spacing of the stream lines and normal lines will correspond to the velocity approximately midway between any two such lines. Therefore, the distance Δs between normal lines at either boundary should indicate the boundary velocity midway between the lines. By measuring successive distances with dividers, the following values of $\Delta s/\Delta s_0$ and of its reciprocal v/v_0 are obtained:

	$\Delta s/\Delta s_0$	v/v_0		$\Delta s/\Delta s_0$	v/v_0
A–1–2	1.00	1.00	A–8–9	0.463	2.16
2–3	1.01	0.99	9–10	0.416	2.40
3–4	1.03	0.97	10–11	0.448	2.23
4–5	1.08	0.93	11–12	0.484	2.06
5–6	1.34	0.75	12–13	0.495	2.02
6–7	1.03	0.97	13–14	0.500	2.00
7–8	0.588	1.70	14–15	0.500	2.00

A plot of the variation in v/v_0 along the developed boundary is shown in the foregoing diagram. It is to be noted that this dimensionless plot, like the flow net from which it was determined, applies to any rate of flow and to any size of flow section, so long as the boundary proportions are not changed.

PROBLEMS

NOTE: In these and subsequent flow-net problems, a sufficiently precise evaluation of the line spacing may be obtained by setting spring dividers between the points in question and then laying off this intercept on an engineer's scale or sheet of graph paper. When the spacing is very small, gross interpolation in scale reading may be avoided by stepping off each intercept as many times as required for ease in reading; unless all distances are thus multiplied by the same factor, of course, each result must be reduced in proportion to the number of steps made with the dividers.

20. Determine and plot to scale the ratio of the centerline velocity v to the velocity of approach v_0 from section 1 to 15 of the flow net shown in Fig. 12.

21. Determine and plot to scale the ratio of the boundary velocity v to the conduit velocity v_0 from section 1 to 12 of the two-dimensional inlet shown in Fig. 15.

22. Determine and plot to scale the ratio of the velocity v to the velocity of approach v_0 along the inner and outer boundaries of the two-dimensional bend of Fig. 13.

23. Determine and plot to scale the variation in the ratio v/v_0 across the midsection of the bend of Fig. 13. Indicate on this plot the limiting values of v/v_0 corresponding to the flow net for the miter bend of Fig. 14.

24. Flow in the two-dimensional draft tube shown in the accompanying sketch may be described by the equations $v_x = v_a x/a$ and $v_y = -v_a y/a$, in which x and y are the coordinates of any point in the flow and v_a is the magnitude of the velocity vector at the distance a from the origin. Construct the corresponding flow net, and check by the given equations the velocity distribution obtained from the net.

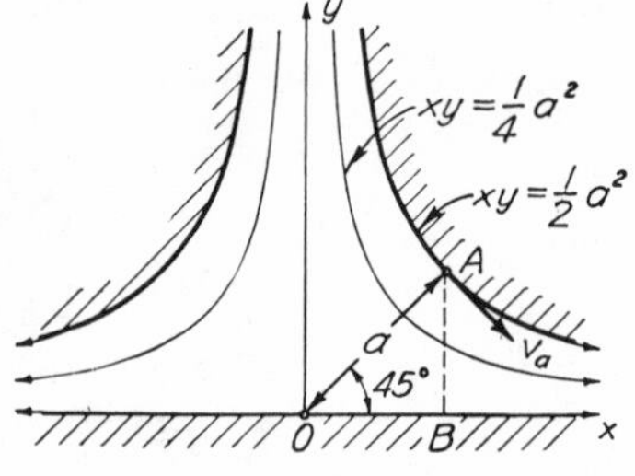

PROB. 24.

25. Show that the rate of flow per unit width between the boundaries of Problem 24 may be obtained by integrating either $v\,dn$ across section OA or $v_x\,dy$ across section BA. Determine the magnitude of q for the conditions that $v_a = 10$ feet per second and $a = 5$ feet.

7. ACCELERATION IN STEADY, NON-UNIFORM FLOW

Velocity variation with time and distance. Any general pattern of fluid motion, such, again, as that made visible by clouds of smoke in the atmosphere, not only will have a different configuration at all points at any instant of observation but also will change continuously in form with time. In other words, on an instantaneous photograph of the cloud no two regions would look alike, nor would any given region have the same form on two successive photographs.

Such variation of the flow pattern with location is termed *non-uniformity* of motion, whereas the variation with time is called *unsteadiness*. To be exact, unsteady flow is that in which the velocity at any fixed point changes from instant to instant; likewise, non-uniform flow is that in which the velocity at any instant changes from point to point along a stream line. Velocity may change, of course, in either magnitude or direction (or both magnitude and direction together), directional changes affecting the shape of the stream lines, and changes in magnitude the stream-line spacing. If a given flow is steady, therefore, the stream-line pattern and the rate of flow will not change with time, regardless of the geometrical shape of the stream lines; if the flow is uniform, on the other hand, all stream lines will be parallel at any instant of observation, regardless of any change which may be occurring in the rate of flow.

Unsteady flow is at best a difficult problem to analyze, except under certain special conditions treated at the end of this chapter. For the present, therefore, it will be assumed that the state of motion under discussion remains constant with time, however involved the stream-line configuration may be; such flow is therefore steady but generally non-uniform. It now remains to investigate the variation in velocity from point to point along typical stream lines for the specific condition that both the rate of flow and the form of the flow pattern remain the same from instant to instant.

Velocity and acceleration. *Velocity* is defined in mechanics as the rate of displacement with respect to time. Since the velocity of the fluid in non-uniform flow is generally not the same at different points along a stream line, the exact mathematical expression of its magnitude must be written in terms of the increment of length ds of the stream line traversed by a fluid element during the infinitesimal time dt; thus,

$$v = v_s = \frac{ds}{dt} \tag{13}$$

The velocity vector is necessarily tangent to ds, so that v_s and v are identical in magnitude; the subscript s is introduced merely for the sake of clarity in the following development. The component of v in any other direction, accordingly, may be expressed simply through substitution of the projection of ds upon the corresponding axis; for instance, with reference to Fig. 18,

$$v_x = v_s \cos\beta = \frac{ds \cos\beta}{dt} = \frac{dx}{dt}$$

Acceleration is defined in mechanics as the rate of change of velocity with respect to time; thus, again in terms of infinitesimal increments,

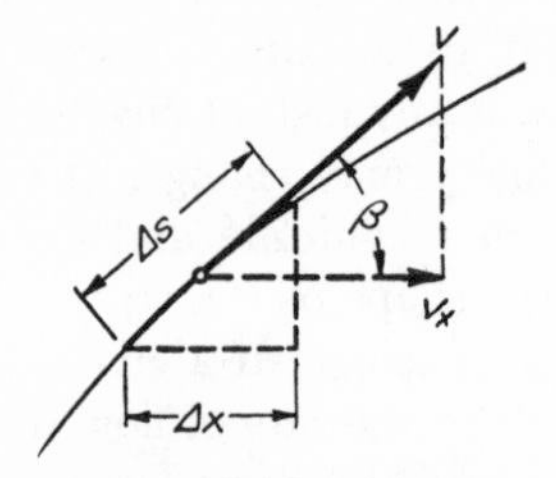

FIG. 18. Geometrical relationship of velocity and displacement.

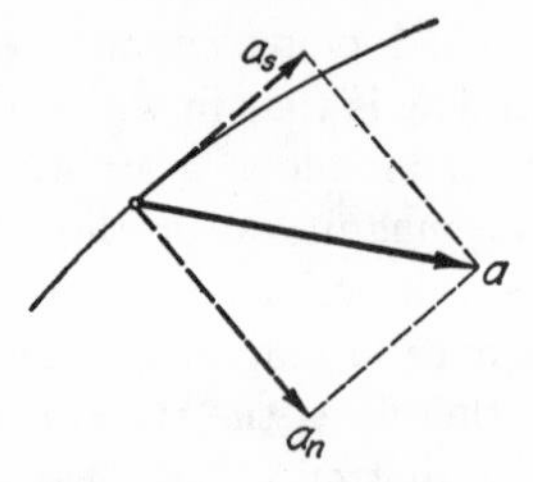

FIG. 19. Tangential and normal components of acceleration.

the acceleration of a fluid element in a direction *tangential* to the stream line will be

$$a_s = \frac{dv_s}{dt} \tag{14}$$

Like velocity, acceleration is a vector quantity. Unlike the velocity vector, however, the acceleration vector has no specific orientation with respect to the stream line. In other words, the vector of acceleration generally has components both tangential and normal to the stream line (see Fig. 19), the *tangential* component a_s embodying the change in the *magnitude* of the velocity, and the *normal* component a_n reflecting the change in *direction:*

$$a_n = \frac{dv_n}{dt} \tag{15}$$

Evidently, only if the stream-line spacing is constant will the tangential component of acceleration be equal to zero; likewise, the normal component of acceleration will be equal to zero only if the stream lines are straight.

Evaluation of tangential and normal components. Consider as a general case, therefore, the stream lines shown in Fig. 20, which are neither straight nor equidistant. During the time dt a fluid element will move the distance ds along the central stream line, its velocity thereby changing slightly in both magnitude and direction. The differential change in magnitude dv_s depends solely upon the change in stream-line spacing from point to point in the s direction and will evidently be equal to the rate of change with distance dv_s/ds times the distance traversed:

$$dv_s = \frac{dv_s}{ds}\,ds$$

Since, from Eq. (13), $ds = v_s\,dt$, it follows that

$$dv_s = \frac{dv_s}{ds}\,v_s\,dt$$

and hence

$$\frac{dv_s}{dt} = v_s\,\frac{dv_s}{ds}$$

which, according to Eq. (14), expresses the tangential component of acceleration.

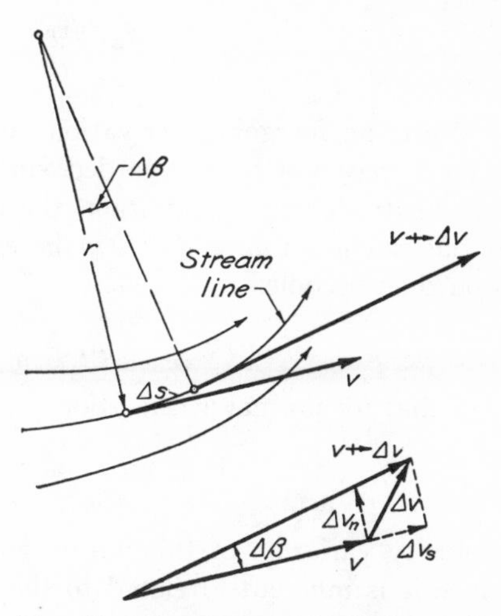

FIG. 20. Components of velocity variation along a stream line.

The velocity vector, obviously, can have no component toward the center of curvature at any time. At the end of the distance ds, nevertheless, the slight change in direction yields the small component dv_n in the *initial* direction n, as shown in Fig. 20. Denoting by dv_n/ds the rate of change of this component with respect to distance traveled in the s direction, it will be seen that the increment dv_n may be expressed by the product of this rate of change and the displacement ds:

$$dv_n = \frac{dv_n}{ds}\,ds$$

Again introducing the relationship $ds = v_s\,dt$ and rearranging terms,

$$\frac{dv_n}{dt} = v_s\,\frac{dv_n}{ds}$$

However, from the similarity of triangles indicated approximately in Fig. 20,

$$\frac{dv_n}{ds} = \frac{v_s}{r}$$

Upon substitution of this value, the normal component of acceleration becomes simply the ratio of v_s^2 to the local radius of curvature r of the stream line:

$$\frac{dv_n}{dt} = \frac{v_s^2}{r}$$

With the foregoing derivation accomplished, the subscript of the term v_s may now be disregarded, since at any point v_s and v are identical. Introducing, in addition, the elementary relationship of differential calculus $v\,dv = d(v^2/2)$, the expression for tangential acceleration then becomes

$$a_s = \frac{1}{2}\frac{d(v^2)}{ds} \tag{16}$$

and that for normal acceleration

$$a_n = \frac{v^2}{r} \tag{17}$$

Velocity variation of this nature is known as *convective* acceleration, since it is inherently related to the convection, or translation, of the fluid through space.

Example 5. The velocity distribution around the front part of a sphere held in the wind closely follows the expression for irrotational flow

$$v = \tfrac{3}{2}\,v_0 \sin\beta$$

in which v_0 is the velocity of the approaching wind and β the angle between the wind direction and the radius to the point in question. Determine the velocity and the acceleration of the air at the 60° point on a sphere 3 feet in diameter for a wind velocity of 50 feet per second.

From direct substitution it is evident that

$$v = \tfrac{3}{2} \times 50 \times 0.866 = 64.9 \text{ fps}$$

the velocity being tangent to the circumference of the longitudinal section shown in the illustration. According to the convergence of the stream lines in this zone, acceleration in the same direction must be expected to occur. Thus, from Eq. (16) the tangential acceleration may be written

$$a_s = \frac{1}{2}\frac{d(v^2)}{ds} = \frac{1}{2}\left(\frac{3}{2}v_0\right)^2 \frac{d(\sin^2\beta)}{ds} = \frac{9v_0^2}{8}\,2\sin\beta\cos\beta\,\frac{d\beta}{ds}$$

or, since $ds = r\,d\beta$,

$$a_s = \frac{9}{8}\frac{v_0{}^2}{r}\,2\sin\beta\cos\beta$$

Introducing the given values of v_0, r and β,

$$a_s = \frac{9}{8} \times \frac{\overline{50^2}}{1.5} \times 2 \times 0.866 \times 0.5 = 1625 \text{ fps}^2$$

In the normal direction, from Eq. (17) it follows at once that

$$a_n = \frac{v^2}{r} = \frac{\overline{64.9^2}}{1.5} = 2810 \text{ fps}^2$$

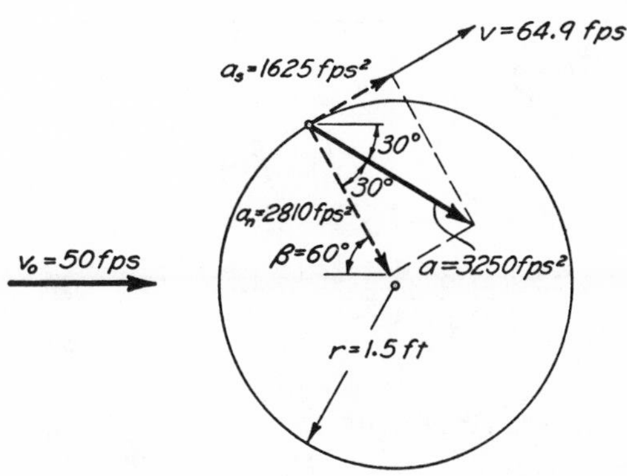

The acceleration vector will therefore have the magnitude

$$a = \sqrt{a_s{}^2 + a_n{}^2} = \sqrt{\overline{1625^2} + \overline{2810^2}} = 3250 \text{ fps}^2$$

or about 100 times the acceleration of gravity. The angle between the vector and the 60° radius will be

$$\theta = \tan^{-1}\frac{1625}{2810} = \tan^{-1} 0.577 = 30°$$

and the vector is hence inclined at an angle of 60° − 30° = 30° to the wind direction, as indicated in the figure.

Example 6. What will be the maximum values of the tangential and the normal acceleration along the boundary of Fig. 12 if $v_0 = 10$ feet per second and $n_0 = 12$ feet?

Since $a_s = \frac{1}{2}d(v^2)/ds$, the maximum value of a_s will occur at the point of steepest slope of the curve of $(v/v_0)^2$ plotted against $s/\Delta s_0$. From the squares of the velocity data of Example 4 the accompanying curve was prepared, from which it is found that, at point A–8 (or G–8),

$$\left(\frac{d(v/v_0)^2}{d(s/\Delta s_0)}\right)_{\max} = \frac{\Delta s_0}{v_0{}^2}\left(\frac{d(v^2)}{ds}\right)_{\max} = 3.75$$

Therefore

$$(a_s)_{\max} = \frac{1}{2}\left(\frac{d(v^2)}{ds}\right)_{\max} = \frac{3.75}{2}\frac{v_0^2}{\Delta s_0} = \frac{3.75}{2} \times \frac{\overline{10}^2}{2} = 93.6 \text{ fps}^2$$

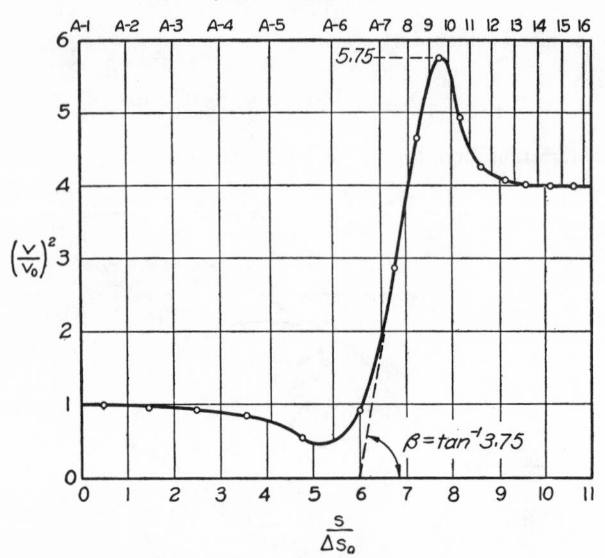

The greatest value of a_n will evidently occur where the ratio v^2/r is a maximum. The greatest magnitude of $(v/v_0)^2$—5.75—is seen to occur between points A–9 and A–10 (or G–9 and G–10), which lie on the boundary curve having the minimum radius of curvature $r = 0.35n_0$; hence

$$(a_n)_{\max} = \left(\frac{v^2}{r}\right)_{\max} = \frac{5.75v_0^2}{0.35n_0} = \frac{5.75 \times \overline{10}^2}{0.35 \times 12} = 137 \text{ fps}^2$$

PROBLEMS

26. A nozzle is so shaped that the velocity of flow along the centerline changes linearly from 5 to 50 feet per second in a distance of 15 inches. Determine the magnitude of the convective acceleration at the beginning and end of this distance.

27. A centrifuge rotates a container of liquid at the constant rate of 1800 revolutions per minute. Compare with the acceleration of gravity the liquid acceleration at a radial distance of 4 inches from the axis of rotation.

28. At what rate is the velocity of the air changing just as it reaches the outlet of the diffusor of Problem 15?

29. The crest of a spillway has a radius of curvature of 5 feet at its highest point. What local velocity v will produce at this point a normal acceleration equal to the acceleration of gravity?

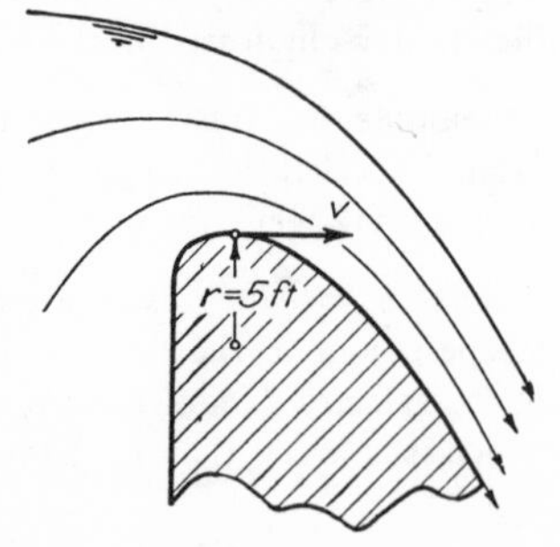

PROB. 29.

30. If $v = 15$ feet per second at the topmost point of the spillway section of Problem 29, and if the water at this point is changing speed at the rate of ½ foot per second per inch of travel, determine and show in a sketch the magnitude and direction of the vector of convective acceleration.

31. The bucket of a spillway has a radius of curvature of 20 feet. When the spillway is discharging 40 cubic feet of water per second per foot length of crest, the average thickness of the sheet of water over the bucket is 15 inches. Compare the resulting centripetal acceleration with the acceleration of gravity.

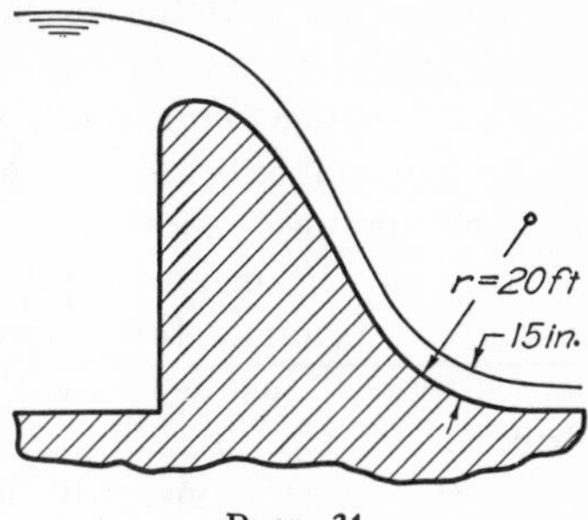

PROB. 31.

32. Water in a reservoir approaches the intake of a pump approximately in accordance with the equation $v = C/R^2$, in which R is the radial distance from the intake along any stream line. If $v = 1.5$ feet per second when $R = 4$ feet, determine the acceleration of the water at distances of 4 feet and 2 feet from the intake.

33. Determine the magnitude of the acceleration at a radial distance of 2 feet upstream from the slot of Problem 16.

34. What will be the maximum difference in normal acceleration between the inside and the outside boundaries of Fig. 13 if $v_0 = 10$ feet per second and $n_0 = 4$ feet?

8. ACCELERATION IN UNSTEADY FLOW

Local, convective, and total acceleration. The boundary profile shown in Fig. 21 might be considered to represent, schematically, either the nozzle at the end of a garden hose or the hydraulic giant used in placer mining, in either of which the rate of discharge and the direction of the jet may be changed at will. Since the stream lines converge through the nozzle, it is evident that the fluid must accelerate tangentially, in accordance with Eq. (16); and, since all stream lines except that at the axis are curved, it is evident that the fluid must also accelerate normally, in accordance with Eq. (17). Such acceleration is purely convective, for it is due entirely to the movement of the fluid from one point to another in a nonuniform zone. In the uniform zones to the left and right of the nozzle, on the other hand, convective acceleration does not occur.

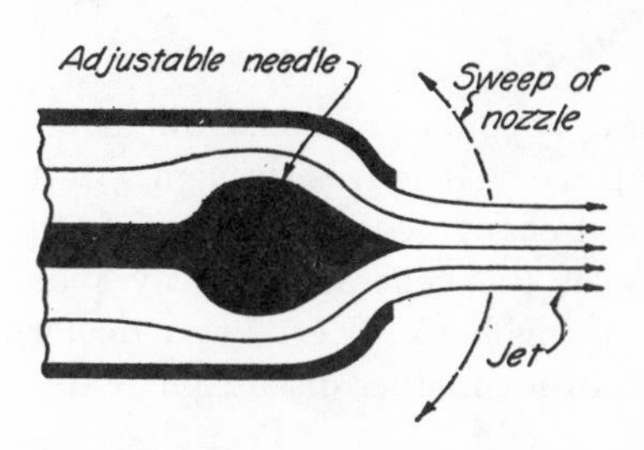

FIG. 21. Flow in which magnitude and direction of velocity vary with time and space.

Assume, now, that the rate of flow through the nozzle is rapidly increased. It is obvious that even in the zones of uniform flow the velocity must increase accordingly, indicating that a type of tangential acceleration not included in the term $\frac{1}{2}d(v^2)/ds$ must be taking place. Assume further that the nozzle is rapidly changed in inclination. It is then obvious that even in the zones of uniform flow the velocity must also change in direction, which indicates a type of normal acceleration not embodied in the term v^2/r. As a matter of fact, such acceleration is purely *local*, as distinct from convective, for it involves variation in the magnitude and direction of the velocity with time at a given locality. Local acceleration is thus the result of unsteadiness of motion, just as convective acceleration is the result of non-uniformity.

For the case of unsteady, non-uniform flow, it therefore appears that Eqs. (16) and (17) are incomplete unless terms describing the local acceleration are introduced. So far as the magnitude of the velocity is concerned, its local rate of change may be written in terms of the tangential component simply as $\partial v_s/\partial t$, the partial derivative indicating that only time, and not displacement, is taken into account. With respect to direction, likewise, the local rate of change may be expressed in terms of the normal component as $\partial v_n/\partial t$. Since the *total* acceleration $a = dv/dt$ must be the sum of the local and convective terms (each of which now has to be written as a partial derivative), it follows that in the tangential and the normal directions

$$a_s = \frac{dv_s}{dt} = \frac{\partial v_s}{\partial t} + \frac{1}{2}\frac{\partial (v^2)}{\partial s} \tag{18}$$

$$a_n = \frac{dv_n}{dt} = \frac{\partial v_n}{\partial t} + \frac{v^2}{r} \tag{19}$$

If such a state of motion is truly general, of course, not only will deformation and rotation of the fluid take place as it moves through space, but also the pattern of stream lines will change continuously in form with time. The obvious complexity of the general case leaves its detailed study well beyond the scope of this book. Certain particular phases of the general case, however, warrant further discussion at this point.

Unsteady flow past fixed boundaries. If unsteady flow between fixed boundaries is also irrotational, it may be fully described by means of the flow net or its three-dimensional counterpart. As in steady flow, the velocity at any instant will vary inversely with the streamline spacing. On the other hand, although the continuity principle is thus generally applicable, the rate of flow can no longer be regarded as constant with time. Consider, for example, the problem of varying discharge through the boundary contraction shown in Fig. 12. The same flow net will indicate the same relative velocity distribution at successive instants, since it is governed in form by the boundary geometry alone; but a continuous increase of velocity at all points requires a continuous increase in the quantity $\Delta q = v_0 \, \Delta n_0 = v \, \Delta n$. In other words, if the magnitude of either q or v_0 and the rate of change $dq/dt = n_0 \, \partial v_0/\partial t$ are known at any instant, the velocity and the acceleration at any point may at once be determined by the methods already discussed.

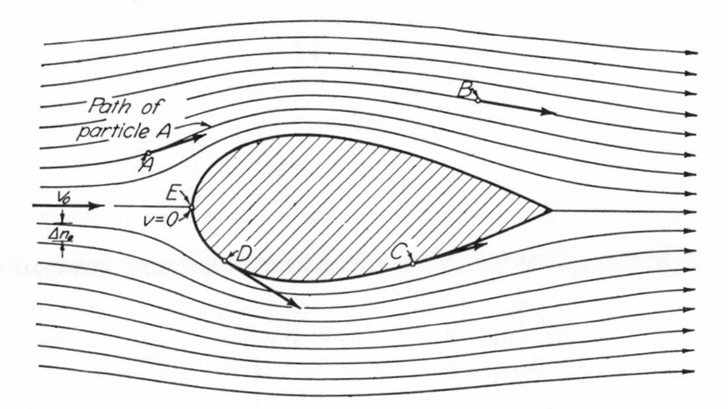

FIG. 22. Pattern of steady flow around a stationary streamlined strut.

A parallel instance is irrotational flow around stationary bodies, such as the streamlined strut shown in Fig. 22. If the flow is steady, the quantity $\Delta q = v_0 \, \Delta n_0$ characterizing the flow net will again remain constant. If the velocity of the approaching fluid increases, however, this quantity will change with time; but, if its value and its rate of change are known at any instant, the corresponding velocity and acceleration may then also be determined at any point in the flow.

Relative motion of boundary and fluid. Somewhat different from these examples of flow past stationary boundaries is the unsteady motion produced by the passage of a body through a fluid. If the fluid is initially at rest, its displacement by the moving body will yield a flow pattern which at any fixed point changes continuously with time. For instance, the movement of the streamlined strut will produce at any instant a series of stream lines which diverge from the nose of the body and close in at the rear, as shown in Fig. 23, the velocity as usual varying inversely with the stream-line spacing. As the body advances through the fluid, however, this system of stream lines must necessarily advance at the same speed, which is greater, evidently, than the velocity of the fluid everywhere except at the very nose of the body. Any fluid particle, therefore, can follow one such stream line for only a brief instant, and its complete path (refer to the illustration) will

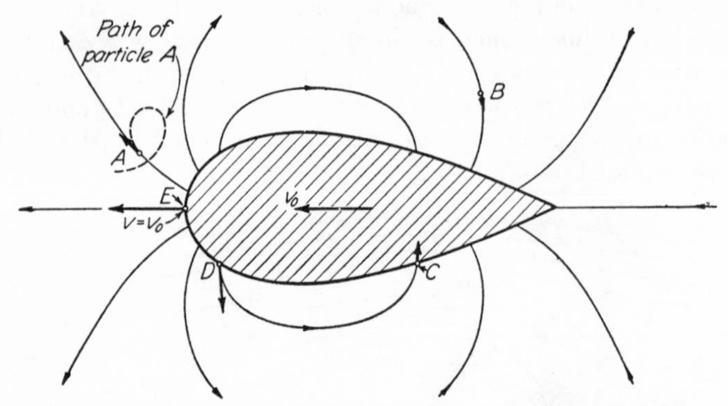

FIG. 23. Pattern of unsteady flow around a moving streamlined strut.

consist of a series of increments of successive stream lines. In this particular type of unsteady motion, and in the general case as well, the stream lines and the actual paths of the particles are therefore not identical; evidently, moreover, the outline of the body is not a stream line. Indeed, the boundary will coincide with a stream line, and the stream lines will represent path lines, only if the boundary remains fixed with respect to the observer.

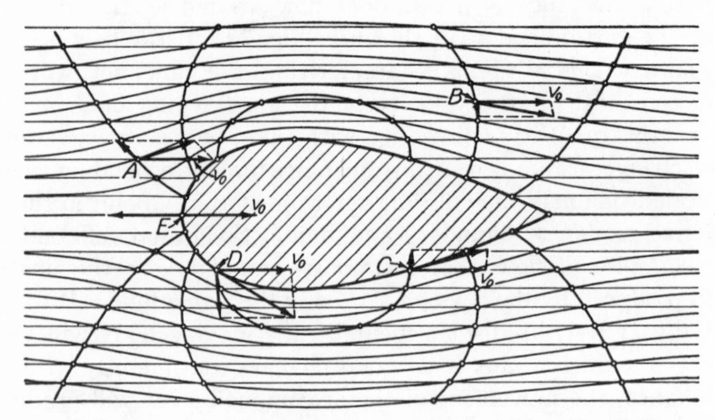

FIG. 24. Interrelationship of the flow patterns of Figs. 22 and 23.

The type of unsteady flow produced by a body moving at constant speed is of particular importance in that the pattern may be transformed into one of steady flow by means of a very useful procedure.

If, instead of watching such a body go past the point of observation, the observer were to move with the body, not only the body itself but also the resulting pattern of stream lines would appear to him to

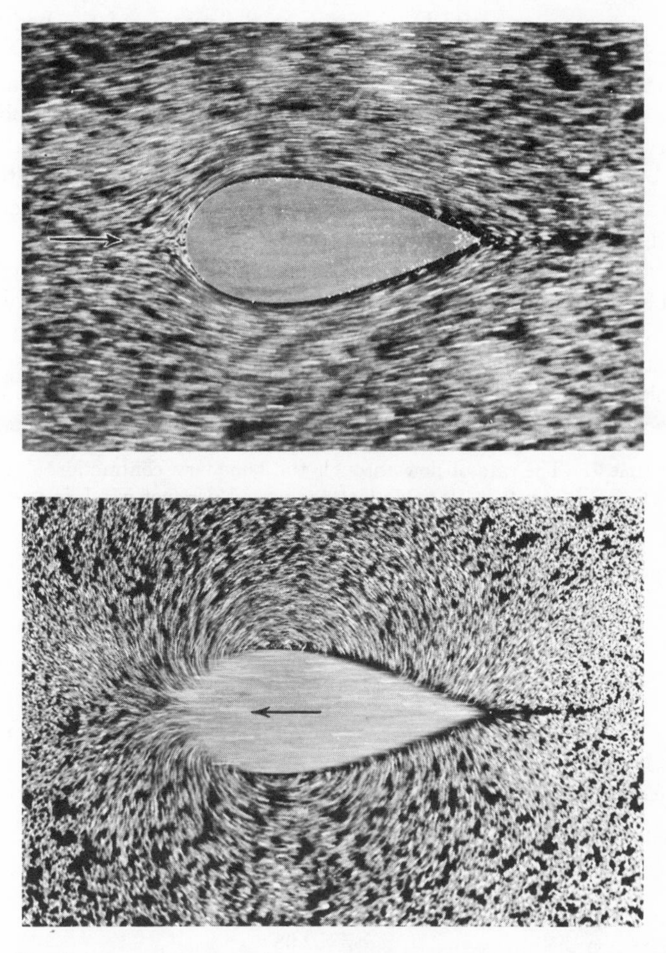

PLATE IV. Patterns of steady and unsteady flow made visible by reflection of light from flakes of aluminum scattered on water surface. Above, both camera and strut are mounted on towing carriage; below, camera is held stationary while strut is towed.

remain stationary. In other words, a man may stand by the roadside and watch cars pass, each of which produces unsteady motion of the dust-laden air; but to the occupants of any car the car appears to be at rest, and the air, and the ground as well, to be moving steadily in the opposite direction. The phenomenon is purely one of relative motion, and Fig. 23 may therefore be transformed as shown in Fig. 24 by adding to the velocity vectors of the unsteady flow a series of velocity vectors opposite in direction to the motion of the body but equal in magnitude to its speed.

Since stream lines represent both the direction and the magnitude of the velocity at all points, such vector addition may be accomplished simply by combining graphically the stream lines for the two velocity distributions. Thus the heavier curves in Fig. 24 correspond to the unsteady flow, and the parallel lines indicate the constant velocity added at every point by translation of the point of observation. The lighter curves drawn through the intersections of these two systems will be seen to yield the pattern of steady flow for this type of body. The significance of this method of transformation from an unsteady to a steady pattern of motion is of more than passing interest, since it completely eliminates from Eqs. (18) and (19) the troublesome terms for local acceleration.

Example 7. The rate of flow through the boundary contraction of Fig. 12 varies linearly from 0 to 50 cubic feet per second per foot width in 4 seconds. Determine the local accelerations at points A–1 and A–9. (Assume $n_0 = 3$ feet.)

From the given data, q may be expressed in terms of time as

$$q = \frac{50}{4} t$$

whence, for each neighboring pair of stream lines,

$$\Delta q = \frac{q}{6} = \frac{50}{6 \times 4} t = 2.08t$$

Therefore, the velocity and local acceleration at any point may be written in terms of time and stream-line spacing:

$$v_s = v = \frac{\Delta q}{\Delta n} = \frac{2.08t}{\Delta n}$$

whence

$$\frac{\partial v_s}{\partial t} = \frac{2.08}{\Delta n}$$

At A–1: $\Delta n = \Delta n_0 = 0.5$ ft and $\dfrac{\partial v_s}{\partial t} = \dfrac{2.08}{0.5} = 4.16$ fps^2

At A–9: $\Delta n = 0.44 \Delta n_0 = 0.22$ ft and $\dfrac{\partial v_s}{\partial t} = \dfrac{2.08}{0.22} = 9.45$ fps^2

Since the stream lines do not change in form or orientation with time, $\partial v_n / \partial t = 0$ at all points.

PROBLEMS

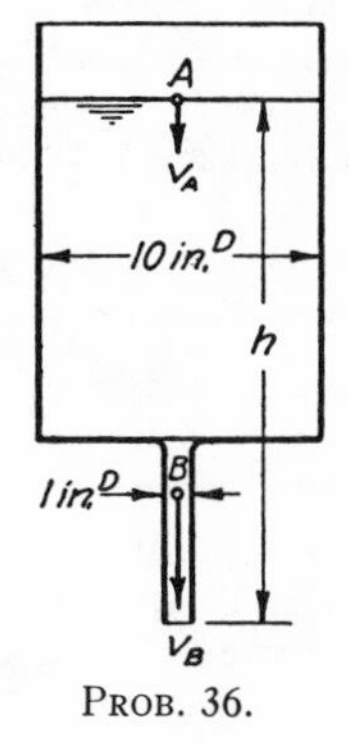

PROB. 36.

35. The rate of flow through a tapered pipe varies linearly from 0 to 5 cubic feet per second in 15 seconds. Determine the local accelerations at normal sections 12, 9, and 6 inches in diameter.

36. The rate at which a tank may be emptied through a bottom outlet is very nearly proportional to the square root of the surface elevation above the outlet (i.e., $Q = C\sqrt{h}$). If, for the tank and outlet dimensions shown, $Q = 4.5$ cubic feet per minute at the instant that $h = 3$ feet, determine the local accelerations at points A and B. (Note that $-dh/dt = v_A$.)

37. The velocity of fluid directly in the path of a moving body may be represented by the equation $v = v_b b/2x$, in which b is a characteristic dimension of the body and v_b is its speed. If the body starts from rest with an acceleration of 3 feet per second per second, what is the fluid acceleration at the point $x = b = 5$ feet: (a) when $t = 0$, and (b) when $t = 2$ seconds?

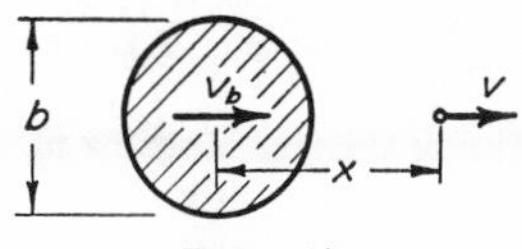

PROB. 37.

38. Assuming that the body of Problem 37 has just attained a speed of 10 feet per second, plot to scale the instantaneous velocity variation along the line of motion (a) relative to the fluid and (b) relative to the body.

39. Let the relative velocity between fluid and profile in Figs. 22–24 be 25 feet per second. Determine the corresponding values of the maximum and minimum fluid velocities at the boundary for both the steady and the unsteady patterns of flow.

QUESTIONS FOR CLASS DISCUSSION

1. Why are Eqs. (10) and (11) valid only if the areas of successive sections are bounded by identical stream lines?
2. What type of velocity distribution invariably characterizes uniform, irrotational flow?
3. Show that a fluid element passing the boundary curve in Fig. 15 displays irrotational characteristics.
4. The flame of a candle may be extinguished far more easily by exhaling than by inhaling; why? In which instance is the motion more nearly irrotational?
5. Why is the nose of the strut shown in Fig. 22 a point of stagnation?
6. How can the flow net be used to design an efficient boundary transition?
7. When is flow in a pipe (a) unsteady, and (b) non-uniform?
8. A common lawn sprinkler consists of a jet at either end of an arm which rotates about a vertical axis. Is the flow from such a sprinkler (a) uniform or non-uniform, and (b) steady or unsteady?
9. Does the surf on a beach represent (a) steady or unsteady flow, and (b) uniform or non-uniform flow? Give reasons for answers.
10. Define and cite examples of tangential and normal acceleration.
11. A wave travels through a long channel of otherwise still water. Is the free surface a stream line? Why cannot Eq. (11) be used conveniently to determine the mean velocity of the water under the wave crest?

12. How may the wave motion of the foregoing question be transformed to steady flow? Could Eq. (11) then be more conveniently used? Can the same transformation be performed in the case of the surf of Question 9?

13. Cite further examples of unsteady flow (*a*) which can, and (*b*) which cannot, be transformed to steady flow by the principle of relative motion.

14. Distinguish between local and convective acceleration. Show that the motion of the air around a propeller illustrates both types. Suggest means by which such flow might be transformed to eliminate variation with time.

SELECTED REFERENCES

DURAND, W. F. "Fluid Mechanics, Part I." *Aerodynamic Theory*, Vol. I, Springer, 1934.

PRANDTL-TIETJENS. *Fundamentals of Hydro- and Aerodynamics*. Engineering Societies Monograph, McGraw-Hill, 1934.

ROUSE, H. *Fluid Mechanics for Hydraulic Engineers*. Engineering Societies Monograph, McGraw-Hill, 1938.

CHAPTER III

PRESSURE VARIATION IN ACCELERATED FLOW

9. PRINCIPLE OF MASS ACCELERATION

Newton's laws of motion. In the foregoing chapter on fluid *kinematics,* attention was paid to the variation of the velocity of flow with time and distance, but no mention was made of the forces necessary to produce such acceleration. The latter phase of the problem is one of *dynamics,* rather than kinematics, and involves the principle of *mass acceleration* formulated in the seventeenth century by Sir Isaac Newton. Newton's three laws of motion may be restated briefly as follows:

1. A body will remain at rest or in a given state of motion until acted upon by an external force.
2. The acceleration of a body takes place in the direction of the force which produces it; it is directly proportional to the magnitude of the force and inversely proportional to the mass of the body.
3. Every action is accompanied by an equal and opposite reaction.

The first law describes a characteristic of matter known as *inertia;* the second law, in the light of the third, states that the *accelerative action* of a force is countered by an *inertial reaction* of the matter acted upon.

If, as has become universally customary, the proportionality constant of the second law is considered to be absorbed both numerically and dimensionally by one or another of the quantities involved, this law may be expressed mathematically by the familiar equation

$$F = Ma \tag{20}$$

In American engineering units, force is measured in *pounds,* and mass in *slugs.* Equation (20) thus states that a force of 1 pound will cause a mass of 1 slug to accelerate 1 foot per second per second. Mass is, of course, a *scalar,* since it is purely a quantitative measure of matter and does not possess direction. Force, however, like acceleration, is a true vector, for it possesses magnitude and direction as well. According to Newton's second law, both the force and the acceleration of Eq. (20) must invariably have the same direction.

Introduction of fluid density. In the case of a solid body, the motion to which Eq. (20) refers is that of its mass center, since the relative positions of all particles of matter within the body are considered to remain unchanged. Fluid particles, on the other hand, are generally engaged in a continuous variation of relative position and form, so that the motion of the particle at the mass center of any fluid body seldom represents the mean motion of the fluid as a whole. It therefore becomes necessary to investigate the forces producing motion at each and every point of a given flow. For this reason, one is not interested in the total mass of a moving fluid so much as in the mass per unit volume at any point of observation. The mass per unit volume is known as the *mass density* of the fluid and is given the symbol ρ (the Greek letter rho). Although the density of any fluid varies to some extent with conditions of flow, such variation is frequently quite inappreciable even in gases. For the present, therefore, only flows with essentially constant density will be considered.

If, now, the mass density is used instead of the mass itself in Eq. (20), the force term must also be referred to the unit fluid volume. Designating by f the vector magnitude of the corresponding force per unit volume, the Newtonian relationship for fluid acceleration thus becomes

$$f = \rho a \tag{21}$$

Example 8. In Example 5 it was found that the air passing around a 3-foot sphere held in a 50-foot-per-second wind would attain an acceleration of 3250 feet per second per second at a point 60° around the circumference. If the density of the air were 0.0025 slug per cubic foot, what force per unit volume would be acting at this point?

From Eq. (21)

$$f = \rho a = 0.0025 \times 3250 = 8.13 \text{ lb/ft}^3$$

PROBLEMS

40. What force would have to be exerted by a piston in an 8-inch pipe to accelerate a 50-foot column of water 3 feet per second per second? (Assume $\rho = 1.94$ slugs per cubic foot.)

41. A drum of oil is rotated about a vertical axis at the constant rate of 240 revolutions per minute. If the oil has a density of 1.7 slugs per cubic foot, evaluate the accelerative force per unit volume at radial distances of 0, ½, and 1 foot from the axis.

42. A water nozzle produces a linear velocity change from 10 to 35 feet per second in a distance of 18 inches. What force per unit volume must act upon the water at the beginning and at the end of the 18-inch distance, if the density is 1.94 slugs per cubic foot?

43. Vanes having a 3-inch radius of curvature are placed in a ventilation duct to guide the air around a corner (see Fig. 135). If the mean air velocity is 35 feet per second, estimate the maximum accelerative force per unit volume exerted by the vanes (let $\rho = 0.0025$ slug per cubic foot).

10. ACCELERATION DUE TO PRESSURE GRADIENT

Intensity of fluid pressure. Fluid *pressure* is defined as the normal force exerted by fluid matter upon any surface—normal, that is, in the sense that the vector for the pressure upon a differentially small portion of even a curved surface must be at right angles to the surface in that locality. The ratio of such a differential force to the differential area over which it acts is known as the *intensity* of pressure p in the given locality; evidently, since the area is of differential magnitude, p represents, in effect, the intensity of pressure at a point on the surface. Such a surface may, of course, be either a solid boundary or an imaginary plane passed through a fluid for purposes of analysis. Under the latter circumstances, p will represent the intensity of pressure at a point within the fluid. For the condition that no tangential stress (shear) exists in the same region, the pressure intensity at any point will necessarily be independent of the orientation of the surface under consideration.

Strictly speaking, the magnitude of the pressure intensity within a fluid should be expressed in terms of pounds per square foot (or per square inch) above absolute zero. It is generally more convenient, however, to use the prevailing intensity of *atmospheric* pressure as a reference, the relative intensity p then representing the difference between the absolute intensity p_{abs} and the atmospheric intensity p_{at}:

$$p = p_{abs} - p_{at}$$

Under normal conditions, $p_{at} = 14.7$ pounds per square inch absolute (psia).

Since the quantity p represents simply a local intensity of stress, it must be regarded as a scalar rather than a vector. Lacking, as it does, all directional characteristics, the pressure intensity within a fluid—regardless of how great its magnitude may be—is therefore no measure of the accelerative forces which may exist. In other words, just as a solid body will change its velocity only if a greater stress is exerted on one side than on the other, fluid pressure can produce acceleration only if the intensity p decreases in some direction through the fluid.

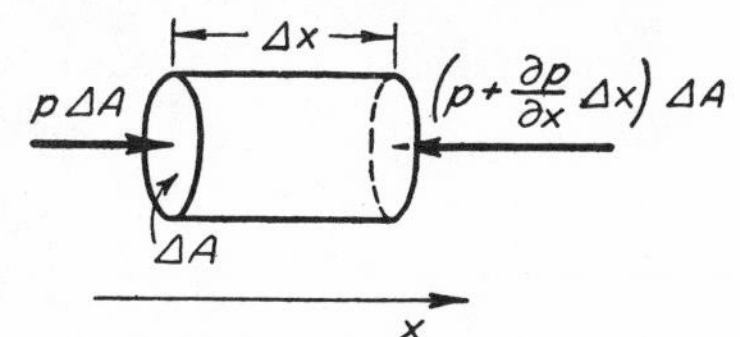

FIG. 25. Pressures on opposite faces of a fluid element.

Force produced by pressure variation from point to point. To determine the magnitude of such an accelerative force in any direction x, consider a small element of fluid (like that in Fig. 25) having the form of a right circular cylinder with faces normal to the x axis. If the rate of change, or *gradient*, of pressure intensity in the x direction is represented by the quantity $\partial p/\partial x$, and if the force upon the left face of the cylinder is $p\,dA$, then the force on the other end, acting in the negative x direction, will be $[p + (\partial p/\partial x)\,dx]\,dA$. The difference between these quantities is the x component of force due to pressure variation with distance:

$$dF_x = p\,dA - \left(p + \frac{\partial p}{\partial x}\,dx\right) dA = -\frac{\partial p}{\partial x}\,dx\,dA$$

Dividing by the infinitesimal volume of the cylinder $dx\,dA$ and letting $dF_x/(dx\,dA) = f_x$, it will be seen that

$$f_x = -\frac{\partial p}{\partial x} \tag{22}$$

Evidently, at any point within a fluid the accelerative force per unit volume in a given direction is equal to the rate of decrease of pressure intensity (i.e., the negative pressure gradient) in that direction.

Equations of acceleration. From the Newtonian laws of motion it follows that the product of the mass per unit volume and the component of acceleration at any point within a moving fluid must be equal to the corresponding component of force per unit volume acting at that point; thus, in any direction x,

$$f_x = -\frac{\partial p}{\partial x} = \rho a_x$$

Expressions for the tangential and normal components of acceleration are already at hand in Eqs. (18) and (19), while the components of force per unit volume in these directions may be expressed simply through substitution of s and n for x in the foregoing equation. By combining these expressions, one obtains the following significant relationships between the variation of pressure intensity with distance, the mass density, and the variation of the velocity with distance and time:

$$-\frac{\partial p}{\partial s} = \rho\frac{\partial v_s}{\partial t} + \frac{\rho}{2}\frac{\partial (v^2)}{\partial s} \tag{23}$$

$$-\frac{\partial p}{\partial n} = \rho\frac{\partial v_n}{\partial t} + \frac{\rho v^2}{r} \tag{24}$$

Two important facts should now be quite apparent. On the one hand, it will be seen that the pressure intensity may vary from instant to instant without producing acceleration, since the accelerative force per unit volume does not depend upon the derivative of p with respect to time. On the other hand, variation of p from point to point must be accompanied by either local or convective acceleration of the fluid; obviously, therefore, in either unsteady or non-uniform flow a pressure gradient must exist.

Integration of the acceleration equation along a stream line. As has already been emphasized, the general problem of unsteady flow is of too complex a nature for more than brief recognition in an elementary course. Nevertheless, in many cases of unsteady motion the flow pattern can be transformed into one of steady motion by translating the point of observation. As in truly steady flow, elimination of the troublesome terms $\partial v_s/\partial t$ and $\partial v_n/\partial t$ will then reduce the foregoing equations of acceleration to relatively simple forms. Equation (23), for instance, becomes for steady flow

$$-\frac{\partial p}{\partial s} = \frac{\rho}{2}\frac{\partial(v^2)}{\partial s} \tag{25}$$

which indicates that the pressure intensity will vary along a stream line in proportion to the *negative* variation of the square of the velocity. Integration of this relationship between successive points on the same stream line then yields the equality

$$p_1 - p_2 = \frac{\rho}{2}(v_2^2 - v_1^2) \tag{26}$$

Evidently, it makes no difference how great the absolute magnitude of the pressure intensity may be within the fluid, for it is simply the difference in p at the two points which must correspond to the product of $\rho/2$ and the negative difference in v^2. It is for this reason that the pressure intensity may change with time without producing acceleration, for it is necessary only that it change at the same rate at both points. Equation (26) therefore permits the evaluation of the variation in pressure intensity along any stream line in steady flow once the density and the variation in velocity along that stream line are known.

Example 9. A nozzle is so shaped that the velocity of flow along the centerline varies linearly from 5 to 50 feet per second in a distance of 15 inches. (*a*) Assuming the fluid discharged to have a density of 1.94 slugs per cubic foot, determine the change in pressure intensity corresponding to the velocity change of 45 feet per second. (*b*) What are the magnitudes of the pressure gradients at the beginning and end of the 15-inch distance?

(*a*) From Eq. (26),

$$p_1 - p_2 = \frac{\rho}{2}(v_2^2 - v_1^2) = \frac{1.94}{2}(50^2 - 5^2) = 2400 \text{ psf}$$

(*b*) Writing the velocity as a function of distance along the centerline,

$$v = 5 + \frac{45}{1.25} s = 5 + 36s$$

whence, from Eq. (25),

$$\frac{\partial p}{\partial s} = -\frac{\rho}{2}\frac{\partial(v^2)}{\partial s} = -\rho v\frac{\partial v}{\partial s} = -1.94(5 + 36s)36$$

Where $s = 0$ ft,

$$\frac{\partial p}{\partial s} = -1.94(5 + 0)36 = -349 \text{ psf/ft}$$

Where $s = 1.25$ ft,

$$\frac{\partial p}{\partial s} = -1.94(5 + 36 \times 1.25)36 = -3490 \text{ psf/ft}$$

Evidently, the pressure intensity not only decreases (as indicated by the negative sign) in the direction of flow, but decreases more and more rapidly from point to point.

PROBLEMS

44. The rate at which water flows through a horizontal 10-inch pipe is increased linearly from 1 to 5 cubic feet per second in 3 seconds. What pressure gradient must exist to produce this acceleration? What difference in pressure intensity will prevail at sections 25 feet apart? (Express p in pounds per square foot.)

45. If a tank car 30 feet in length is completely filled with oil ($\rho = 1.6$ slugs per cubic foot), what average difference in pressure intensity in pounds per square inch must exist between the two ends as the car is given a horizontal acceleration of 2 feet per second per second?

46. A slender tube which is 18 inches long and closed at one end is half full of mercury ($\rho = 26.4$ slugs per cubic foot). If the tube is rotated about the open end in a horizontal plane at the constant rate of 180 revolutions per minute, what pressure intensity in pounds per square inch will prevail at the closed end? (Note that the axis of the tube is not a stream line.)

47. Mercury completely fills a closed tube 2 feet long. If the tube is rotated at a constant speed of 90 revolutions per minute about a vertical axis 6 inches in from end A, and if the pressure intensity of the mercury at end A is then 5 pounds per square inch, determine the intensity at end B. (Note that the two ends do not lie on the same stream line.)

48. Just upstream from the test section of a wind tunnel the centerline velocity is changed from 50 to 250 miles per hour by convergence of the tunnel walls. Assuming an air density of 0.0025 slug per cubic foot, what will be the accompanying change in pressure intensity?

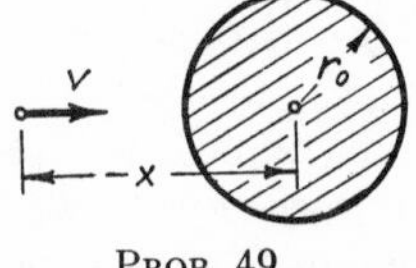

PROB. 49.

49. Fluid approaching an immersed cylinder with axis at right angles to the flow decreases in velocity according to the expression $v = v_0(1 - r_0^2/x^2)$, in which r_0 is the radius of the cylinder and $-x$ is the distance upstream from the axis. What intensities of pressure in pounds per square foot will exist at points 5, 10, and 20 feet ahead of the axis of a chimney 10 feet in diameter during a 50-mile-per-hour wind?

50. During the motion of a body through still water at a constant speed of 20 feet per second, the water velocity at a point 3 feet ahead of the nose of the body has a magnitude of 12 feet per second. Determine the difference in pressure intensity in pounds per square foot between the nose and the point 3 feet ahead.

11. PRESSURE DISTRIBUTION IN STEADY, IRROTATIONAL FLOW

Significance of the pressure equation. Although Eq. (26) is perfectly valid for all points on the same stream line, in the general case it is not possible to compare points on different stream lines by means of this relationship. In other words, adding and subtracting the term $\frac{1}{2}\rho\,\partial(v_s^2)/\partial n$, and recalling from Fig. 20 that $v/r = \partial v_n/\partial s$, it will be seen that for steady flow Eq. (24) may be written in the form

$$-\frac{\partial p}{\partial n} = \frac{\rho}{2}\frac{\partial(v^2)}{\partial n} + \rho v\left(\frac{\partial v_n}{\partial s} - \frac{\partial v_s}{\partial n}\right) \tag{27}$$

which obviously cannot be integrated to yield a simple expression like Eq. (26). It so happens, however, that the last quantity within parentheses represents the speed of rotation at a point within the fluid. As long as steady flow is rotational, therefore, Eq. (26) will apply only to points on the same stream line, and a difference in velocity between neighboring stream lines (as, for example, in Fig. 10) will not necessarily indicate a difference in pressure intensity. For conditions of irrotational motion, however, the local angular velocity $\partial v_n/\partial s - \partial v_s/\partial n$ will everywhere be equal to zero, and Eq. (27) may then be integrated to yield an expression which is valid between points on the same normal line and hence on different stream lines:

$$p_1 - p_2 = \frac{\rho}{2}(v_2^2 - v_1^2) \tag{28}$$

Since this expression for points along a normal line is identical with Eq. (26) for points along a stream line, it obviously applies to all points, regardless of location, in a region of steady, irrotational flow.

Under such conditions of motion it has already been shown that the velocity distribution for any boundary conditions may be determined by means of the flow net. It now follows that application of Eq. (28) will permit, in addition, the evaluation of the pressure distribution throughout the field of motion. Assume, for example, that the mass density of the fluid and the velocity and intensity of pressure at some point in the flow are known; the velocity at any other point may at once be determined from the stream-line configuration and the principle of continuity, and application of Eq. (28) will then yield the corresponding intensity of pressure.

Dimensionless representation of pressure distribution. Since the flow net represents a general solution of steady, irrotational motion for any boundary profile, it has been seen that the dimensionless ratio of

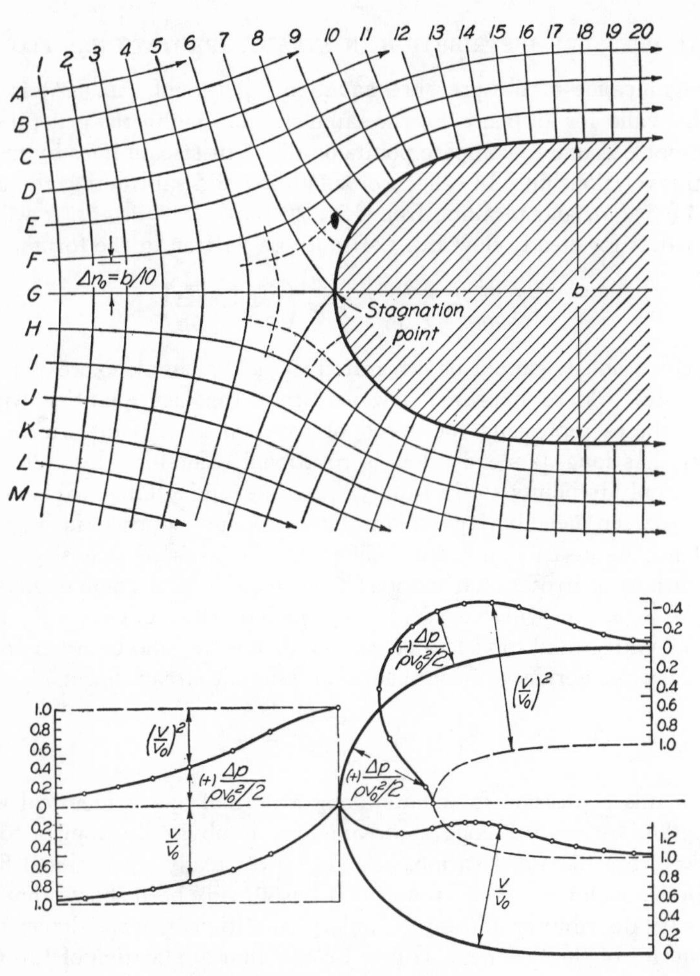

FIG. 26. Flow net and relative distribution of velocity and pressure intensity at the front of a pier or strut.

the velocity at any point to that in some reference zone will be independent of the rate of flow. By means of the following modification of Eq. (28), it will be found that the variation in pressure intensity as well may be studied conveniently in the same dimensionless manner

Thus, letting p_0 represent the pressure intensity in a zone of uniform motion where the velocity is v_0, and p the intensity corresponding to the velocity v at any other point,

$$p - p_0 = \frac{\rho}{2}(v_0^2 - v^2)$$

If one then indicates by Δp the difference $p - p_0$ and divides both sides by $\rho v_0^2/2$, the following dimensionless expression will be obtained:

$$\frac{\Delta p}{\rho v_0^2/2} = 1 - \left(\frac{v}{v_0}\right)^2 \tag{29}$$

The variation of the ratio v/v_0 is at once at hand from the configuration of the flow pattern (see, for instance, the plotted values in Example 4). Evidently, the corresponding variation of $\Delta p/(\rho v_0^2/2)$ may be found simply by subtracting $(v/v_0)^2$ from unity at every point. The resulting general solution is illustrated in Fig. 26 for the nose of a streamlined strut or pier, the plotted curves showing at a glance the relative variation in pressure intensity ahead of and around the boundary of the body. Owing to the dimensionless nature of all terms in Eq. (29), the plotted values will be seen to be independent of the size of the body and the absolute magnitudes of p_0, v_0, and ρ.

Measurement of pressure intensity and velocity. As is indicated by Fig. 26, the pressure intensity in the flow around such a body differs considerably from that of the originally uniform motion, tending to rise above this reference value in the immediate vicinity of the leading edge, where the velocity is reduced, and to drop below it where the velocity is increased. The maximum pressure intensity is reached at the very nose of the body, where the velocity is zero. This is aptly called a *point of stagnation* and is invariably characterized by the condition $\Delta p/(\rho v_0^2/2) = 1$; the accompanying rise in pressure intensity is evidently equal to $\rho v_0^2/2$, the *stagnation pressure* therefore having the magnitude

$$p_{st} = p_0 + \frac{\rho v_0^2}{2}$$

If a small hole, known as a *piezometer* orifice, is drilled through the flow boundary at an angle of 90°, and if the edges of the hole are carefully finished, a pressure gage connected with this opening will register the actual pressure intensity $p = p_0 + \Delta p$ of the flow at the location of the opening. The magnitude of Δp will evidently equal the difference between gage readings for the point in question and a point in the region of uniform flow. If the opening is located at a point of stagnation, moreover, the differential measurement will be equal to $\rho v_0^2/2$ from which the velocity of the approaching flow may be determined. Indeed, this principle is utilized in the *Pitot tube* (see Fig. 27), a small cylindrical instrument which produces a point of stagnation wherever it is introduced. The tube must be so proportioned as to disturb the general flow to a negligible degree, in order that the pressure intensity along the tube a short distance from the tip will be essentially that of the undisturbed flow. The difference between the stagnation reading

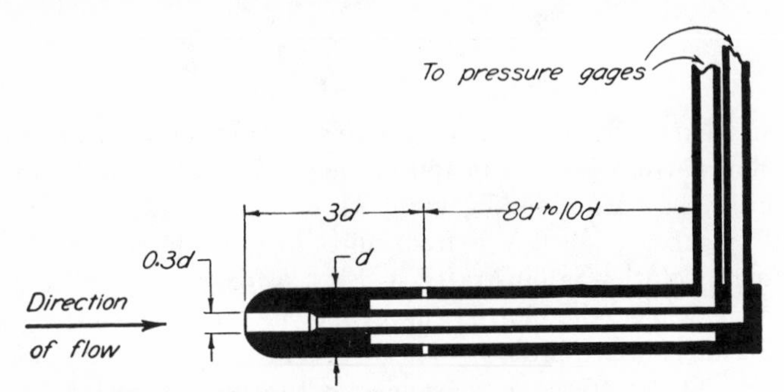

FIG. 27. Pitot-tube detail, after Prandtl.

p_{st} at the tip piezometer and the local intensity p at the side piezometer will then depend upon only the fluid density and the local velocity of the flow:

$$p_{st} - p = \frac{\rho v^2}{2}$$

This velocity may therefore be computed from the relationship

$$v = \sqrt{2(p_{st} - p)/\rho} \tag{30}$$

Example 10. A two-dimensional constriction having the profile shown in Fig. 12 is placed in a rectangular conduit for the purpose of determining the rate of flow. Pressure gages are connected to piezometer orifices located at points A–6 and A–9. Assuming steady, irrotational motion, evaluate the magnitude of the numerical factor C in the discharge equation

$$q = Cn_0\sqrt{2(p_6 - p_9)/\rho}$$

From Eq. (28):

$$p_6 - p_9 = \frac{\rho}{2}(v_9^2 - v_6^2)$$

or, dividing by $\rho v_0^2/2$,

$$\frac{p_6 - p_9}{\rho v_0^2/2} = \left(\frac{v_9}{v_0}\right)^2 - \left(\frac{v_6}{v_0}\right)^2$$

From the plot of $(v/v_0)^2$ accompanying Example 6,

$$\left(\frac{v_9}{v_0}\right)^2 = 5.5 \quad \text{and} \quad \left(\frac{v_6}{v_0}\right)^2 = 0.5$$

so that

$$\frac{p_6 - p_9}{\rho v_0^2/2} = 5.5 - 0.5 = 5.0$$

But from Eq. (7), $q = \int v\, dn = v_0 n_0$, whence

$$q = \frac{1}{\sqrt{5}}\, n_0 \sqrt{2(p_6 - p_9)/\rho}$$

and

$$C = \frac{q}{n_0\sqrt{2(p_6 - p_9)/\rho}} = \frac{1}{\sqrt{5}} = 0.447$$

Example 11. A pier in the submerged entrance of a conduit 8 feet high has the form shown in Fig. 26, the dimension b having the magnitude of 3 feet. What longitudinal force will be exerted upon the upstream face of the pier as a result of the accompanying changes in water velocity, if the velocity of approach is 12 feet per second?

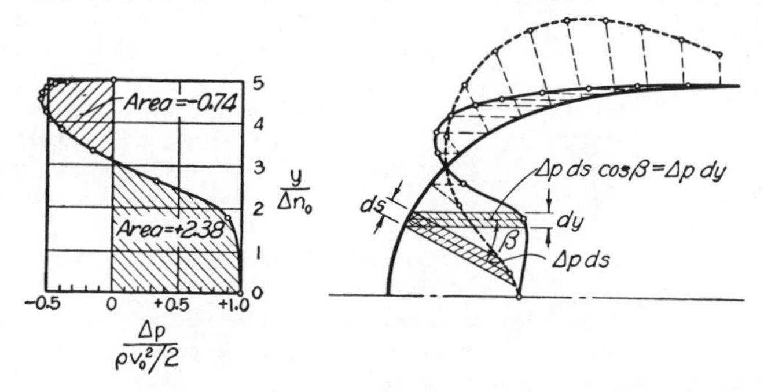

The longitudinal force (see the accompanying figure) is evidently the integral

$$F_x = 8 \int (\Delta p\, ds) \cos \beta$$

taken around the entire boundary curve. But

$$(\Delta p\, ds) \cos \beta = \Delta p\, (ds \cos \beta) = \Delta p\, dy$$

Hence the force F_x may be found from the area under the curve of $\Delta p/(\rho v_0^2/2)$ plotted horizontally from either the boundary itself or its vertical projection; thus, for one half the symmetrical profile

$$\frac{F_x}{2} = 8\, \frac{\rho v_0^2}{2}\, \Delta n_0 \int_0^5 \frac{\Delta p}{\rho v_0^2/2}\, d\left(\frac{y}{\Delta n_0}\right) = 8 \times \frac{1.94 \times \overline{12}^2}{2} \times \frac{3}{10} \times (2.38 - 0.74)$$

$$= 550 \text{ lb}$$

whence

$$F_x = 2 \times 550 = 1100 \text{ lb}$$

PROBLEMS

51. If the duct of Example 10 has a cross section 6 feet wide and 6 feet deep just before the constriction, what rate of air flow would be indicated by a difference in gage readings of 0.5 pound per square inch?

52. What percentage error in the computed rate of flow would result if it were assumed for purposes of rough approximation that the pressure gages in Example 10 were located in zones of uniform flow?

53. The velocity of irrotational flow past a stationary cylinder follows the equation $v/v_0 = 2 \sin \theta$ around the forward portion of the boundary. What pressure intensities in pounds per square foot should prevail at angles of 0°, 30°, and 60° around the front of a smokestack 6 feet in diameter during a wind of 60 miles per hour?

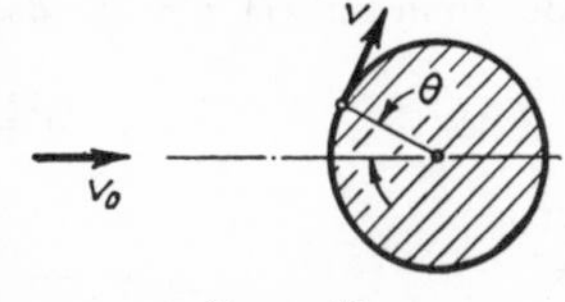

PROB. 53.

54. What increase in force on the pier of Example 11 could be expected if the velocity of flow were increased 50 per cent?

55. The wind velocity near the center of a cyclone may be assumed to vary inversely with distance from the center. If the velocity is 10 miles per hour 30 miles from the center, what pressure gradient should obtain at this point? What reduction in barometric pressure should occur over a radial distance of 25 miles from this point toward the storm center?

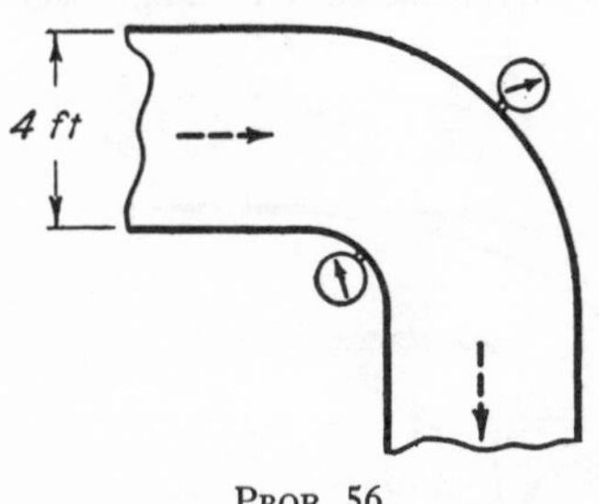

PROB. 56.

56. If a 90° bend in a 4-by-4-foot ventilation duct has the profile shown in Fig. 13, what rate of flow would be indicated by a pressure difference of 3 pounds per square foot between the outside and the inside of the bend? Assume the air to have a density of 0.0025 slug per cubic foot.

57. A Pitot tube is mounted on an airplane to indicate the relative speed of the plane. What differential pressure intensity in pounds per square inch will the instrument register when the plane is traveling at a speed of 165 miles per hour into a 40-mile-per-hour head wind?

58. A cylindrical tube used as a velocity meter contains three piezometer openings placed at 30° intervals around the circumference of a given cross section. The tube is held at right angles to the flow and rotated until pressure gages connected to the two outer piezometer openings indicate the same intensity; the center opening should then be located at the point of stagnation and the outer openings at points for which $\Delta p = 0$ (refer to Problem 53). What velocity of (*a*) air, and (*b*) water, would yield a differential pressure intensity of 0.5 pound per square inch between the stagnation

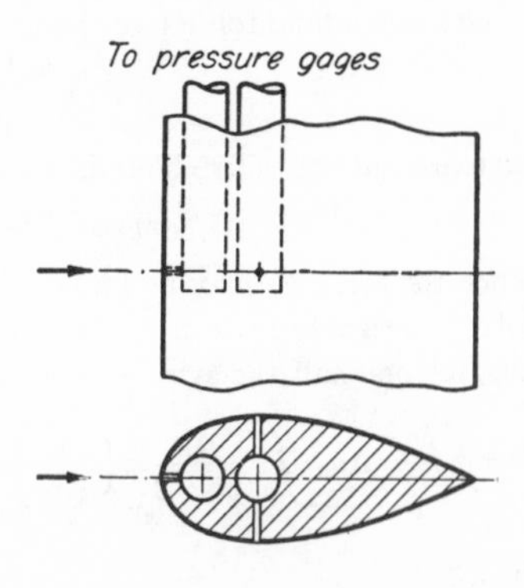

PROB. 59.

opening and the two side openings? (Use as densities of air and water, respectively, 0.0025 and 1.94 slugs per cubic foot.)

59. For purposes of measuring the velocity distribution within a pump, a slender tube having the cross section shown in Fig. 22 is arranged to move axially across the desired section; as indicated in the accompanying sketch, pressure gages are connected to piezometer leads from points of maximum and minimum intensity at a given section of the tube. If the fluid being pumped is gasoline having a density of 1.4 slugs per cubic foot, what velocity will be indicated by a difference in pressure intensity of 1 pound per square inch?

12. APPLICATION OF THE PRESSURE EQUATION TO PROBLEMS OF EFFLUX

Boundary characteristics of jets. If a fluid is discharged from a conduit into the atmosphere through an orifice of any form, the pressure intensity along the *free surface* of the jet (see Fig. 28) must be equal to that of the atmosphere; hence both the pressure intensity and the velocity must be constant at all such points. Within the conduit, of course, not only is the velocity lower than that of the jet at every point, as is evident from the equation of continuity, but the pressure intensity is higher than that of the jet at every point, in accordance with the equations of acceleration. In other words, the pressure distribution within the approaching flow, along the orifice plate, and within the jet must be such as to produce the required change in the velocity of the fluid as it emerges into the atmosphere.

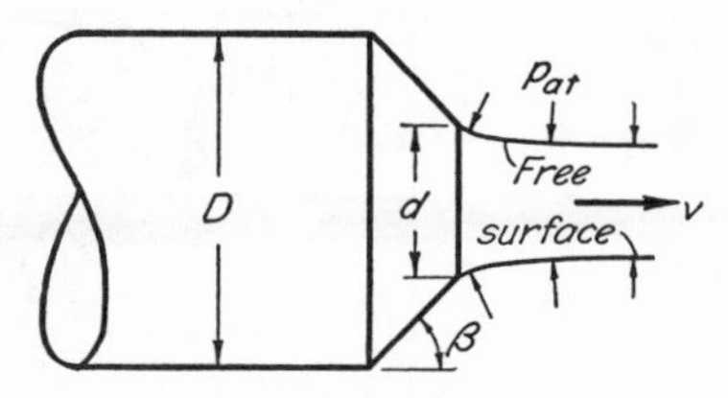

FIG. 28. Efflux from an orifice at the end of a pipe.

Consider, for instance, the two-dimensional boundary profile shown in Fig. 29. The problem is similar to that of flow around or between fixed boundaries, in that the assumption of steady, irrotational motion leads to the unique solution represented by the flow net for any given boundary form. In the present case, although the outermost stream lines of the jet are not guided by boundary walls, they are completely governed in form and position by the fact that the velocity—and hence the spacing of the normal lines of the flow net—must be the same at all points along the free surface, for only then can a constant intensity of pressure prevail at all points in contact with the atmosphere. The flow net and the equations of continuity and acceleration therefore provide a means of simultaneous solution for the jet profile and the distribution of velocity and pressure throughout the flow. Characteristic curves of such distribution along the centerline and the boundary are shown in dimensionless form in the illustration, for the given boundary proportions.

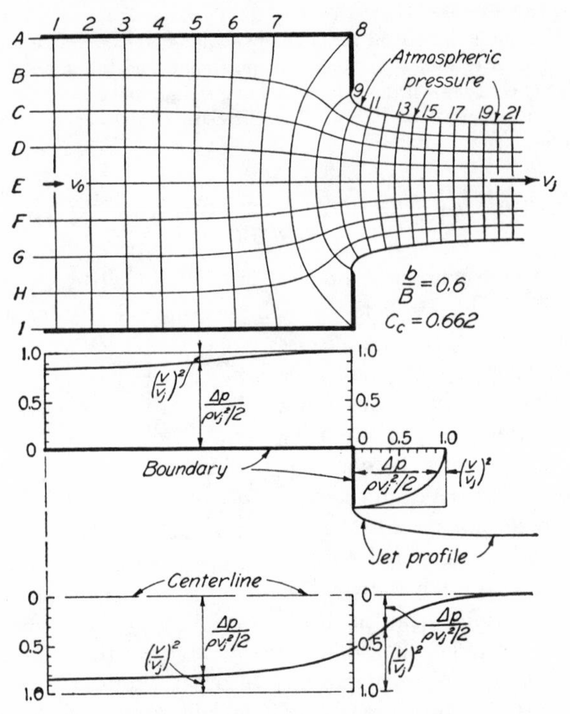

FIG. 29. Flow net and relative distribution of velocity and pressure intensity at a two-dimensional orifice.

Coefficients of jet contraction. Of particular concern in problems of efflux is the degree of jet contraction—the ratio C_c of the final cross-sectional area of the jet to that of the boundary opening—a parameter which is known as the *coefficient of contraction*. If only normal stresses act, the magnitude of this coefficient will depend entirely upon the boundary geometry; C_c may thus be evaluated as a first approximation by means of the flow net, for only the correct jet profile will yield a net in which the normal lines are equally spaced along the free surface. This is, however, a very

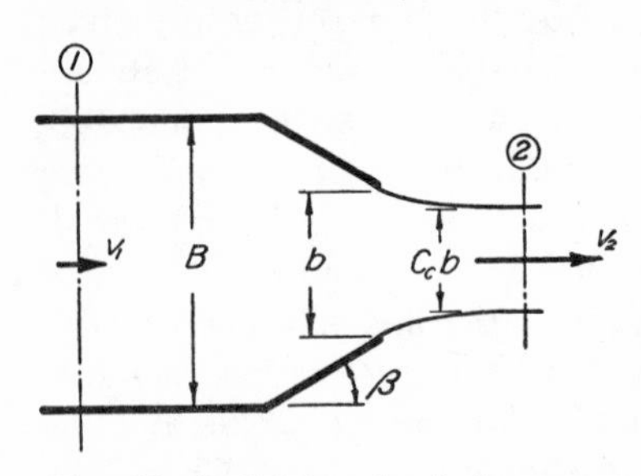

FIG. 30. Definition sketch for two-dimensional analysis of efflux.

tedious process, and not overly precise, owing to the graphical inaccuracies involved. On the other hand, the mathematical principles of which the flow net is a graphic representation permit an exact determination of C_c for many two-dimensional boundary forms. Indeed, it was shown by Kirchhoff nearly a century ago that the contraction coefficient for irrotational efflux from a plane orifice in an extremely large container would have the magnitude $\pi/(\pi + 2) = 0.611$; the values for C_c given in the accompanying table were determined more recently by von Mises for the boundary characteristics indicated schematically in Fig. 30.

TABLE I

COEFFICIENTS OF JET CONTRACTION FOR THE TWO-DIMENSIONAL BOUNDARY CHARACTERISTICS OF FIG. 30

$\frac{b}{B}$	$\beta = 45°$ C_c	$\beta = 90°$ C_c	$\beta = 135°$ C_c	$\beta = 180°$ C_c
0.0	0.746	0.611	0.537	0.500
0.1	0.747	0.612	0.546	0.513
0.2	0.747	0.616	0.555	0.528
0.3	0.748	0.622	0.566	0.544
0.4	0.749	0.631	0.580	0.564
0.5	0.752	0.644	0.599	0.586
0.6	0.758	0.662	0.620	0.613
0.7	0.768	0.687	0.652	0.646
0.8	0.789	0.722	0.698	0.691
0.9	0.829	0.781	0.761	0.760
1.0	1.000	1.000	1.000	1.000

Equations of orifice discharge. Once the magnitude of the contraction ratio is known for given boundary conditions, it is a simple matter to compute the rate of discharge through the orifice, for any given density and intensity of pressure, through use of the equations of pressure and continuity. Writing these equations between sections 1 and 2 of Fig. 30,

$$p_1 - p_2 = \Delta p = \frac{\rho}{2}(v_2^2 - v_1^2)$$

$$v_1 B = v_2 C_c b$$

Simultaneous solution of these expressions will yield the efflux velocity in terms of the density, the pressure change, and the geometrical characteristics of the boundaries:

$$v_2 = \frac{1}{\sqrt{1 - C_c^2(b/B)^2}} \sqrt{2\Delta p/\rho} \tag{31}$$

Since $q = v_2 C_c b$, the rate of efflux may at once be written in the form

$$q = \frac{C_c}{\sqrt{1 - C_c^2 (b/B)^2}} b\sqrt{2\Delta p/\rho} \tag{32}$$

or, more simply,

$$q = C_d \, b\sqrt{2\Delta p/\rho} \tag{33}$$

in which C_d is a *coefficient of discharge*, depending, like C_c, only upon the form of the boundaries; that is,

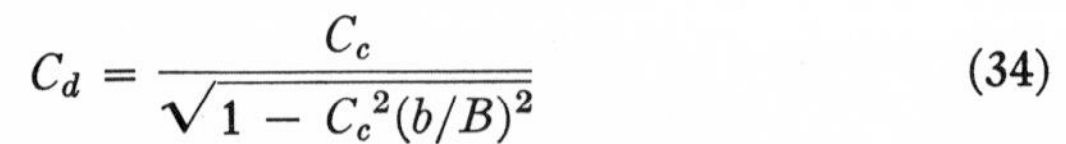

$$C_d = \frac{C_c}{\sqrt{1 - C_c^2 (b/B)^2}} \tag{34}$$

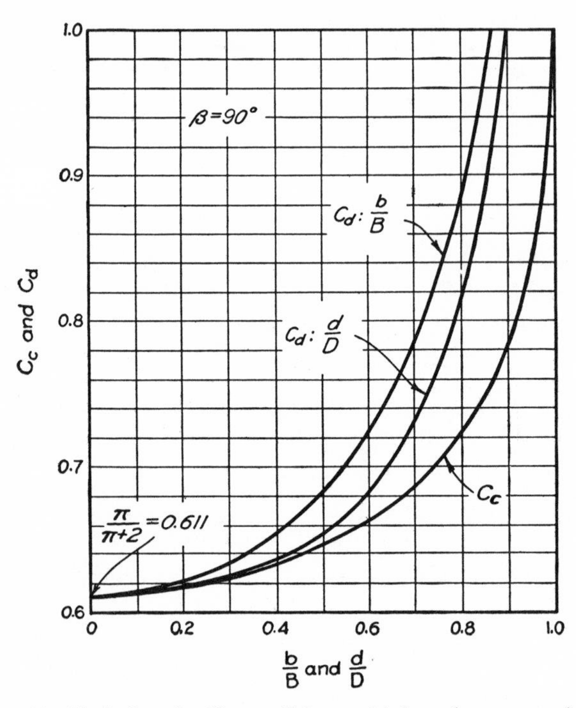

FIG. 31. Variation of orifice coefficients with boundary proportions.

Evidently, the discharge coefficient is independent of the absolute magnitudes of q, Δp, and b. Its variation with the boundary proportions is indicated in Fig. 31 by the curve of C_d as a function of b/B for $\beta = 90°$.

Similar considerations apply to the three-dimensional counterpart of such two-dimensional flow—the efflux of fluid from a circular orifice at the end of a cylindrical conduit, as shown in Fig. 28. Although it has not yet been possible to evaluate analytically the coefficient of contraction for the three-dimensional case, experiments indicate that the values of C_c listed in Table I apply with good approximation to the three-dimensional case as well if b/B is replaced by d/D, the ratio of the orifice diameter to the diameter of the conduit. Since the coefficient of contraction was defined as the ratio of the jet and orifice *areas*, however, it will be evident that the development leading to Eqs. (31)–(34) must be modified as follows for the three-dimensional case:

$$p_1 - p_2 = \Delta p = \frac{\rho}{2}(v_2^2 - v_1^2)$$

$$v_1 \frac{\pi D^2}{4} = v_2 C_c \frac{\pi d^2}{4}$$

The jet velocity then becomes

$$v_2 = \frac{1}{\sqrt{1 - C_c^2(d/D)^4}} \sqrt{2\Delta p/\rho} \tag{35}$$

and the rate of efflux

$$Q = \frac{C_c}{\sqrt{1 - C_c^2(d/D)^4}} \frac{\pi d^2}{4} \sqrt{2\Delta p/\rho} \tag{36}$$

In the simpler form

$$Q = C_d \frac{\pi d^2}{4} \sqrt{2\Delta p/\rho} \tag{37}$$

the coefficient of discharge C_d again depends, like C_c, only upon the boundary geometry:

$$C_d = \frac{C_c}{\sqrt{1 - C_c^2(d/D)^4}} \tag{38}$$

Though the same values of C_c may be used for equal ratios of b/B and d/D, the fact cannot be too strongly emphasized that the inherent differences in cross-sectional area between a slot and a circular opening yield quite different values of C_d for the same linear boundary proportions. This will be evident from comparison of Eqs. (34) and (38), and from the curves of C_d against b/B and d/D plotted in Fig. 31 for the condition that $\beta = 90°$.

Although the phenomenon of efflux is most frequently illustrated by the discharge of a liquid into the atmosphere, the foregoing basic considerations apply quite as well to the efflux of liquids into liquids and gases into gases. In all such instances, the difference Δp must represent the change in pressure intensity experienced by the fluid as it emerges from the conduit, a factor which, for a given rate of flow, is evidently independent of the absolute intensity of pressure.

Example 12. Natural gas having a density of 0.0014 slug per cubic foot is discharged at the rate of 10 cubic feet per second through a 3-inch circular orifice at the end of a 6-inch pipe. Determine the diameter of the contracted jet and the intensity of pressure of the uniform flow within the conduit relative to that of the jet.

From Table I or Fig. 31, $C_c = 0.644$ when $\beta = 90°$ and $d/D = 0.5$. Since

$$C_c = \frac{\pi d_j^2/4}{\pi d^2/4} = \frac{d_j^2}{d^2} = 0.644$$

it follows at once that

$$d_j = 3\sqrt{0.644} = 2.41 \text{ in.}$$

From Fig. 31 or Eq. (38), $C_d = 0.653$ for the given boundary form; therefore, from Eq. (37),

$$\Delta p = \frac{Q^2 \rho}{2C_d^2(\pi d^2/4)^2} = \frac{\overline{10}^2 \times 0.0014}{2 \times \overline{0.653}^2 \times \left[\frac{\pi}{4} \times \left(\frac{3}{12}\right)^2\right]^2} = 68.0 \text{ psf}$$

PROBLEMS

60. Figure 29 may be assumed to represent the flow pattern for the efflux of air ($\rho = 0.0025$ slug per cubic foot) into the atmosphere through a long, narrow slot. If the jet velocity is 120 feet per second, determine the velocity and intensity of pressure (*a*) at either juncture of the horizontal and vertical boundaries, (*b*) at either edge of the slot, and (*c*) at the centerline in the plane of the slot.

61. Water discharges from a 4-inch plate orifice at the end of an 8-inch pipe at the rate of 1.5 cubic feet per second. What is the pressure intensity of the approaching flow within the pipe?

62. What is the highest intensity of pressure which can exist at any point on the upstream face of an orifice plate at the end of a pipe discharging oil ($\rho = 1.7$ slugs per cubic foot) if the mean velocity of approach is 10 feet per second and the orifice diameter is ⅔ that of the pipe?

63. If a 5-inch orifice at the end of a 9-inch pipe produces a jet 4 inches in diameter, what will be the corresponding coefficients of contraction and discharge?

64. A nozzle at the end of a pipe is conical in shape, the apex angle of the walls being 90°. If the pipe is 10 inches and the nozzle outlet 4 inches in diameter, what pressure in the pipe would be required to produce a discharge of 5 cubic feet of air per second?

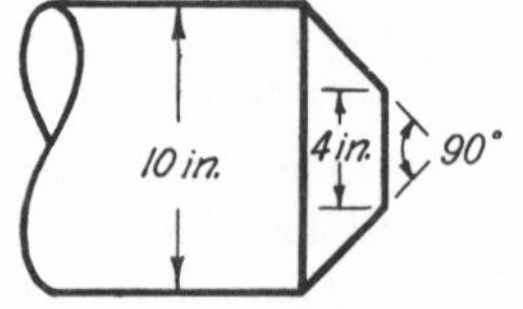

PROB. 64.

65. The outlet of a ventilation duct is protected by a series of 1-inch metal strips 2 inches on center. If the mean velocity in the duct is 25 feet per second, what back pressure will the strips produce?

66. A fire nozzle at the end of a 3-inch hose yields a jet of water 1¼ inches in diameter. In order to determine the rate of flow, a stagnation tube made of a hypodermic needle is connected to a pressure gage and held in the jet. If the gage reading is 25 pounds per square inch, what is the rate of flow? What is the accompanying intensity of pressure within the hose?

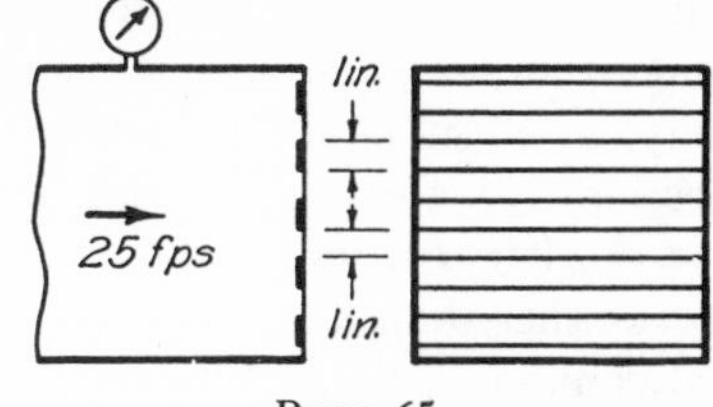

PROB. 65.

67. In a certain phase of chemical manufacture a liquid having a density 1.5 times that of water discharges through a long slot ½ inch wide in the bottom of a large tank. If the pressure intensity within the tank at the level of the slot is 2 pounds per square inch, what rate of efflux per unit length of slot should obtain?

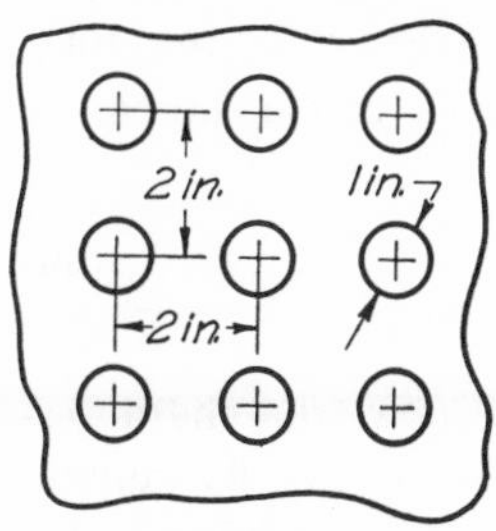

PROB. 68.

68. Estimate the rate of flow of water through a plate perforated with 1-inch holes 2 inches on center when the mean pressure drop across the plate is 3 pounds per square inch. Base the flow rate on a unit cross-sectional area of plate.

69. Estimate the rate at which natural gas would escape through a ¼-inch hole in the side of a large tank if the tank is under a pressure of 5 pounds per square inch. Assume the gas to have a density of 0.0014 slug per cubic foot.

70. The flow of air from a duct 2 feet square in cross section is regulated by means of two 1-by-2-foot plates hinged at the end of the duct. If the duct pressure is 2 pounds per square foot when the plates are turned upwind at an angle of 45°, what will be the corresponding rate of flow?

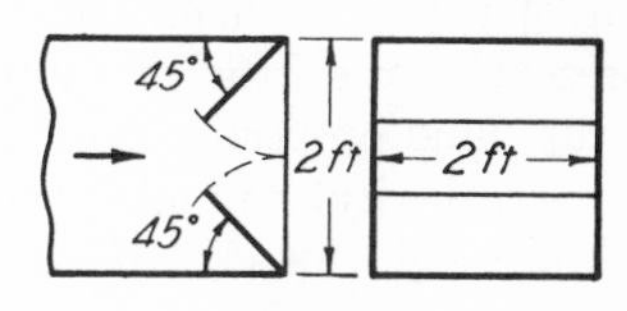

PROB. 70.

13. SIGNIFICANCE OF THE EULER NUMBER

Formulation of a basic flow parameter. Considerable attention has been paid in this chapter to the special case of steady, irrotational flow in which the only accelerative force is provided by a change in the intensity of fluid pressure from point to point. Such attention is warranted not only because of the relative simplicity of the analysis but also because of the convenient basis it provides for evaluating the effects of other forces than fluid pressure in the remaining chapters of this book.

So long as only the pressure gradient, the density, and the acceleration are considered, the assumption of steady, irrotational flow has been seen to yield a unique solution for the distribution of velocity and pressure around any given form of boundary. In other words, a particular value of the dimensionless ratio $\Delta p/\rho(v_0^2/2)$ then characterizes the flow at each point in the flow pattern, regardless of the absolute magnitudes of p_0, v_0, ρ, and the boundary scale. Since such a quantity may, in effect be considered to represent the ratio of a typical unit accelerative force ($f \approx \Delta p/L$) to a typical unit inertial reaction ($\rho a \approx \rho V^2/2L$), its constancy is seen to result from the fact that no force other than fluid pressure is assumed to act.

Leonhard Euler, a Swiss mathematician of the eighteenth century, was the first to realize the true role played by pressure in fluid motion, Eqs. (18) and (19) being special forms of the original Eulerian equations of acceleration. For this reason, a parameter of the form $\Delta p/(\rho V^2/2)$, or, more conveniently, $V/\sqrt{2\Delta p/\rho}$, is called herein as the *Euler number:*

$$\mathbf{E} = \frac{V}{\sqrt{2\Delta p/\rho}} \tag{39}$$

So long as no other fluid property influences the flow, the characteristic Euler number for any boundary form must necessarily remain constant, like, for instance, the discharge coefficient of a given orifice; indeed, comparison with Eq. (37) will show that this parameter is identical with the orifice coefficient (or, for that matter, with the coefficient of any flow meter, such as that of Example 10), the term V representing the ratio of the rate of flow to the nominal flow area. If other fluid properties cause the flow pattern to change appreciably, however, the Euler number should at once reflect this change. Thus, as fluid weight, or gravity, deflects the jet emerging from an orifice, one should expect $\mathbf{E}$ (i.e., the discharge coefficient) to vary. Likewise, as viscous shear gives rise to fluid rotation, as the surface tension of a liquid jet causes it to cling to the orifice plate, or as the compressibility of a gaseous jet becomes too great to be ignored, the Euler number should provide a quantitative measure of each such influence upon the basic flow pattern.

QUESTIONS FOR CLASS DISCUSSION

1. Define and contrast the following terms: inertia, mass, acceleration, density, pressure, force, reaction.

2. If flow in a uniform conduit is rotational, the velocity will vary across any normal section. Why should the pressure intensity nevertheless be constant across the section?

3. Assume Fig. 26 to show in dimensionless form the distribution of pressure in steady flow around the front of the strut of Fig. 22. In what respect would the

distribution curves differ for the corresponding pattern of unsteady flow shown in Fig. 23?

4. In evaluating pressure distribution from a pattern of stream lines, why is it necessary to distinguish between steady and unsteady flow?

5. If the rate of flow through a uniform pipe increases at a constant rate, in which direction must the pressure intensity increase in order to accomplish the acceleration?

6. Why do the insects which strike the windshield of an automobile give a false picture of the motion of the air?

7. The velocity distribution in a cyclone is similar as a first approximation to that of the irrotational vortex (i.e., $v = C/r$). Show whether the pressure intensity must increase or decrease toward the storm center.

8. Under what conditions of flow will a pressure gage connected to the stagnation opening of a Pitot tube yield the same reading for all points of a moving fluid?

9. A long cylindrical tube with two piezometer openings 30° apart may be used as a velocity meter in two-dimensional flow if held at right angles to the plane of motion so that one opening coincides with the local point of stagnation. How might such an instrument be used to determine the direction as well as the magnitude of the velocity vector?

10. When a car passes a truck or bus at high speed, a pronounced side thrust is experienced by the occupants of the car. Explain this phenomenon.

11. Compare the jet dimensions for efflux from a slot and a circular orifice having identical coefficients of contraction.

12. If the flow through the boundary profile shown in Fig. 29 were not irrotational, where would separation probably occur? What effect would this have upon the stagnation pressure and the contraction coefficient?

13. Explain why the curves for C_d in Fig. 31 appear to approach infinity as a limit.

14. If the flow pattern for efflux from a given orifice is known, how may it be used to analyze the efflux from an orifice which has the same proportions but is ten times as great in size?

15. Flow through geometrically similar boundary profiles is said to be dynamically similar if the Euler number is the same for each. Under what circumstances may it be expected to vary?

SELECTED REFERENCES

LAMB, H. "Mathematical Theory of Fluid Motion." *The Mechanical Properties of Fluids*, Chapter II, Blackie, 1925.

PRANDTL, L. "The Flow of Liquids and Gases—Part I." *The Physics of Solids and Fluids*, Chapter V, Blackie, 1928.

CHAPTER IV

EFFECTS OF GRAVITY ON FLUID MOTION

14. MASS AND WEIGHT

Gravitational attraction; specific weight. That property of matter known as mass is evidenced mechanically in two distinct ways. According to the Newtonian laws of motion, as discussed in the foregoing chapter, the mass property gives rise to an *inertial reaction* of matter to any accelerative force. Another, but quite different, mass characteristic of a substance is the *attractive force* which it exerts upon other substances in its vicinity, the magnitude of the mass attraction between two bodies of matter varying directly with the product of their masses and inversely with the square of the distance between their mass centers. Evidently, mass attraction is only one of several kinds of force, whereas inertia is a mass tendency to resist any force causing acceleration. Mass attraction, therefore, may produce other effects than mass acceleration, and mass acceleration may be due to other causes than mass attraction. In a word, these two characteristics of matter are related to one another only to the extent that each is proportional to the mass magnitude.

Gravitational attraction is a specific case of mass attraction, the *weight* of any body of matter being a measure of the attractive force exerted upon that body by the earth. As the mass of the earth may be considered constant, the weight of a given body should be directly proportional to its mass and inversely proportional to the square of its distance from the center of the earth. This proportion is generally written in the form $W = gM$, the coefficient of proportionality g being a function (refer to the Appendix) of elevation and geographical location. Since $g = W/M$, g will be seen to represent the force per unit mass exerted by the earth upon any substance, and hence the magnitude of the *gravitational acceleration* in the case of a freely falling body.

Just as the mass density, or mass per unit volume, was shown to be a significant mass characteristic in the study of fluid motion, an equally significant weight characteristic is found in the weight density, or weight per unit volume, a quantity generally called *specific weight* and having the symbol γ (gamma). The ratio of the weight per unit volume and the mass per unit volume of any substance is obviously equal to the acceleration of gravity,

$$\frac{\gamma}{\rho} = g \tag{40}$$

from which it is apparent that the specific weight of a given substance $\gamma = \rho g$ will vary not only with conditions of pressure and temperature, as does mass density, but also with elevation and geographical location, as does the gravitational acceleration. However, the variation of g is relatively small, and it is sufficient for most engineering calculations to use the approximate value of $g = 32.2$ feet per second per second for the ratio of specific weight to density. It is to be noted, in passing, that the so-called *specific gravity* of a substance is merely a relative term representing the ratio of its density or specific weight to that of water.

15. ACCELERATION DUE TO PRESSURE GRADIENT AND FLUID WEIGHT

Forces normal to surfaces of constant p and z. Since weight, a vector quantity, is the attractive force exerted by the earth upon a substance, its direction of action is evidently toward the center, and therefore normal to the surface, of the earth. It follows that the force per unit volume due to fluid weight will invariably act vertically downward, regardless of the direction of flow. On the other hand, the force per unit volume due to pressure gradient, which is not necessarily dependent upon gravitational attraction, must generally be considered to act in another direction. Nevertheless, since either vector may be resolved in any three coordinate directions, it should not be difficult to obtain an expression for the combined effect of pressure gradient and weight in any direction desired.

From the foregoing chapter it will be recalled that the force per unit volume due to pressure variation will have a component in any arbitrary direction equal to the negative gradient of p in that direction. Along a surface of constant pressure intensity (for instance, the free surface of a liquid), the tangential force component will evidently be zero, since p cannot vary in this direction. At right angles to such a surface, accordingly, p will have its maximum rate of change, from which it follows that the resultant force per unit volume f_p due to pressure gradient (see Fig. 32) must be normal to the surface $p =$ *constant* passing through the point in question. In the arbitrary direction s shown in the figure, the component of the vector will be simply

$$(f_p)_s = -\frac{\partial p}{\partial s}$$

With regard to the force per unit volume due to fluid weight, $f_w = \gamma$, it will be seen that this vector must invariably be normal to

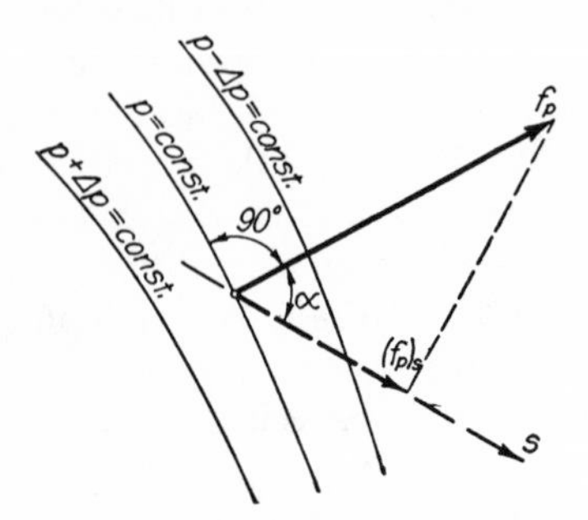

FIG. 32. Orientation of unit force due to pressure gradient.

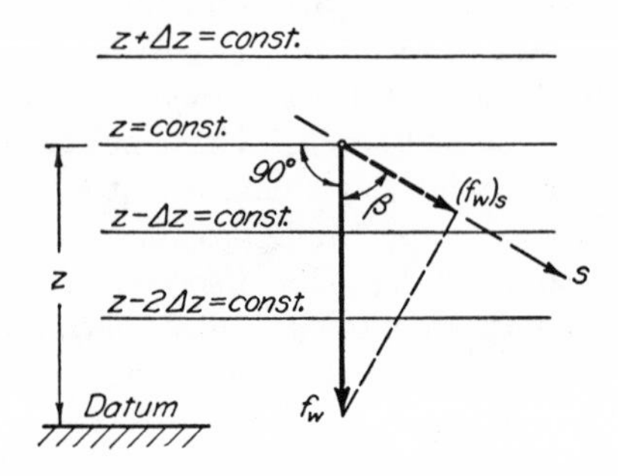

FIG. 33. Orientation of unit force due to fluid weight.

a plane of constant elevation z above an arbitrary geodetic datum (Fig. 33). The component of this force in the arbitrary direction s may at once be written as

$$(f_w)_s = \gamma \cos \beta$$

With reference to Fig. 34, however, it will be apparent that the angle β between the s and z axes determines the change in elevation $-\Delta z$ corresponding to a displacement Δs; the cosine of β hence represents the rate of change of elevation in the s direction:

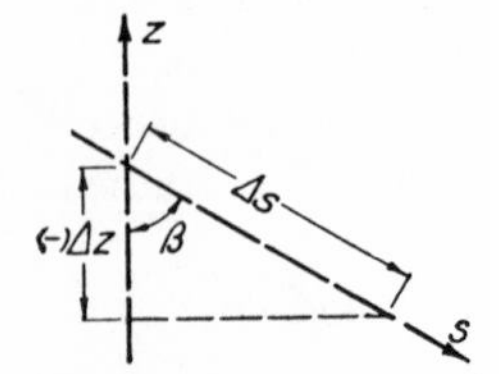

FIG. 34. Variation in elevation along an arbitrary axis.

$$\cos \beta = -\frac{\partial z}{\partial s}$$

Therefore,

$$(f_w)_s = -\gamma \frac{\partial z}{\partial s}$$

Addition then yields the desired expression

$$f_s = (f_p)_s + (f_w)_s = -\frac{\partial p}{\partial s} - \gamma \frac{\partial z}{\partial s} = -\frac{\partial}{\partial s}(p + \gamma z) \qquad (41)$$

which states that at any point in a fluid the component of force per unit volume due to pressure gradient and weight is equal to the rate of decrease in the sum $p + \gamma z$ in the corresponding direction.

Evidently, therefore, if the quantity $p + \gamma z$ varies from point to point in a fluid, and if the fluid is acted upon by no other type of force, acceleration will take place at every point in proportion to and in the direction of the maximum negative gradient of this quantity. Since the component of force per unit volume in any direction must equal the mass per unit volume times the corresponding component of total acceleration, then in the arbitrary direction s

$$-\frac{\partial}{\partial s}(p + \gamma z) = \rho a_s \tag{42}$$

This expression differs from that derived in the foregoing chapter only in the addition of the term $\gamma\, \partial z/\partial s$. Although $\partial z/\partial s$ will approach zero as the s axis approaches the horizontal, γ will be small only in the case of a gas. It is therefore apparent that the relationships of Chapter III are restricted to the motion of gases in which differences in elevation are not excessive, or to the motion of liquids in horizontal planes.

Significance of surfaces of constant h. In order to study the general case of flow under gravitational influence, it will be necessary to integrate expressions for fluid acceleration in the tangential and normal directions. In the case of bodies of liquid which undergo acceleration as a unit, however, Eq. (42) is at once applicable, particularly when written in the following form, obtained by dividing all terms by γ:

$$-\frac{\partial}{\partial s}\left(\frac{p}{\gamma} + z\right) = \frac{a_s}{g} \tag{43}$$

The quantity p/γ is commonly known as the *pressure head*, since it, like z, has the dimension of length. The sum of these two linear quantities may often conveniently be replaced by the single term h, which (for reasons to be discussed in the following section) is called the *piezometric head*. Equation (43) then becomes

$$-\frac{\partial h}{\partial s} = \frac{a_s}{g} \tag{44}$$

which states that the rate at which h decreases in any direction s is equal to the ratio of the corresponding component of total acceleration to the acceleration of gravity. By the same process of reasoning which led to Figs. 32 and 33, it may be seen that the force vector (and hence the acceleration vector) must be at right angles to the surface $h = constant$ passing through the point in question, as indicated in Fig. 35. If, as is customary, the pressure intensity of a liquid is measured with respect to that of the atmosphere, then the free surface of a liquid is a surface of zero pressure intensity. Therefore, as shown in the following examples, the magnitude of h along any surface $h = constant$ is equal to the elevation z of the intersection of this surface with the free surface $p = 0$ of the liquid.

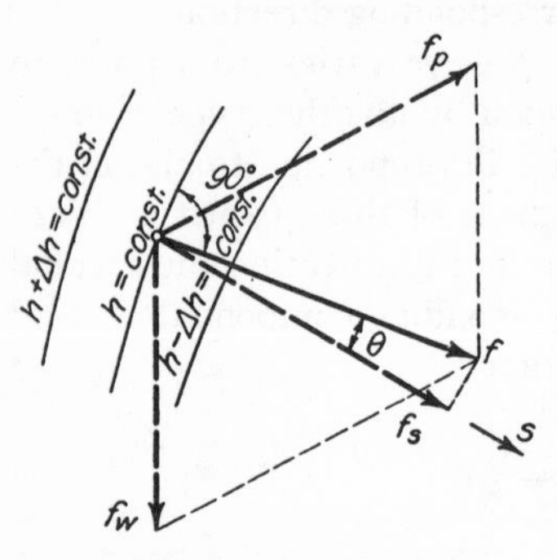

FIG. 35. Orientation of unit force due to pressure gradient and fluid weight.

Example 13. Determine the slope of the free surface of oil in a tank car during a constant horizontal acceleration of 5 feet per second per second.

Since h represents the elevation of the free surface above an arbitrary horizontal datum plane, the term $\partial h/\partial s = \partial h/\partial x$ is the tangent of the angle θ between the free surface and the horizontal. Therefore

$$\tan\theta = \frac{\partial h}{\partial x} = -\frac{a_x}{g} = -\frac{5}{32.2} = -0.155$$

$$\theta = \tan^{-1} - 0.155 = -8^\circ\, 49'$$

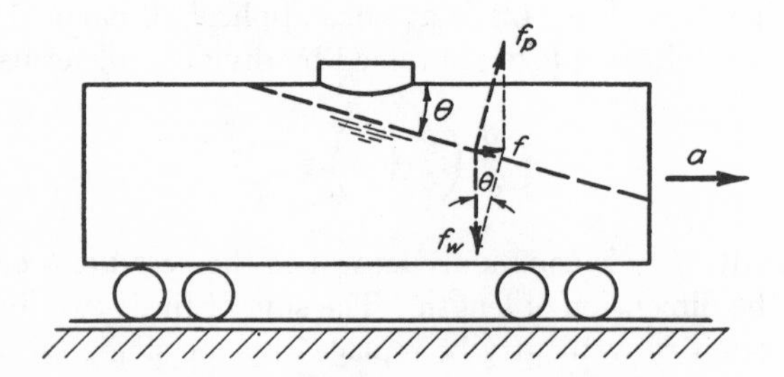

Example 14. A cylindrical tank partly filled with liquid is rotated at the constant angular velocity ω about its vertical axis. What form will the free surface assume? How will the pressure intensity vary along the bottom of the tank?

Since the acceleration is entirely normal, $\partial h/\partial s = 0 = \partial h/\partial z$; in other words, the surfaces of constant h are coaxial cylinders, as shown. Then, since the surfaces of constant z are invariably horizontal, surfaces of constant p will all be similar to the free surface, and the same vector diagram will apply to all points of the same radius r.

From Eq. (17), $a_n = v^2/r$, and since $v = \omega r$

$$-\frac{\partial h}{\partial n} = \frac{a_n}{g} = \frac{\omega^2 r}{g}$$

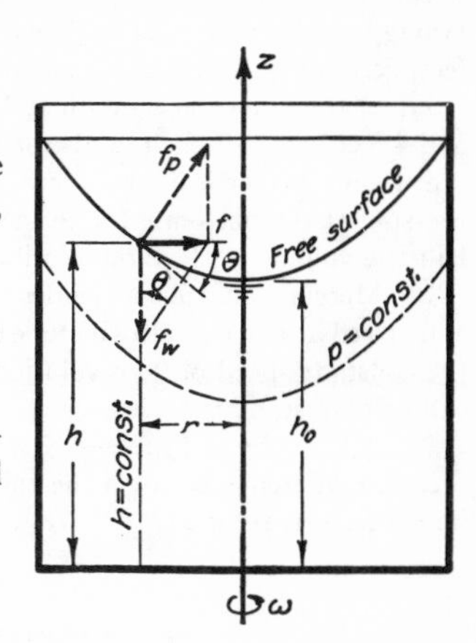

As n is measured *toward* the center of curvature and r is measured *from* the center of curvature, $\partial h/\partial n = -\partial h/\partial r$; hence

$$\int_0^r \frac{\partial h}{\partial r}\,dr = \int_0^r \frac{\omega^2 r}{g}\,dr$$

The integral of the left side is simply the difference between the value of h at the radius r and its value h_0 at the axis of rotation; thus

$$h - h_0 = \frac{\omega^2 r^2}{2g}$$

which indicates that the free surface (and hence every surface of constant p) is a paraboloid of revolution. Finally, since $h = p/\gamma + z$, the pressure intensity along the bottom, where $z = 0$, will be

$$p = \gamma\left(\frac{\omega^2 r^2}{2g} + h_0\right)$$

PROBLEMS

71. A construction hoist carries a tank of fresh mortar to the top of a scaffold, changing velocity at the rate of 10 feet per second per second at the beginning and end of its trip. If the mortar weighs 170 pounds per cubic foot, determine the pressure gradient within the mortar (*a*) as the hoist starts and (*b*) as it stops. If the mortar depth is 40 inches, determine by integration the pressure intensity at the bottom of the tank (*c*) as the hoist starts and (*d*) as it stops.

72. An open tank of water slides down an inclined plane. Show that the free surface will be (*a*) horizontal if the velocity is constant, and (*b*) parallel to the plane if the acceleration is equal to the component of g in the direction of motion—i.e., if the tank slides without frictional resistance.

73. Oil stands within 2 feet of the top of an open tank car 30 feet long, 8 feet wide, and 7 feet deep. What is the greatest horizontal acceleration which the car may be given without spilling the oil?

74. Determine the slope of the water surface in a small container which is mounted on a turntable 6 feet from the axis of rotation, when the angular velocity of the turntable is 30 revolutions per minute.

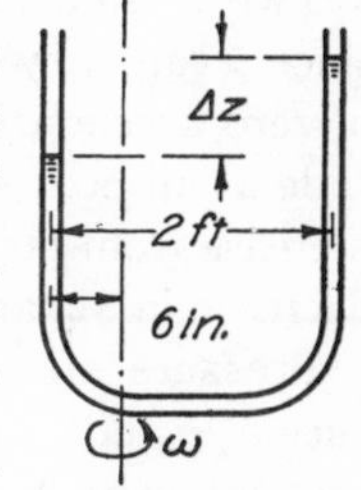

PROB. 75.

75. A U-tube having vertical legs 2 feet on center is partly filled with carbon tetrachloride (specific gravity = 1.6) and rotated about a vertical axis 6 inches in from one leg. What will be the difference in elevation of the two free surfaces when the angular velocity is 100 revolutions per minute?

76. If the angular velocity of one's arm is sufficiently great, a glass of water may be swung through a vertical circle without spilling. Assuming an arm length of $r = 2.5$ feet, determine its safe minimum angular velocity as the glass passes its highest point.

77. Water stands at a depth of 2 feet in an open cylindrical tank 3 feet in diameter and 4 feet high. If the tank is rotated at the constant rate of 90 revolutions per minute about its vertical axis, determine the maximum and minimum intensities of pressure at the bottom. (Note that the volume of a paraboloid of revolution is one-half the volume of the circumscribing cylinder.)

78. Mercury half fills a slender tube which is ¼ inch in diameter, 15 inches long, and closed at one end. If the tube is rotated in a vertical plane about its open end at the constant speed of 90 revolutions per minute, what is the maximum force which will be exerted by the mercury upon the closed end?

79. If the tube of Problem 47 were rotated in a vertical plane, what would be the pressure difference between the ends as the tube attained (*a*) its uppermost, and (*b*) its lowermost, position?

16. PRINCIPLES OF HYDROSTATICS

Pressure variation in steady, uniform flow. Under conditions of zero acceleration in the arbitrary direction s, Eq. (42) will reduce to the form

$$\frac{\partial}{\partial s}(p + \gamma z) = 0$$

If the acceleration is zero in every other direction, moreover, it follows that in no direction can the sum $p + \gamma z$ vary. In other words, in the case of steady, uniform flow,

$$p + \gamma z = C \tag{45}$$

in which the factor C is a constant of integration, indicating that the sum $p + \gamma z$ will be the same at all points within a fluid which undergoes no acceleration.

Two related facts are at once apparent from this equation: First, since p can vary only if γz also varies, it follows that under conditions of zero acceleration the pressure intensity will have the same magnitude at all points of equal elevation in the same fluid. Second, p will increase from point to point by an amount equal to the corresponding decrease in γz, and vice versa.

Pressure measurement by liquid piezometers. Such interdependence of p and γz is particularly significant in the measurement of pressure intensity by means of *liquid piezometers* rather than mechanical pressure gages. Consider, for example, the case of a conduit carrying fluid, in which it is necessary to know the pressure intensity at some boundary point a, which may or may not be in a zone of acceleration.

A piezometer orifice (see Fig. 36) is drilled through the boundary at this point and connected to a transparent tube having a vertical leg open to the atmosphere. If the fluid in the conduit is a liquid, a portion of this liquid will fill the piezometer tube to a height depending only upon the specific weight of the liquid and the pressure intensity at point a relative to that of the atmosphere. In other words, since $p = 0$ at the free surface in the tube, since there is no acceleration of the liquid column, and since points a and b lie at the same elevation in the same fluid,

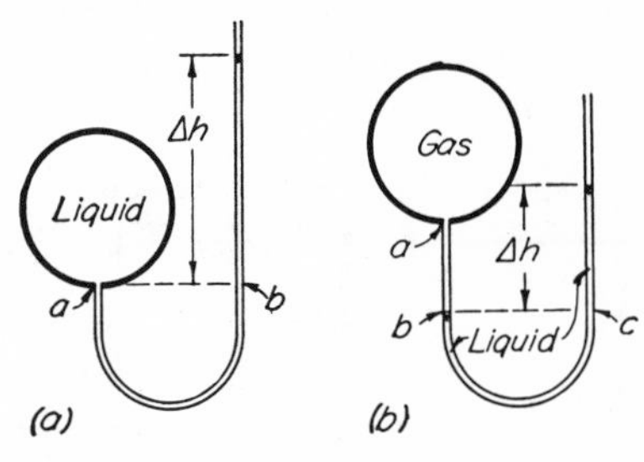

FIG. 36. Piezometers for liquids and gases.

$$p_b = 0 + \gamma \Delta h = p_a$$

If the conduit transmits a gas, it is only necessary to fill the lower part of the tube with a suitable gage liquid of specific weight γ', whereupon (ignoring the weight of the gas column a–b)

$$p_c = \gamma' \Delta h = p_b \approx p_a$$

Evidently, Δh will be positive or negative in each case depending upon whether the absolute pressure intensity at point a is greater or less than atmospheric.

In many devices, such as the Pitot tube, it is necessary to know only a pressure difference, a *differential manometer* of the type shown in Fig. 37 then being especially useful. Assuming that a Pitot tube is to be used in water, the piezometer leads (see Fig. 37) may be connected to an inverted U-tube partially filled with a gage liquid of specific weight γ' which is somewhat lower than that of water. The height of the gage above (or below) the Pitot tube is evidently immaterial, since the pressure changes from the point of measurement to points a and b are identical. Points c and d must be under the same pressure, as both points are at the same level in the same continuous body of fluid. But $p_c = p_a - \gamma \Delta h$ and $p_d = p_b - \gamma' \Delta h$, whence, since $p_c = p_d$,

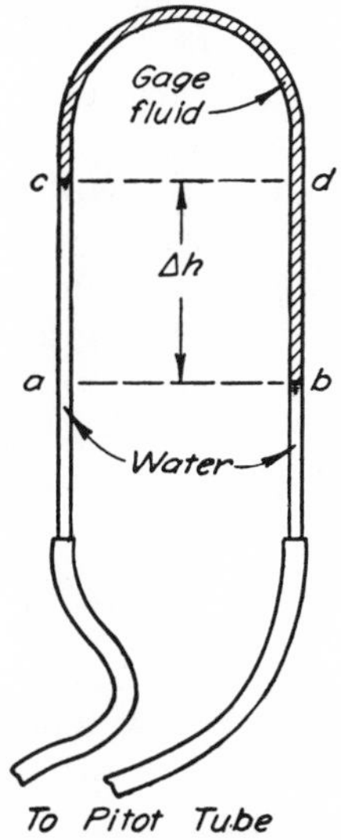

FIG. 37. Differential manometer.

$$p_a - p_b = \Delta h(\gamma - \gamma')$$

Evidently, the smaller the difference between γ and γ', the more sensitive the manometer will be—i.e., the greater the reading Δh for a given differential pressure.

Hydrostatic distribution of pressure; piezometric head. In the development of Eq. (42), only gravitational attraction and the normal stresses exerted upon a fluid element were taken into account, tangential stresses (i.e., viscous resistance to deformation) thereby being ignored. Since tangential stresses are truly zero only in the event that the fluid undergoes no deformation, variation in pressure intensity according to the relationship $p + \gamma z = C$ requires not only that the acceleration be equal to zero but also that the velocity be everywhere the same. Such conditions are generally realized only if the velocity and acceleration are both equal to zero, and hence the expression $p + \gamma z = C$ is, strictly speaking, an equation of *hydrostatics*, the study of fluids at rest. Nevertheless, under many conditions of motion both the acceleration and the tangential stresses are small in comparison with other factors; under such circumstances the assumption of *hydrostatic pressure distribution* permits a relatively simple approximate analysis of the actual state of motion and of the accompanying forces exerted by the fluid upon the boundaries.

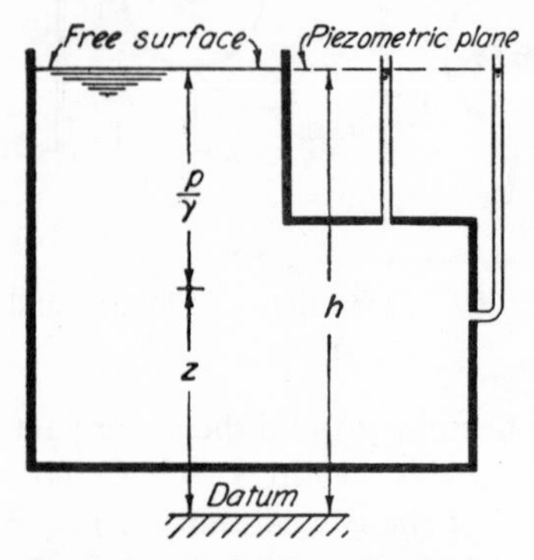

FIG. 38. Definition sketch for pressure head and piezometric head.

Under conditions of liquid flow with hydrostatic pressure distribution, it is convenient to divide the several terms of Eq. (45) by γ, so that each will have the dimension of length:

$$\frac{p}{\gamma} + z = \frac{C}{\gamma}$$

Since the symbol h was introduced in the foregoing section to represent the sum $p/\gamma + z$, and since this sum is constant under conditions of hydrostatic pressure distribution (see Fig. 38),

$$h = \frac{p}{\gamma} + z = \text{constant} \tag{46}$$

At the free surface of a liquid, p—and hence p/γ—has the magnitude of zero, so that the constant h in this equation must represent the surface elevation. It follows at once that in flow with hydrostatic pressure distribution the pressure head at any point is equal to its depth below the free surface. If the flow itself does not possess a free surface (for instance, liquid moving in a closed conduit), h nevertheless represents the height to which the liquid would rise in an open tube connected to a piezometer orifice at any point in the boundary. For this reason, the sum h of pressure head and elevation is known as the *piezometric head.*

In the case of a gas, it must be noted, the piezometric head has no such direct significance, since a stable free surface cannot then exist. The pressure intensity of a gas, moreover, changes only imperceptibly with appreciable changes in elevation because of the very small magnitude of γ. However, it is common practice to express the pressure "head" of a gas in terms of that of a liquid at the same pressure intensity, the pressure intensity of the gas then being equal to the product of the specific weight of the liquid and the corresponding liquid head.

Example 15. A vertical conduit carrying crude oil (sp. gr. = 0.85) contains, for purposes of flow measurement, a reduction in cross-sectional area. Piezometer orifices are located at points A and B, in zones of essentially parallel flow, to which is connected a differential mercury gage as shown. Determine the difference in (*a*) pressure intensity and (*b*) piezometric head when $\Delta h_{Hg} = 3$ inches of mercury.

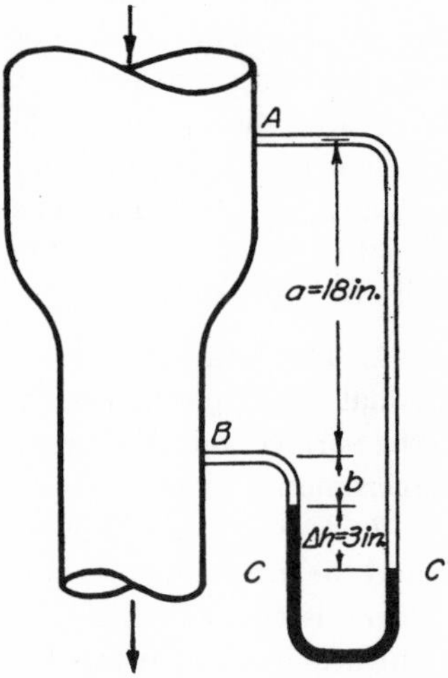

A
a=18in.
B
b
Δh=3in.
C
C'

(*a*) Since points C and C' lie at the same elevation in the same continuous body of liquid (i.e., mercury), $p_C = p_{C'}$. But $p_C = p_B + \gamma b + \gamma' \Delta h_{Hg}$ and $p_{C'} = p_A + \gamma(a + b + \Delta h_{Hg})$; therefore

$$p_B + \gamma b + \gamma' \Delta h_{Hg} = p_A + \gamma(a + b + \Delta h)$$

or

$$p_A - p_B = (\gamma' - \gamma)\Delta h_{Hg} - \gamma a$$

Evidently, the distance b to the U-tube is of no importance as long as both columns are filled with the same fluid Introducing the given values,

$$p_A - p_B = (13.6 - 0.85) \times 62.4 \times \tfrac{3}{12} - 0.85 \times 62.4 \times \tfrac{18}{12} = 119.4 \text{ psf}$$

(*b*) From Eq. (46)

$$h_A - h_B = \frac{p_A - p_B}{\gamma} + z_A - z_B = \frac{119.4}{0.85 \times 62.4} + \frac{18}{12} = 2.25 + 1.5 = 3.75 \text{ ft}$$

Hydrostatic pressure on boundary surfaces. The determination of resultant forces due to fluid pressure in general is based upon two fundamental relationships of mechanics:

I. Any component of the resultant normal force exerted upon a surface of area A is equal to the integral (see Example 11) of the corresponding component of the differential normal force $p\,dA$:

$$F_x = \int (p\,dA)_x \tag{47}$$

II. The product of this force component and the normal distance y_0 of its line of action from a parallel reference plane is equal to the integral of the first moment of the differential normal force with respect to the reference plane:

$$F_x\,y_0 = \int y(p\,dA)_x \tag{48}$$

Evidently, both the resultant force and the location of its line of action may be found from the foregoing equations written for each of the coordinate directions.

These general relationships must always be followed in the computation of boundary, or free-body, forces, regardless of whether the pressure distribution has to be determined by means of the flow net for accelerated motion, as in the following section of this chapter, or by the elementary relationship for hydrostatic variation in motion which is steady and uniform. The latter case, however, permits the use of special forms of these relationships, as embodied in the following principles of hydrostatics:

1. The total force due to liquid pressure upon a plane surface is equal to the product of the specific weight of the liquid, the area of the surface, and the depth of its centroid below the free surface or piezometric plane.
2. The distance between the center of pressure of a plane surface and the intersection of this plane with the free surface or piezometric plane is equal to the ratio of the second and first moments of the surface area about this line of intersection.
3. The horizontal component of liquid pressure upon any surface has the same magnitude and line of action as the horizontal force computed for the projection of this surface upon a vertical plane.
4. The vertical component of liquid pressure upon any surface is equal in magnitude to the weight of a column of the liquid extending vertically from the surface in question to the free surface or piezometric plane, its line of action passing through the center of gravity of this liquid column.

Verification of computation principles. The first of these principles may be verified by means of Eq. (47) written for the vector itself; that is, with reference to Fig. 39 (in which the pressure head or depth below the free surface is indicated by the symbol h_p), the total force

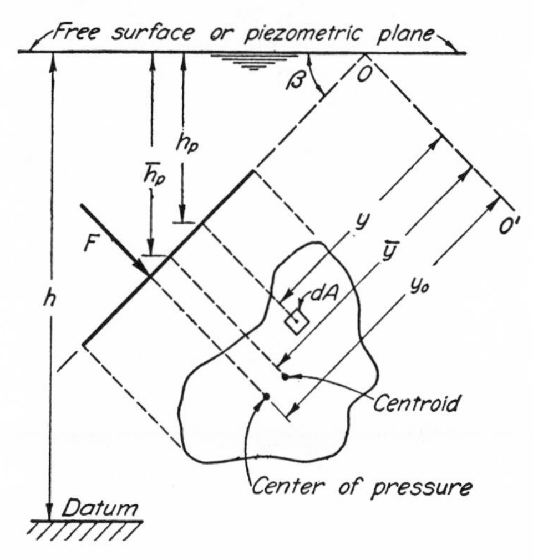

FIG. 39. Evaluation of fluid pressure upon a plane surface.

upon the given surface is the integral of the product of specific weight, pressure head, and the increment of area:

$$F = \int p \, dA = \int (\gamma h_p) \, dA = \gamma \sin \beta \int y \, dA$$

But the quantity $\int y \, dA$ is known from mechanics to be the moment M_0 of the area about the line O–O', or the product of the area and the distance $\bar{y}$ from the line to its centroid; therefore,

$$F = \gamma \sin \beta \, \bar{y} A = \gamma \bar{h}_p A \tag{49}$$

the product of γ and the pressure head $\bar{h}_p$ at the centroid evidently corresponding to the mean pressure intensity $\bar{p}$ upon the surfacc.

The second principle may be verified from the same figure by means of Eq. (48) written in terms of the force vector; thus, designating the point of intersection of the action line and the surface as the center of pressure, which lies the distance y_0 from the reference line O–O',

$$F y_0 = \int y p \, dA = \int y(\gamma h_p) \, dA = \gamma \sin \beta \int y^2 \, dA$$

Solving for y_0 and replacing F by its equivalent integral,

$$y_0 = \frac{\gamma \sin \beta \int y^2\, dA}{\gamma \sin \beta \int y\, dA} = \frac{\int y^2\, dA}{\int y\, dA}$$

But $\int y^2\, dA$ is known from mechanics to represent the second moment I_0 of the area about the line O–O'. This in turn is equal to

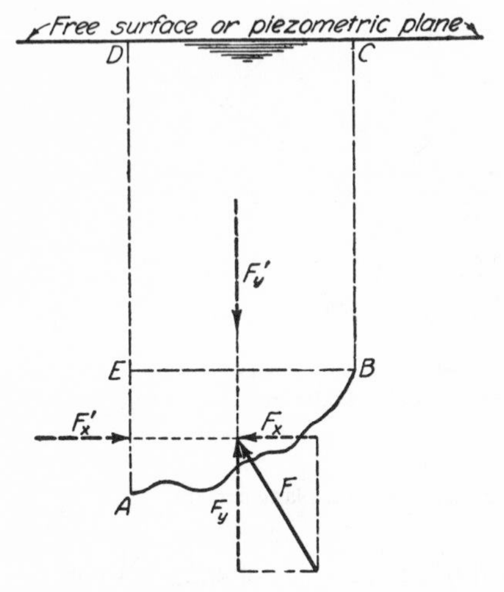

FIG. 40. Evaluation of pressure components upon any surface.

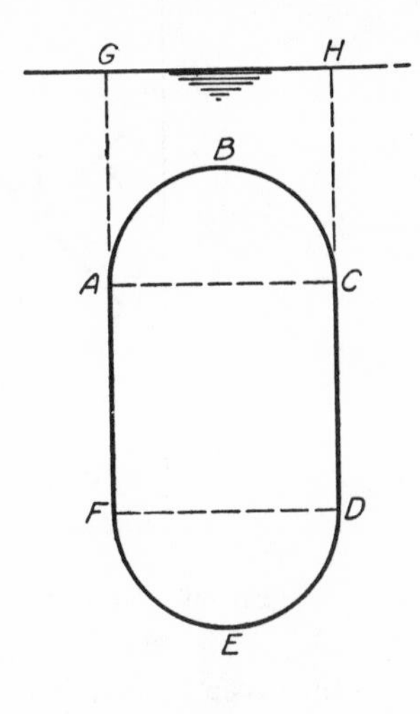

Fig. 41. Definition sketch for fluid buoyancy.

$\bar{I} + \bar{y}^2 A = (\bar{k}^2 + \bar{y}^2)A$, in which $\bar{k}$ is the radius of gyration of the area about a centroidal axis parallel to O–O'. Therefore,

$$y_0 = \frac{I_0}{M_0} = \bar{y} + \frac{\bar{I}}{\bar{y}A} = \bar{y} + \frac{\bar{k}^2}{\bar{y}} \tag{50}$$

Principle 3 may be checked by considering the horizontal forces acting upon the free body of liquid ABE in Fig. 40, from which it will be seen at once that the horizontal component of force exerted upon the curved surface AB must be in equilibrium with the normal force upon the vertical projection AE. Similarly, considering the free body $ABCD$, it is apparent that the vertical component of force exerted upon the surface AB must be in equilibrium with the weight of the superposed liquid, as stated in principle 4. It should be noted at this point that principle 4 explains the phenomenon of *buoyancy*, in which, as discovered by Archimedes two thousand years ago, the vertical hydrostatic force exerted upon an immersed body by the surrounding fluid is equal to the weight of the fluid which the body displaces. Thus, with reference to Fig. 41, according to principle 4 the downward force upon the top of body *ABCDEF* is equal to the weight of the liquid column *ABCHG*, and the upward force of the liquid upon the bottom is equal to the weight of a liquid column having the dimensions *DEFGH*. The net buoyant force is obviously the difference of these two and hence equal to the weight of liquid displaced by the body. If this is equal to the weight of the body itself, a state of static equilibrium must obviously prevail.

Example 16. An open conduit having the cross section shown in the accompanying figure carries water at a uniform depth of 6 feet. Determine the magnitude, direction, and line of action of the force per unit length of conduit exerted upon sections *AB*, *BC*, and *CD* of the boundary.

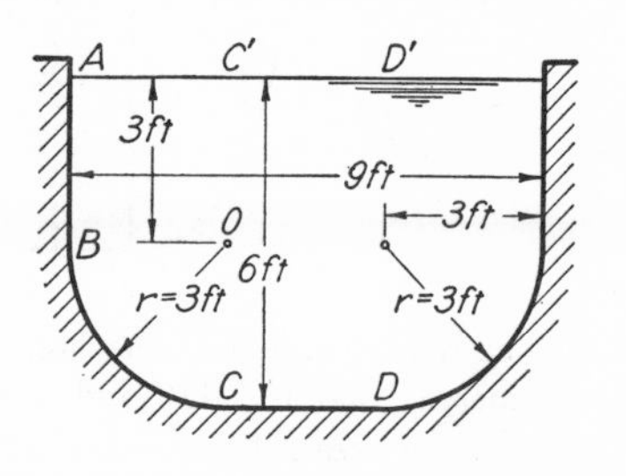

As in Example 11, the required forces may be found by plotting the pressure distribution around the boundary and determining the areas under the curves (see sketch *a*). However, since the pressure varies linearly with the depth, the forces may be computed more conveniently as follows:

The total force per unit length on the horizontal bottom is evidently equal to the weight of the water directly above it:

$$\frac{F_{CD}}{L} = 6 \times 3 \times 62.4 = 1124 \text{ lb/ft}$$

and its line of action passes vertically through the centerline of the conduit as shown in sketch *b*.

Since the mean pressure head on the section *AB* is 1.5 feet, from Eq. (49)

$$\frac{F_{AB}}{L} = 62.4 \times 1.5 \times 3 = 281 \text{ lb/ft}$$

From Eq. (50)

$$(y_0)_{AB} = 1.5 + \frac{\dfrac{L \times 3^3}{12}}{1.5 \times 3 \times L} = 2 \text{ ft}$$

The line of action therefore lies 2 feet below the water surface.

Since the mean pressure head on the vertical projection of *BC* is 4.5 feet,

$$\frac{(F_{BC})_x}{L} = 62.4 \times 4.5 \times 3 = 843 \text{ lb/ft}$$

and since the weight of water above BC must equal the vertical component of force,

$$\frac{(F_{BC})_y}{L} = 62.4\left(3 \times 3 + \frac{\pi \times 3^2}{4}\right) = 1003 \text{ lb/ft}$$

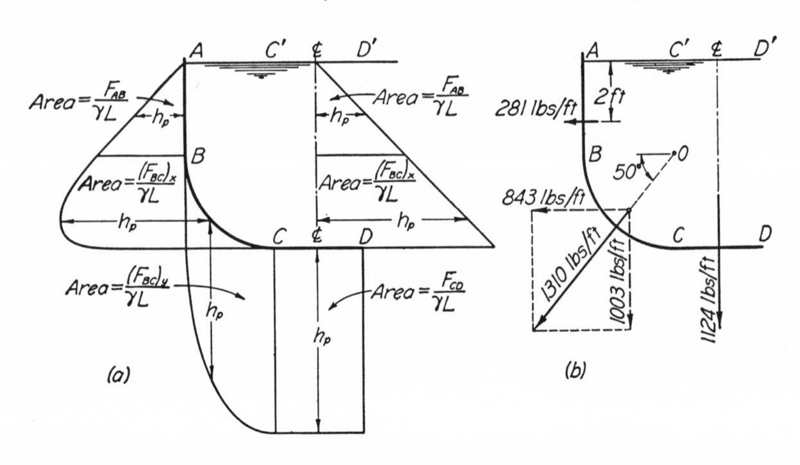

The line of action of the horizontal component may be found in the same manner as that for AB, and the vertical component must pass through the centroid of area $ABCC'$. But since their resultant is the integral of a *normal* force around the circular arc CD, it must evidently pass through the center O (see sketch b). Its slope will be

$$\tan \beta = \frac{(F_{BC})_y}{(F_{BC})_x} = \frac{1003}{843} = 1.19 \quad (\text{i.e., } \beta = 50°)$$

and its magnitude

$$F_{BC} = \sqrt{843^2 + 1003^2} = 1310 \text{ lb/ft}$$

PROBLEMS

80. An upright U-tube partly filled with mercury (specific gravity = 13.6) is connected to piezometers on pipes A and B to measure the pressure difference at various rates of flow. Determine the differential pressure $p_A - p_B$ (a) if both pipes carry a gas having a specific weight of 0.01 pound per cubic foot, and (b) if both pipes carry water.

81. A closed tank is partly filled with oil under pressure, piezometer tubes being arranged as shown to indicate oil level and pressure. If the specific gravity of the oil is 0.75, what is the pressure intensity of the air above the oil?

82. The differential head for a flow meter in a pipe carrying water is measured by means of the inverted U-tube shown in the accompanying sketch. What is the difference in piezometric head between points A and B (a) if the space above the water columns is filled with air, and (b) if the space is filled with a liquid having a specific gravity of 0.8?

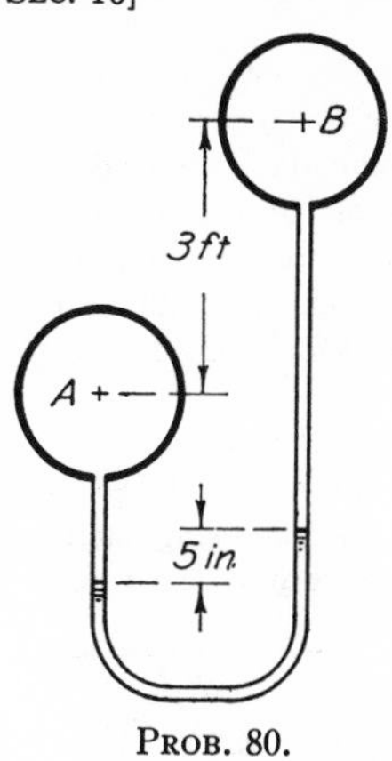

PROB. 80.

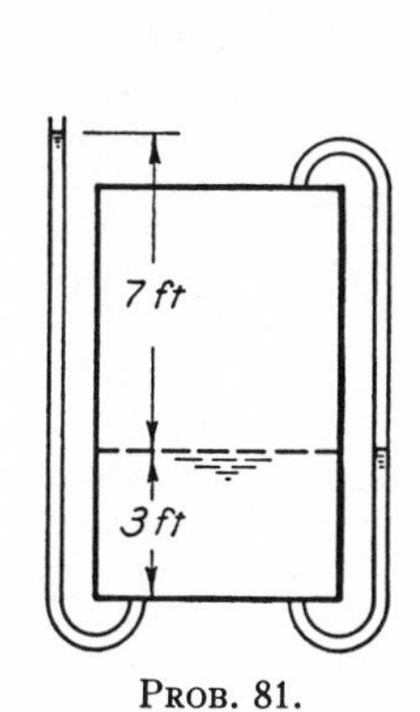

PROB. 81.

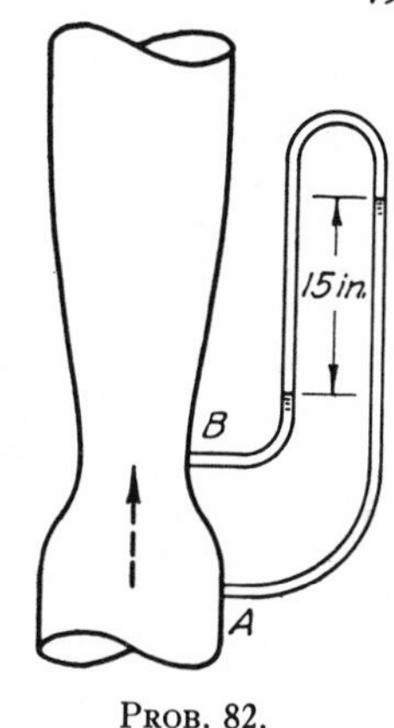

PROB. 82.

83. The tank shown in the accompanying sketch will discharge liquid at a constant rate so long as the liquid surface within the tank lies above the bottom of the air inlet. Determine the liquid level in piezometers A and B under the conditions shown. What is the pressure intensity of the air in the tank if the specific weight of the liquid is 85 pounds per cubic foot?

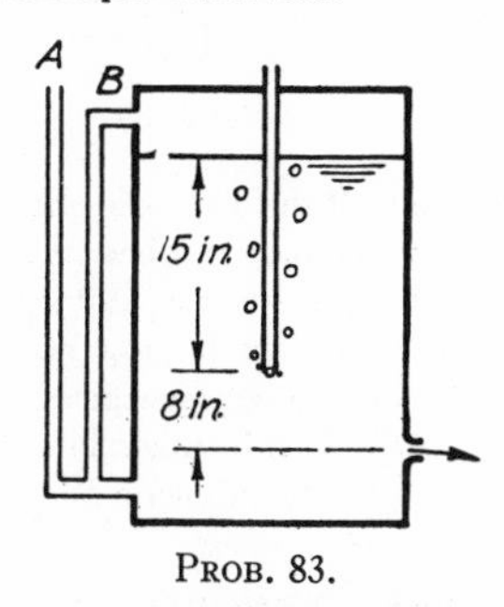

PROB. 83.

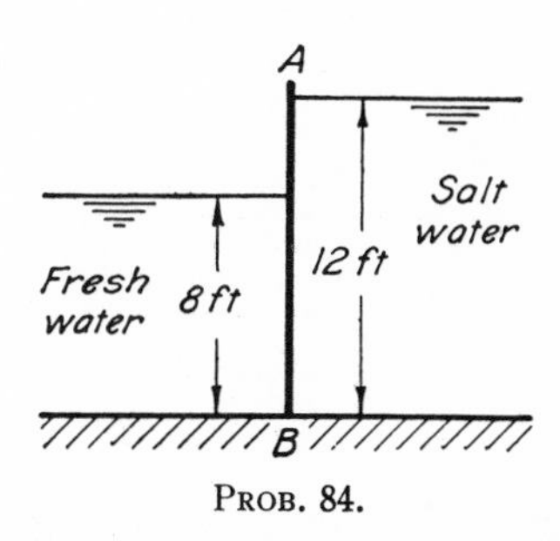

PROB. 84.

84. Fresh water stands at a depth of 8 feet on one side of vertical sheet piling AB, and salt water at a depth of 12 feet on the other side. Assuming the salt water to have a specific gravity of 1.03, determine the moment about B of the resultant pressure per foot of piling.

85. The upstream face of a dam is inclined at an angle of 60° to the horizontal. If the free surface of the water in the reservoir lies 25 feet above the base of the dam, determine the horizontal and vertical components and the line of action of the resultant force of the water per unit length of dam.

86. A circular butterfly gate 10 feet in diameter is pivoted about a horizontal axis passing through its center. What force F applied at the bottom is required to hold the gate in position if water stands 2 feet above the gate on one side and the other side is completely exposed to the air?

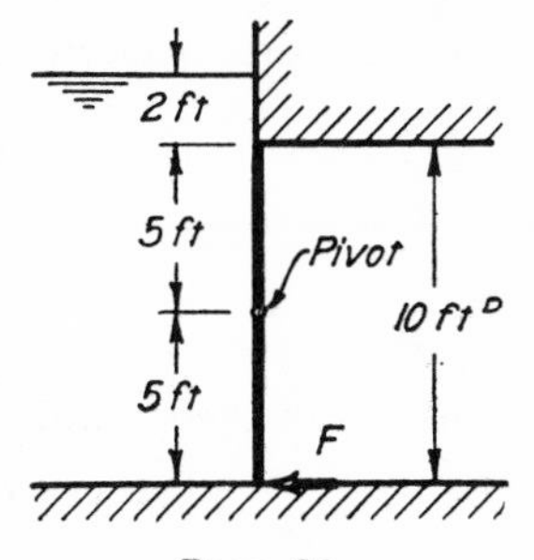

PROB. 86.

87. A cylindrical control weir (which may be raised on racks not shown in the figure) has a diameter of 8 feet and a length of 15 feet. Determine the magnitude, direction, and location of the horizontal and vertical components of water pressure upon the weir under the conditions illustrated. Show that the resultant force passes through the axis of the cylinder.

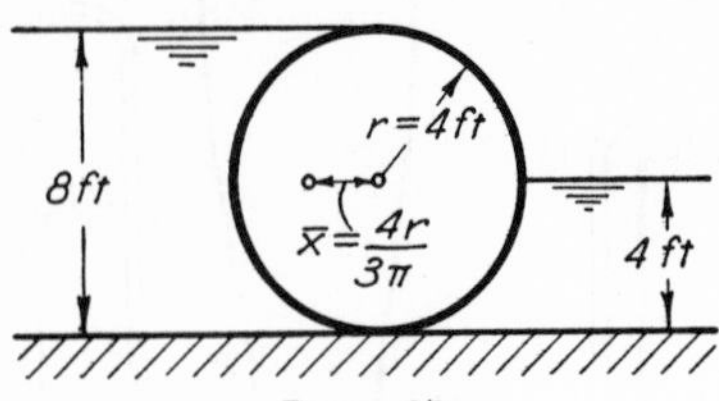

PROB. 87.

88. Oil having a specific gravity of 0.85 half fills a cubic tank measuring 3 feet on a side, the air above the oil being under a pressure of 1 pound per square inch. Determine the total pressure on the top, the bottom, and one side of the tank, and locate the center of pressure in the last case.

89. A tainter gate having a length of 10 feet is mounted on a spillway in the position shown. When water stands at the top of the gate, what are the magnitude and the direction of the resultant pressure upon the gate? (Note that the line of action of the resultant must pass through the gate axis.)

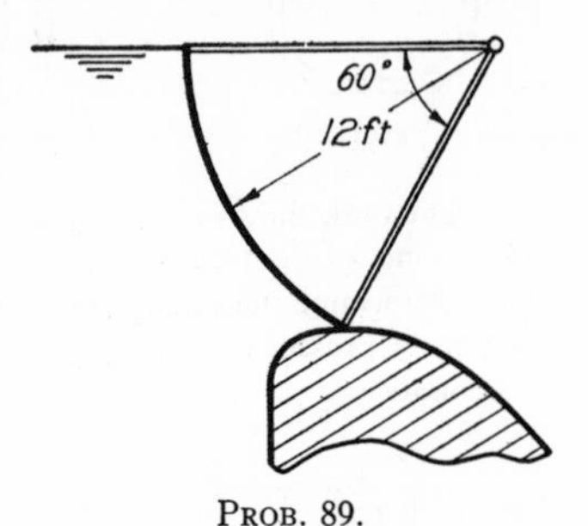

PROB. 89.

90. Determine completely the forces exerted per unit length on the sides and bottom of the barge shown in cross section in the accompanying sketch.

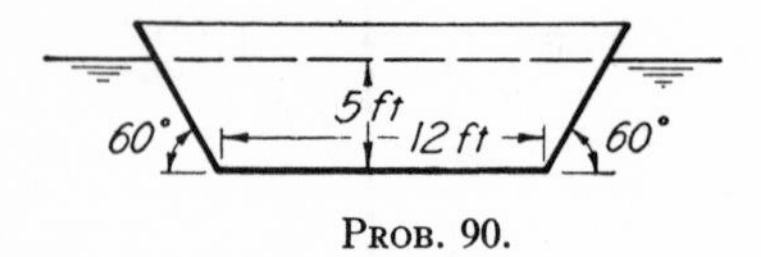

PROB. 90.

91. If the displacement of a 600-ton submarine when cruising at the ocean surface is 90 per cent of its total volume, how many cubic feet of water (specific gravity = 1.025) must be taken into its tanks in order to submerge?

17. DERIVATION AND SIGNIFICANCE OF THE BERNOULLI THEOREM

Integration of the Euler equation along a stream line. If Eq. (43) is written for the tangential and normal directions, and if the quantities a_s and a_n are replaced by their equivalent values from Eqs. (18) and (19), the Eulerian equations of acceleration will take the form

$$-\frac{\partial}{\partial s}\left(\frac{p}{\gamma} + z\right) = \frac{1}{g}\frac{\partial v_s}{\partial t} + \frac{1}{2g}\frac{\partial(v^2)}{\partial s} \tag{51}$$

$$-\frac{\partial}{\partial n}\left(\frac{p}{\gamma} + z\right) = \frac{1}{g}\frac{\partial v_n}{\partial t} + \frac{v^2}{gr} \tag{52}$$

The first of these relationships may also be written as

$$-\frac{\partial}{\partial s}\left(\frac{v^2}{2g} + \frac{p}{\gamma} + z\right) = \frac{1}{g}\frac{\partial v_s}{\partial t} \tag{53}$$

which states that the sum $v^2/2g + p/\gamma + z$ must vary along a stream line if the velocity varies with time. If the flow is steady, on the other hand (i.e., if $\partial v_s/\partial t = 0$), integration of this equation between any two points on the same stream line will yield a relationship known, in honor of a Swiss scientist Daniel Bernoulli (1700–1782) living at the same time as Euler, as the *Bernoulli theorem:*

$$\frac{v_1^2}{2g} + \frac{p_1}{\gamma} + z_1 = \frac{v_2^2}{2g} + \frac{p_2}{\gamma} + z_2 \tag{54}$$

In the preceding sections it was shown that the terms p/γ and z represent lengths, or heads, the sum of which (i.e., the piezometric head h) indicates the elevation of the liquid surface in an open piezometer tube connected to a small boundary orifice at the point in question. Only in uniform flow, of course, it is possible to assume a constant magnitude of h at all points of flow, since according to Eq. (44) h will vary in the direction of acceleration. In the measurement of boundary pressure, therefore, it is necessary that the piezometer orifice be so formed that the flow is in no way disturbed, since any disturbance tending to produce local acceleration will have a corresponding effect upon the piezometric head. Indeed, the disturbance produced by the introduction of a Pitot tube—the formation of a point of stagnation—causes the piezometric head to increase at this point by the amount $v^2/2g$, a quantity aptly termed the *velocity head.* The sum $v^2/2g + p/\gamma + z = H$ is therefore indicated by a stagnation tube and is known as the *total head* of the flow at the point in question. Equation (53) thus states that the total head will change in the direction of motion if local acceleration occurs (i.e., if the flow is unsteady), and Eq. (54) that the total head will have the same instantaneous magnitude at all points along any stream line if the flow is steady.

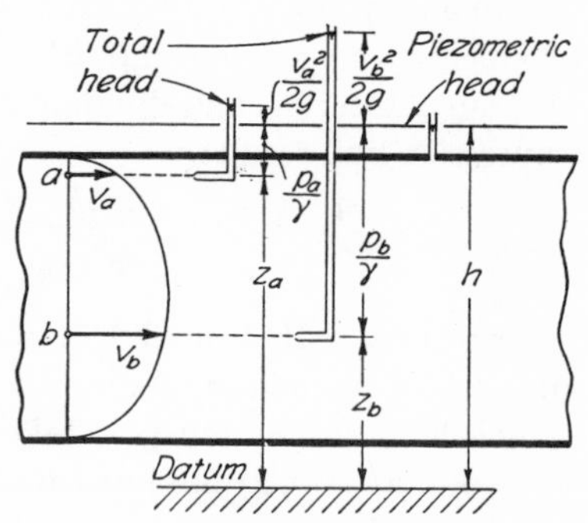

FIG. 42. Variation of total head in rotational flow.

The Bernoulli equation for steady, irrotational flow. If the flow is rotational, it will be found that the total head will vary from one stream line to the next, however constant it may remain along any one such line. Consider, for instance, the steady, uniform, rotational motion indicated in Fig. 42; the piezometric head will necessarily be the same at all points, owing to the absence of acceleration, but the variation in velocity, and hence in velocity head, across the flow will yield a different total head for each stream line, as shown by the several stagnation tubes. This may be seen for the general case (compare with the corresponding operation in Section 11) by adding the quantity $-\partial(v^2/2g)/\partial n$ to both sides of Eq. (52) for the case of steady flow:

$$-\frac{\partial}{\partial n}\left(\frac{v^2}{2g}+\frac{p}{\gamma}+z\right)=\frac{v}{g}\left(\frac{\partial v_n}{\partial s}-\frac{\partial v_s}{\partial n}\right)$$

Evidently, only if $\partial v_n/\partial s = \partial v_s/\partial n$ will the flow be irrotational, and only then may this equation be integrated in the n direction to yield the counterpart of Eq. (54) along lines normal to the stream lines. Thus, only in the case of steady, irrotational flow will the total head have the same instantaneous magnitude throughout the moving fluid, the Bernoulli equation then applying between any two arbitrary points:

$$\frac{v_1^2}{2g}+\frac{p_1}{\gamma}+z_1=\frac{v_2^2}{2g}+\frac{p_2}{\gamma}+z_2 \tag{55}$$

Variation of piezometric head in enclosed flow. If such conditions of motion are essentially fulfilled, the flow net and the equation of continuity may be used as before to evaluate the velocity distribution, and Eq. (55) will then permit determination of the pressure distribution in any desired region. It is noteworthy, in this connection, that in liquid flow which is fully confined by solid boundaries the variation in piezometric head will be directly proportional to the variation in the quantity Δp studied in the preceding chapter. That is, upon introducing the relationship $\Delta h = h - h_0$ in the Bernoulli equation, rearranging terms, and dividing by $v_0^2/2g$, the following dimensionless expression will result:

$$\frac{\Delta h}{v_0^2/2g}=1-\left(\frac{v}{v_0}\right)^2 \tag{56}$$

Comparison with Eq. (29) of the last chapter will indicate that, for any closed boundary profile, the plot of $\Delta h/(v_0^2/2g)$ for conditions in which fluid weight must be taken into account will be identical with that of $\Delta p/(\rho v_0^2/2)$ obtained for flow with negligible weight effects.

The significance of this similarity is as follows: In confined flow, the stream-line pattern (and hence the velocity distribution) for steady, irrotational motion is governed entirely by the geometry of the boundaries. If the fluid is a gas, or if the motion is in the horizontal plane, the variation in $\Delta p(\rho v_0^2/2)$ will in turn be governed solely by the boundary geometry. If the fluid is a liquid, however, and if the motion involves vertical displacement, the pressure intensity will also vary owing to differences in elevation. But, since the boundary geometry completely determines the form of the stream lines, gravitational effects can have no influence upon the velocity distribution in flow which is wholly enclosed by fixed or moving boundaries. The sole effect of fluid weight in confined flow is therefore embodied in the

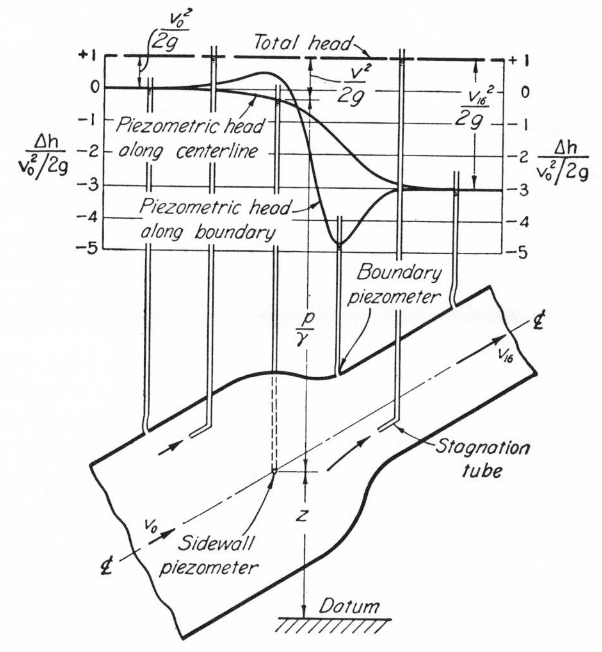

FIG. 43. Variation of velocity head and piezometric head for the flow net of Fig. 12.

pressure relationship $p = \gamma(h - z)$, for h is a function only of the velocity distribution.

Such conditions are clearly illustrated by liquid flowing through the two-dimensional profile of Fig. 12, the axis being inclined at some arbitrary angle as shown in Fig. 43. Assuming steady, irrotational motion, the total head will be the same at all points, and hence may be represented by a horizontal line showing the elevation to which liquid will rise in a piezometer column connected to a stagnation tube held at any point in the flow. Columns connected to boundary orifices, on the other hand, will yield a line of piezometric head of variable elevation, as plotted in the figure for the upper boundary; since the velocity distribution (and hence the variation in velocity head) is the same along either boundary, the piezometric heads will necessarily be the same at corresponding boundary points. Along the centerline, however, the difference in velocity distribution will cause the line of piezometric head to differ from that for either boundary, as indicated in the illustration.

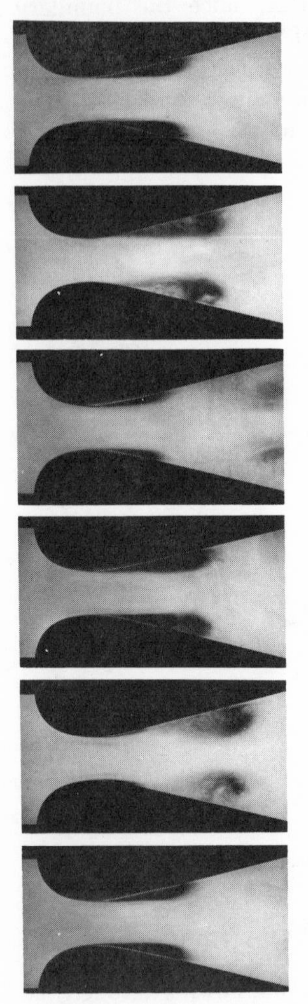

PLATE V. Cavitation at a two-dimensional boundary constriction; successive frames from a motion picture, showing formation and collapse of vapor pockets (camera speed: 64 frames per second).

Variation of pressure intensity; cavitation. Once the piezometric head is known for any point, it is obviously a simple matter to subtract therefrom the elevation of the point to obtain the pressure head. Indeed, since the gradient of piezometric head can change only with the rate of flow, it will be seen that an arbitrary change in pressure head throughout the system will raise or lower a line of piezometric head the same distance at all points, its form thereby remaining constant. A change in the rate of flow, on the other hand, will change only the absolute magnitudes of Δh, the relative proportions of the curve being unaffected. In brief, the ratio of Δh to the velocity head $v_0^2/2g$ at any point will be independent of both the overall pressure load and the rate of flow, in accordance with Eq. (56); the dimensionless scale for $\Delta h/(v_0^2/2g)$ at either side of the figure is therefore generally characteristic of this particular boundary profile and is necessarily identical to the scale for $\Delta p/(\rho v_0^2/2)$ as evaluated in Chapter III for gaseous flow, or for liquid flow in the horizontal plane.

Although the Bernoulli equation indicates the existence of neither a maximum velocity nor a minimum pressure intensity for either liquid or gaseous motion, it must be noted that any liquid will begin to boil once a value of p equal to its vapor pressure is reached at some point in the field of motion. Any tendency to reduce the pressure intensity below this limiting value will result in a discontinuity of the flow due to the rapid formation and collapse of vapor cavities in the liquid, a phenomenon aptly known as *cavitation*. From the Bernoulli equation it will be evident that the reduction of pressure intensity at any point may be caused by a reduction in the pressure load upon the system as a whole, or by an increase in either the elevation or the velocity of flow. Cavitation generally results from a combination of these several influences, and it should be avoided by proper design of boundary form and regulation of operating conditions.

Example 17. Let Fig. 12 represent the profile of a two-dimensional constriction in a 4 by 4-foot rectangular conduit. If the axis of the conduit is vertical and the direction of motion is downward, what difference in intensity will be indicated by pressure gages at points A–6 and A–9 as 200 cubic feet of water per second pass through the conduit?

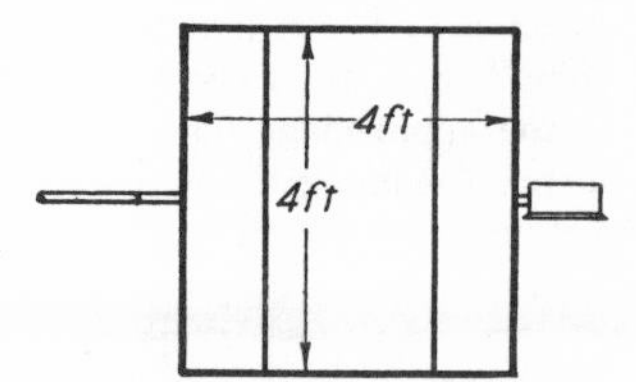

Since $Q = v_0 A_0$

$$v_0 = \frac{200}{16} = 12.5 \text{ fps}$$

From the plot of $(v/v_0)^2$ accompanying Example 6,

$$\left(\frac{v_6}{v_0}\right)^2 = 0.5 \quad \text{and} \quad \left(\frac{v_9}{v_0}\right)^2 = 5.5$$

Therefore

$$v_9^2 - v_6^2 = v_0^2(5.5 - 0.5)$$

$$= \overline{12.5}^2 \times 5.0 = 781 \text{ (fps)}^2$$

From Eq. (55)

$$\frac{p_6 - p_9}{\gamma} = \frac{v_9^2 - v_6^2}{2g} + z_9 - z_6$$

Since, by scaling, $z_9 - z_6 = -0.29 n_0 = -0.29 \times 4 = -1.16$ ft,

$$p_6 - p_9 = 62.4\left(\frac{781}{64.4} - 1.16\right) = 684 \text{ psf} = 4.75 \text{ psi}$$

It is to be noted that $(v_9^2 - v_6^2)/2g$ corresponds to the difference in piezometric head between boundary points 6 and 9. Since this difference is read directly on a differential manometer using air as the gage fluid (see figure),

$$\Delta h = h_6 - h_9 = \frac{781}{64.4} = 12.1 \text{ ft}$$

PROBLEMS

92. By means of open glass columns connected to a wall piezometer and a stagnation tube in a pipe carrying oil, the following data were obtained: when $y = 1$ inch, $h = 2$ inches; when $y = 2$ inches, $h = 3$ inches. If the specific gravity of the oil is 0.85, what velocities are indicated?

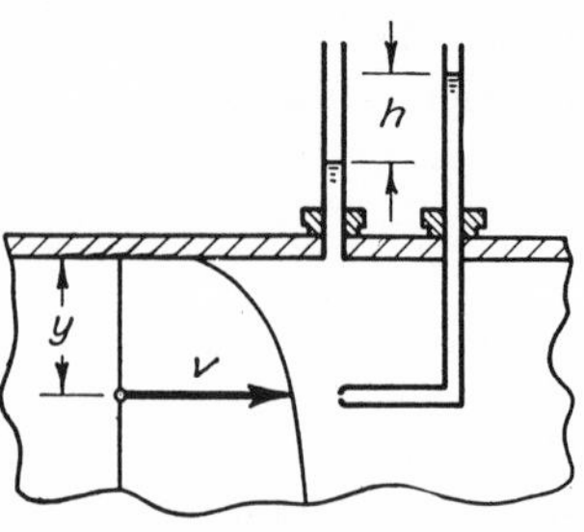

PROB. 92.

93. If the orifice of Problem 83 has a diameter of ½ inch and is so rounded that the coefficient of contraction is unity, what will be the rate of efflux under the head indicated?

94. Cavitation is found to occur at an orifice meter in a 4-inch pipe at a rate of flow of 0.5 cubic foot of water per second when the pressure in the approaching flow is 6 pounds per square inch. Assuming the vapor pressure of the water to be 0.5 pound per square inch absolute, estimate the maximum change in water velocity at the orifice.

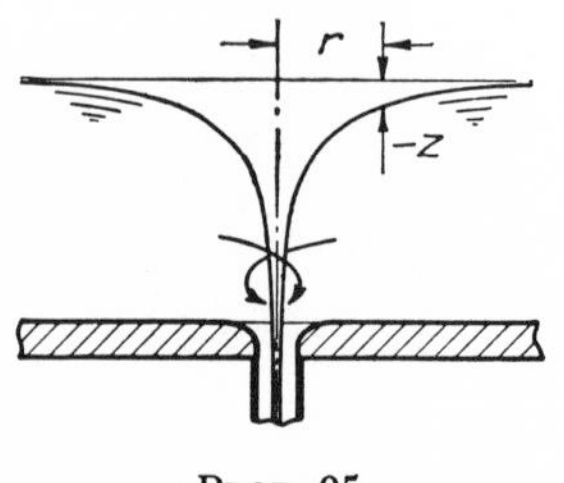

PROB. 95.

95. The irrotational vortex or "whirlpool" which tends to form above an open drain in a relatively shallow tank is characterized by a tangential velocity which varies inversely with radial distance from a vertical axis through the drain. If the tangential velocity of a given vortex is 3 inches per second 15 inches from the axis, what will be the decrease in surface elevation (*a*) at this distance, and (*b*) at a distance of 1 inch from the axis?

96. Assuming the conduit transition of Example 17 to be horizontal with points A–6 and A–9 along the upper boundary, compute the piezometric head at point A–6 which will indicate the onset of cavitation at point A–9 for the given rate of flow and a vapor pressure of 0.5 pound per square inch.

97. The outlet in the bottom of a tank is so formed that the velocity at point A is 1.5 times the mean velocity within the outlet pipe. If the depth of water within the tank is 3 feet, what is the greatest length L of pipe which may be used without producing cavitation? (Assume a vapor pressure of 0.5 pound per square inch absolute.)

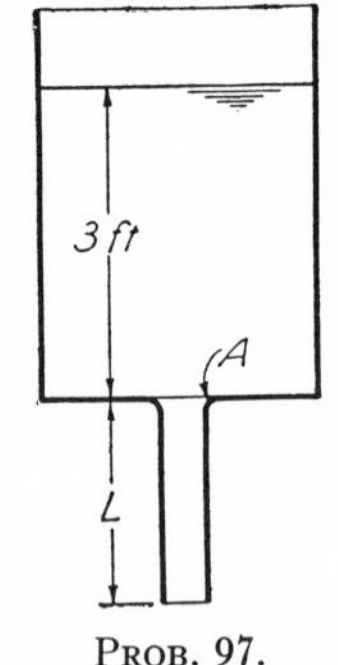

PROB. 97.

98. Let the profile of Fig. 26 represent a vertical section through a horizontal guide vane at the entrance to a power plant. Assuming the velocity of the approaching water to be 20 feet per second and the vertical thickness of the vane to be 3 feet, determine the difference in (*a*) piezometric head, and (*b*) pressure intensity, between the point of stagnation and the point of minimum pressure intensity.

99. The intake to a hydroelectric installation has the form of the upper half of Fig. 15. Assuming the intake to be of sufficient width for the flow to be essentially two-dimensional, estimate the minimum intensity of pressure which will prevail along the curved boundary under the conditions illustrated.

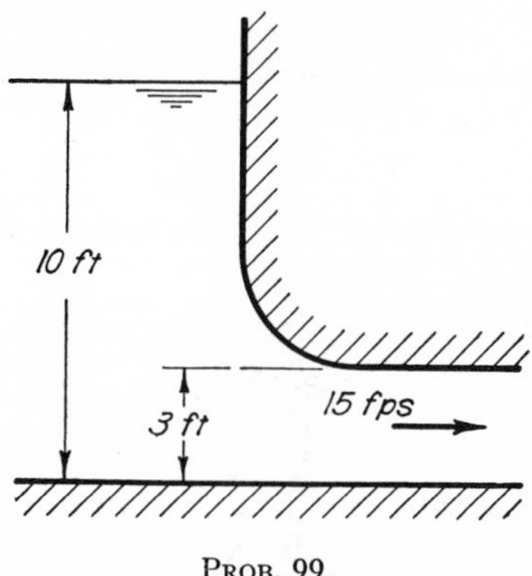

PROB. 99.

18. CURVILINEAR FLOW WITH A FREE SURFACE

Constancy of pressure intensity over a free surface. If in any zone a moving fluid comes into contact with a fluid of different density, the pattern of motion as a whole will no longer be governed in form solely by the geometry of fixed boundaries, for gravitational attraction may now have an appreciable effect upon the acceleration of the fluid. The form of the surface of contact is still uniquely determined, however, by the fact that the pressure intensity at every point along any such surface must be the same in both fluids. In the case of a liquid in contact with the atmosphere, the pressure intensity along the outermost stream lines—the free surface—will then be constant, the configuration of the entire flow pattern being such as to satisfy this boundary condition in addition to the general equations of motion.

Perhaps the most familiar illustration of curvilinear flow with a free surface is found in the "whirlpool" which tends to form as liquid is discharged through an outlet in the bottom of a shallow tank. The velocity distribution of such flow closely approximates that of the irrotational vortex, in which the tangential velocity varies inversely with radial distance from the axis (i.e., $v = C/r$). Since the pressure head will be equal to zero at all points on the free surface, the surface elevation must decrease with increasing velocity head; in other words, the surface will fall away more and more rapidly as the vortex axis is approached, with the result that a funnel-shaped filament of air will seem to descend into the body of the liquid. It is to be noted, however, that the accompanying liquid acceleration is entirely in the horizontal direction (that is, toward the center of curvature of the circular stream lines), so that the pressure distribution is still hydrostatic in the vertical direction. One must therefore seek beyond the irrotational vortex for flow in which gravity is even partly responsible for the fluid acceleration.

PLATE VI. Characteristic surface profile of the irrotational vortex, produced by circulation of water above an outlet.

Flow of this nature is illustrated by Fig. 44, in which will be seen the pattern of two-dimensional, steady, irrotational flow of a liquid through a sharp-edged slot in a vertical plate at the end of a channel. In the zone of essentially uniform flow to the left of the figure, the lack of appreciable acceleration in any direction indicates that the line of piezometric head h_0 must coincide with the free surface. The horizontal line of constant total head must therefore lie the distance $v_0^2/2g$ above the free surface in this zone, the quantity v_0 being the velocity of the approaching uniform flow. At the two points of stagnation the velocity is reduced to zero and hence the piezometric head must increase by the amount $v_0^2/2g$, corresponding to an increase in pressure head at the lower corner and an increase in elevation at the free surface. As the liquid passes through the slot, the velocity along the outermost stream lines must increase to such an extent that, when the pressure intensity becomes equal to that of the atmosphere, the velocity head will equal the distance between the total-head line and either free surface (i.e., when $p/\gamma = 0$, $H = v^2/2g + z$). But, since the two free surfaces lie at different elevations, it is evident that the velocity (and thus the stream-line spacing) must differ accordingly at the

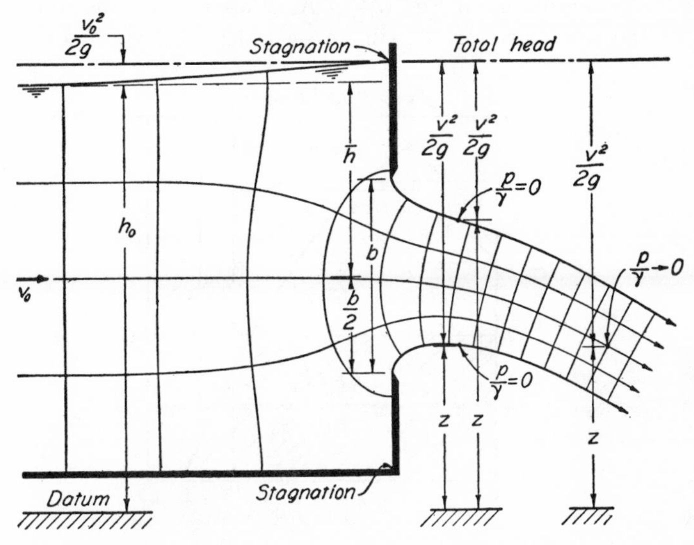

FIG. 44. Flow net for efflux from a slot under the influence of gravity.

two edges of the slot, and hence the deformation of the meshes of the flow net due to gravitational effects must already be noticeable before the slot is reached (compare Fig. 44 with Fig. 29). As the fluid emerges from the slot, the jet as a whole will be subject to full gravitational acceleration. The pressure intensity within the jet, however, will be reduced to that of the atmosphere only as the convergence of the stream lines approaches a minimum well beyond the slot. Within the latter zone of freely falling liquid the velocity head will be equal to the vertical distance below the line of constant total head at all points.

Discharge equation for a horizontal slot. So long as the average head $\bar{h}$ (i.e., the piezometric head of the approaching flow referred to the slot axis) is relatively great, the mean velocity of the contracted jet will almost exactly equal the quantity $\sqrt{2g\bar{h}}$. The rate of efflux may then be computed according to the methods of Section 12, the quantity Δp being equal to the product of $\bar{h}$ (which is simply the average Δh) and the specific weight γ of the liquid, and the coefficient of contraction of the flow (for a symmetrically located slot) being taken from Table I. Thus for large values of $2h/b$ (i.e., B, b),

$$q = \frac{C_c}{\sqrt{1 - C_c^2(b/2\bar{h})^2}} b\sqrt{2g\bar{h}} = C_d\, b\sqrt{2g\bar{h}} \tag{57}$$

C_d varying according to Eq. (34). As $\bar{h}$ approaches the dimension $b/2$ of the slot, however, it will be seen that such methods should

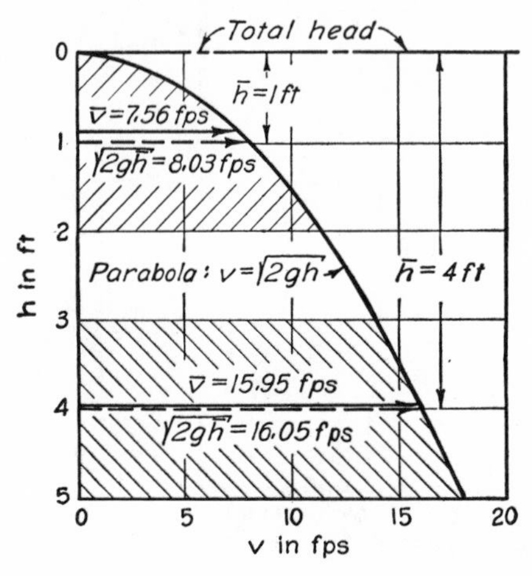

FIG. 45. Approximation of mean velocities from mean velocity heads.

become less and less exact, owing to the increasing difference between $\sqrt{2g\bar{h}}$ and the average velocity across the jet; in other words (see Fig. 45), the mean velocity can no longer be evaluated from the mean velocity head, since the square root of the mean of several values and the mean of their square roots are not numerically the same. Under the latter circumstances, the rate of efflux may be more closely approximated by integration. From the Bernoulli equation, the velocity at any point in the contracted jet will have the magnitude

$$v = \sqrt{2g\left(\frac{v_0^2}{2g} + \bar{h} - z\right)}$$

Indicating the increment of cross section per unit length of jet by the product of a local contraction factor c_c and the height increment dz, across a section of zero pressure intensity the differential rate of efflux per unit length of slot will be

$$dq = v\, c_c\, dz = c_c \sqrt{2g\left(\frac{v_0^2}{2g} + \bar{h} - z\right)}\, dz$$

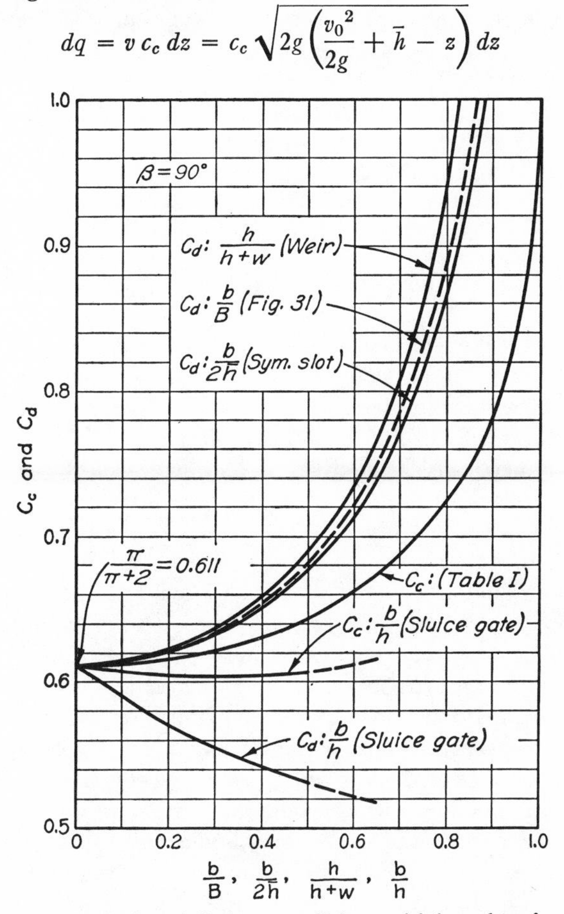

FIG. 46. Variation of discharge coefficients with boundary form.

The integral of this expression will involve both c_c and z as variables. However, if c_c is replaced by a mean value C_c, and if z is assumed to vary from $-b/2$ to $+\,b/2$, integration will yield

$$q = \tfrac{2}{3} C_c \sqrt{2g}\left[\left(\frac{v_0^2}{2g} + \bar{h} + \frac{b}{2}\right)^{3/2} - \left(\frac{v_0^2}{2g} + \bar{h} - \frac{b}{2}\right)^{3/2}\right] \qquad (58)$$

Despite the deflection of the jet, the values of C_c in Table I still provide a close approximation to the coefficient of this equation. However, owing to the troublesome factor $v_0^2/2g$, which in itself depends upon q, evaluation of the rate of efflux from Eq. (58) must proceed according to the method of successive approximation. That is, v_0 is first assumed equal to zero and the corresponding q is evaluated; the value $v_0 = q/2\bar{h}$ is then used to obtain a more exact magnitude of q, and so forth.

On the other hand, Eq. (58) may be written in the simpler form, corresponding to Eq. (33) of Chapter III,

$$q = C_d\, b\sqrt{2g\bar{h}} \tag{59}$$

in which

$$C_d = \tfrac{2}{3}C_c\sqrt{\frac{b}{\bar{h}}}\left[\left(\frac{v_0^2}{2gb} + \frac{\bar{h}}{b} + \frac{1}{2}\right)^{3/2} - \left(\frac{v_0^2}{2gb} + \frac{\bar{h}}{b} - \frac{1}{2}\right)^{3/2}\right] \tag{60}$$

Thus, although the relationship as a whole is still as tedious to evaluate, the fact that C_d is dimensionless and depends only upon the boundary proportions permits its determination once and for all in tabular or graphical form. Figure 46 contains such a graph of C_d versus $b/2\bar{h}$ for the symmetrically located slot. The extent of the gravitational effect may be judged through comparison with the neighboring curve from Fig. 31.

Needless to say, the contraction coefficients of Table I are applicable without modification only if the slot is symmetrically located with respect to the uniformly approaching flow. The foregoing discussion would therefore be merely of academic interest were it not for the fact that it serves as an introduction to two practically important types of free-surface transition—flow over a *weir* and flow under a *sluice gate.*

Flow over a sharp-crested weir. If $\bar{h}$ in Fig. 44 becomes equal to $b/2$, the upper edge of the slot will no longer be effective, and the lower portion of the boundary will become what is known as a *sharp-crested weir*—that is, a thin wall over which liquid flows (Fig. 47). Noting that the dimension b of Eq. (58) is now equal to the head h (i.e., the elevation of the free surface of the *uniform* flow with respect to the weir crest),

$$q = \tfrac{2}{3}C_c\sqrt{2g}\left[\left(\frac{v_0^2}{2g} + h\right)^{3/2} - \left(\frac{v_0^2}{2g}\right)^{3/2}\right] \tag{61}$$

in which C_c may again be taken from Table I by placing $b/B = h/(h + w)$. Again introducing a discharge coefficient,

$$q = \tfrac{2}{3}C_d\sqrt{2g}\,h^{3/2} \tag{62}$$

in which

$$C_d = C_c\left[\left(\frac{v_0^2}{2gh} + 1\right)^{3/2} - \left(\frac{v_0^2}{2gh}\right)^{3/2}\right] \tag{63}$$

C_d will as before be found to depend only upon the boundary proportions, as plotted in Fig. 46 for a vertical weir. The latter curve, it may be noted, is closely approximated by the empirical formula

$$C_d = 0.611 + 0.075 \frac{h}{w}$$

Since the foregoing relationships were based upon the assumption of atmospheric pressure on both sides of the *nappe* or sheet of falling liquid, it is evident that they will be applicable only if this assumption is strictly fulfilled. In other words, if the liquid impinges upon the

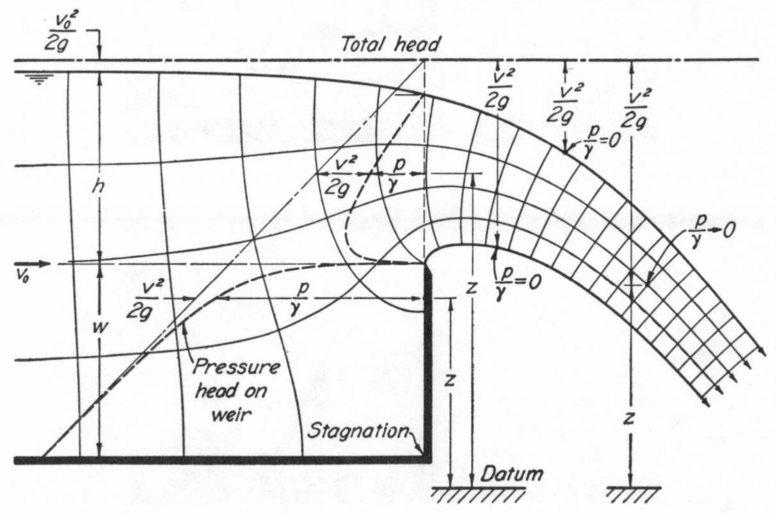

FIG. 47. Flow net for gravity discharge over a sharp-crested weir.

channel floor beyond the weir, any air originally enclosed by the nappe will gradually be removed by the flow; as the pressure intensity is thereby reduced to an appreciable degree, the same discharge will occur under a considerably lower head. If such a weir is to be used as a flow meter, means should therefore be provided for complete *ventilation* of the zone below the nappe.

Flow under a sluice gate. The sluice gate (Fig. 48) differs somewhat from the other limiting condition for flow from a slot in that the jet is generally not free, but guided by a horizontal floor; as a result, the final jet pressure is not atmospheric but hydrostatically distributed. Writing the continuity and Bernoulli equations between sections 0 and 1 of Fig. 48,

$$v_0 h = v_1 C_c b \qquad \frac{v_0^2}{2g} + h = \frac{v_1^2}{2g} + C_c b$$

and solving for v_1,

$$v_1 = \sqrt{2g \frac{h - C_c b}{1 - C_c^2 (b/h)^2}} = \frac{\sqrt{2gh}}{\sqrt{1 + C_c b/h}}$$

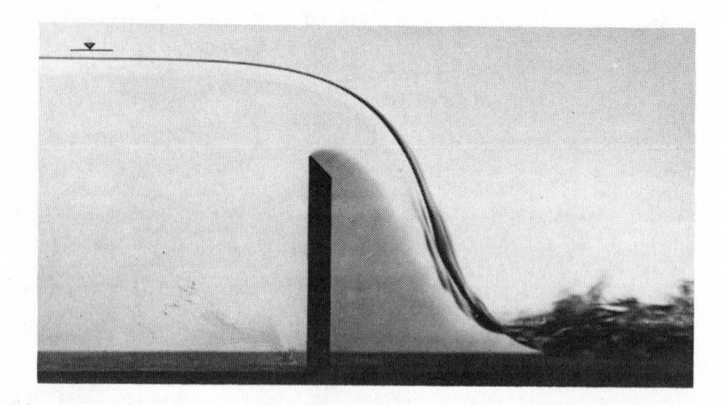

Plate VII. Top: profile of flow over a ventilated sharp-crested weir. Center: ventilation pipe withdrawn; pressure under nappe is gradually reduced as air is carried away by flow. Bottom: removal of air complete; eddy has been colored by dye. Note successive changes in head and deflection of nappe.

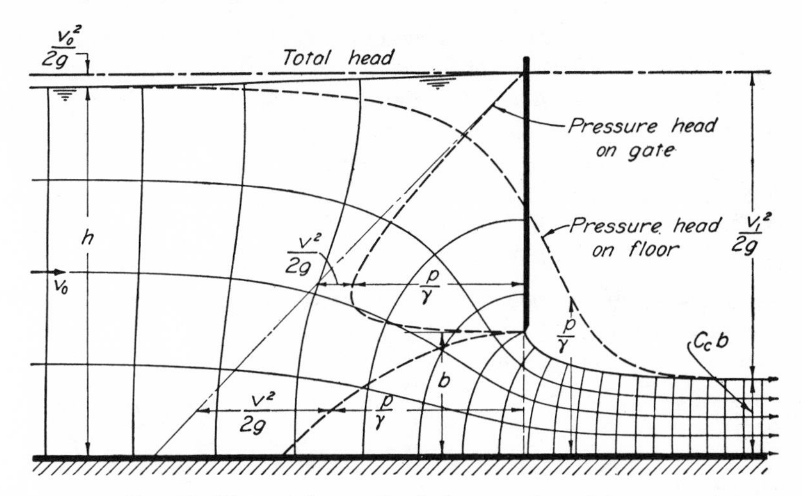

FIG. 48. Flow net for gravity discharge under a sluice gate.

Since $q = v_1 C_c b$,

$$q = \frac{C_c}{\sqrt{1 + C_c b/h}} b\sqrt{2gh} \tag{64}$$

Again introducing a discharge coefficient,

$$q = C_d\, b\sqrt{2gh} \tag{65}$$

in which

$$C_d = \frac{C_c}{\sqrt{1 + C_c b/h}} \tag{66}$$

Thus C_d, like C_c, again depends only upon the boundary proportions. Owing to the different pressure conditions within the contracted jet,

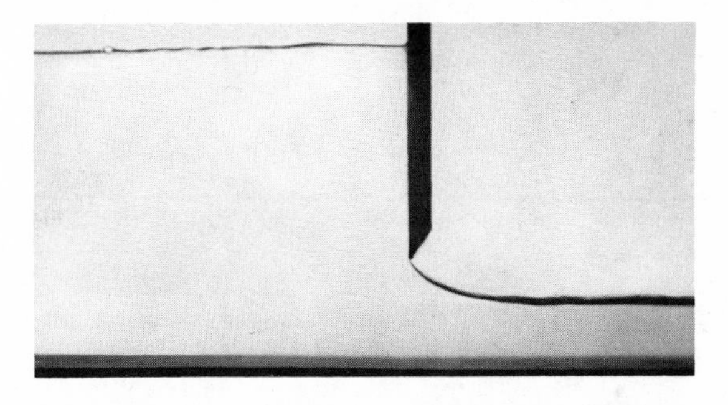

PLATE VIII. Profile of flow under a vertical sluice gate.

however, the contraction coefficients of Table I are not applicable here. That is, the degree of jet contraction remains nearly constant over a considerable range of variation of b/h, as shown in Fig. 46, yielding the curve of $C_d : b/h$ for a vertical sluice gate which is included in this diagram. As will be seen from Fig. 48, the head h will deviate more and more from the velocity head of the jet (where the pressure variation is hydrostatic) as b/h increases, thus causing the downward trend of C_d in Fig. 46.

Example 18. Water discharges at a 4-foot head over a vertical sharp-crested weir 4 feet high and 10 feet long. Determine (*a*) the rate of flow and (*b*) the total force exerted by the water upon the weir.

(*a*) From Fig. 46, $C_d = 0.691$ for $h/(h + w) = 0.5$. Therefore, according to Eq. (62)

$$q = \tfrac{2}{3} \times 0.691 \times \sqrt{2 \times 32.2} \times 4^{3/2} = 29.6 \text{ cfs/ft}$$

whence

$$Q = 10q = 296 \text{ cfs}$$

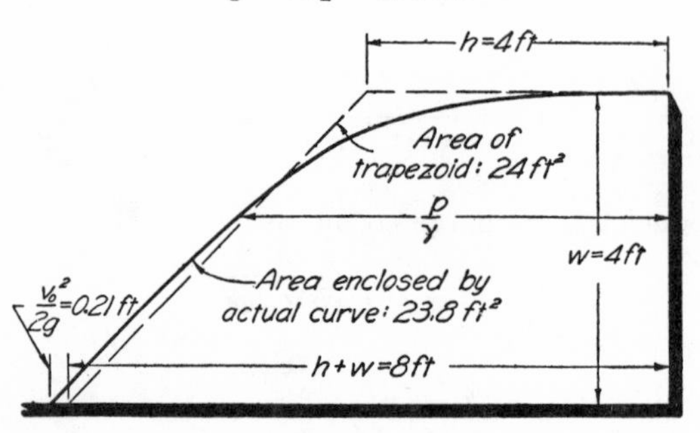

(*b*) After plotting the pressure-distribution curve of Fig. 47 to scale, as shown, the total normal force may be determined by multiplying the area under the curve by the specific weight and the length of weir:

$$F = 23.8 \times 62.4 \times 10 = 14{,}800 \text{ lb}$$

According to the diagram, this value may be approximated by assuming hydrostatic conditions:

$$F \approx \frac{4 + 8}{2} \times 4 \times 62.4 \times 10 = 15{,}000 \text{ lb.}$$

Evidently, however, the true center of pressure will be lower than that indicated by this assumption.

Example 19. A 1-foot sill with upstream face inclined at an angle of 45° is installed at the end of an open channel of rectangular cross section. If the channel is 8 feet wide and 5 feet deep, and if a minimum freeboard of 6 inches is specified, estimate the greatest rate of flow which the channel may carry.

Owing to lack of a plot of C_d for a sloping weir, q must be found from Eq. (61) by successive approximation. Thus, since

$$\frac{h}{h+w} = \frac{3.5}{4.5} = 0.778$$

the corresponding value of C_c may be obtained from Table I by interpolation; that is, when $b/B = 0.778$ and $\beta = 45°$, $C_c = 0.784$. Then first assuming that $v_0 = 0$,

$$q \approx \tfrac{2}{3} C_c \sqrt{2g}\, h^{3/2} = \tfrac{2}{3} \times 0.784 \times \sqrt{2 \times 32.2} \times \overline{3.5}^{3/2} = 27.5 \text{ cfs/ft}$$

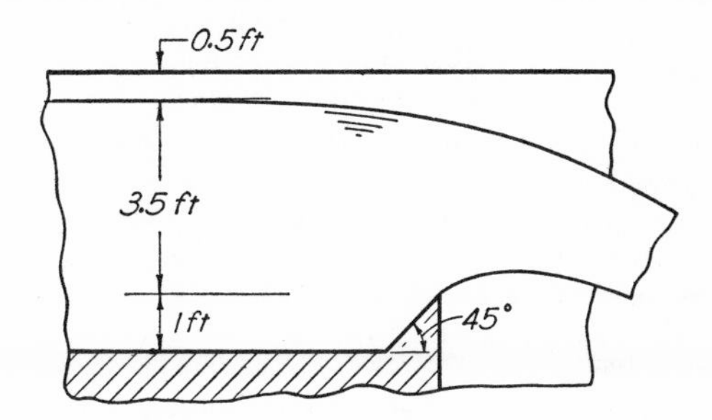

For this approximate value of q,

$$v_0 \approx \frac{q}{h+w} = \frac{27.5}{4.5} = 6.11 \text{ fps} \quad \text{and} \quad \frac{v_0^2}{2g} \approx \frac{\overline{6.11}^2}{64.4} = 0.58 \text{ ft}$$

wherewith a closer approximation of q may now be obtained:

$$q \approx \tfrac{2}{3} \times 0.784 \times 8.02 \times [(0.58 + 3.5)^{3/2} - (0.58)^{3/2}] = 32.7 \text{ cfs/ft}$$

A third approximation will then yield the value

$$q \approx 34.6 \text{ cfs/ft}$$

from which

$$Q \approx 8 \times 34.6 = 277 \text{ cfs}$$

PROBLEMS

100. Water discharges under a head of 8 feet from a sharp-edged horizontal slot 15 inches high in the side of a reservoir. If the guide vanes shown in broken lines were installed just upstream from the slot, what would be the percentage change in (*a*) velocity of efflux, and (*b*) rate of efflux?

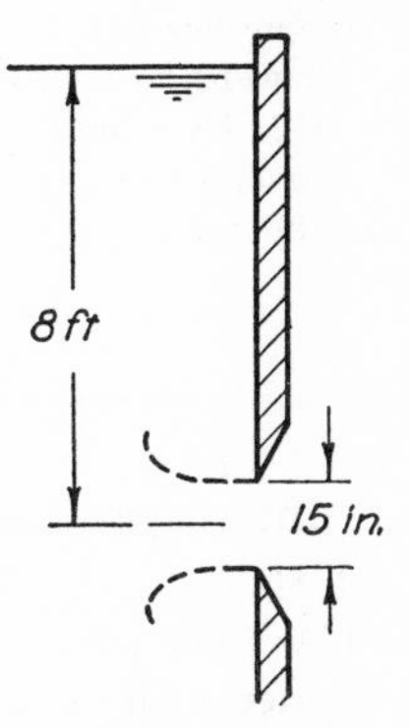

PROB. 100.

101. A slot such as that shown in Fig. 44 is 6 inches in height, the depth of the approaching flow being 18 inches and the mean head 9 inches. Determine (*a*) the rate of flow per unit length of slot, (*b*) the maximum elevation attained by the liquid surface, and (*c*) the intensity of pressure at the lower stagnation point.

102. A vertical, sharp-crested weir 3 feet high must discharge 50 cubic feet of water per second without exceeding a head of 9 inches. What length of crest is required?

103. How high above the crest will water rise in a piezometer column connected to a stagnation tube held in the nappe of flow over a 1-foot vertical weir under a head of 2 feet?

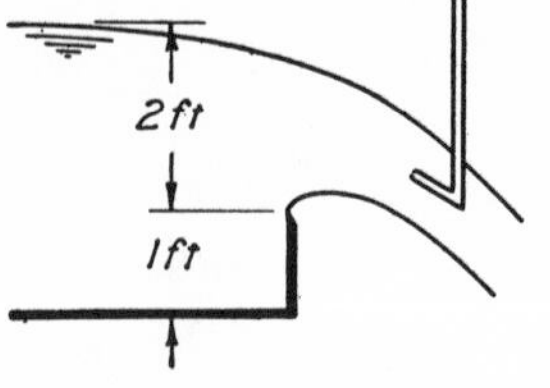

PROB. 103.

104. Water flows at the rate of 10 cubic feet per second per foot width in a rectangular channel 6 feet deep. A vertical weir is to be installed at the end of the channel to bring the free surface 6 inches from the top of the channel wall. Estimate the required weir height, assuming full ventilation.

105. From the distribution of velocity head through the nappe directly above the weir crest shown in Fig. 47, determine as in Example 3 the rate of flow $q = \int v_x \, dy$ over the weir under a 1-foot head; compare this result with the value indicated by Eq. (62).

106. Estimate the unit rate of flow over a ventilated weir inclined downstream at an angle of 30° from the vertical under a head of 2 feet, the weir crest having an elevation of 3 feet above the channel floor.

107. Estimate the force per unit length of crest exerted by the flowing water upon the weir of Problem 106.

108. What force will be exerted upon a vertical sluice gate 8 feet wide when discharging water under a head of 9 feet, if the gate opening is 3 feet (refer to Fig. 48)?

109. Determine the rate of flow per unit width under the sluice gate of Problem 108. What is the maximum height to which water will rise on the upstream side of the gate?

110. From the distribution of velocity head in the plane of the sluice gate shown in Fig. 48, determine as in Example 3 the rate of flow $q = \int v_x \, dy$ for the condition that $h = 6$ feet; compare this result with the value indicated by Eq. (65).

111. Assuming that submergence of a sluice-gate opening does not affect appreciably the form of the outflow pattern, estimate the rate of flow per unit width under the conditions shown.

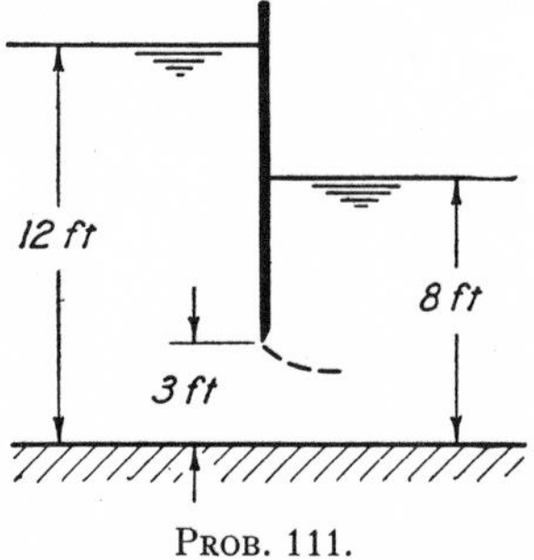

PROB. 111.

112. Determine the change in head due to a 10 per cent increase in unit discharge (*a*) over a 1-foot vertical weir, and (*b*) through a 1-foot sluice-gate opening, the depth of the approaching flow being 2 feet in each case.

19. GEOMETRY OF LIQUID JETS

Interdependence of velocity head and elevation. A particular case of steady, curvilinear flow of a liquid with a free surface is that in which the pressure intensity is zero (i.e., atmospheric) at every point. Under such circumstances the term for pressure head disappears from the Bernoulli equation, and, in the event that no external forces other than gravitational attraction influence the flow, the sum of velocity head and elevation will at all points be the same. Flow of this nature was mentioned briefly in the foregoing section in connection with the efflux of liquid from a horizontal slot and the related case of flow over a weir. It now remains to investigate the pattern of motion satisfying these particular conditions.

If the arbitrary reference plane from which elevation is measured is assumed to pass through the point at which the velocity of a free jet has the magnitude v_0, it follows from the Bernoulli equation that at any other point

$$\frac{v^2}{2g} + z = \frac{{v_0}^2}{2g} \tag{67}$$

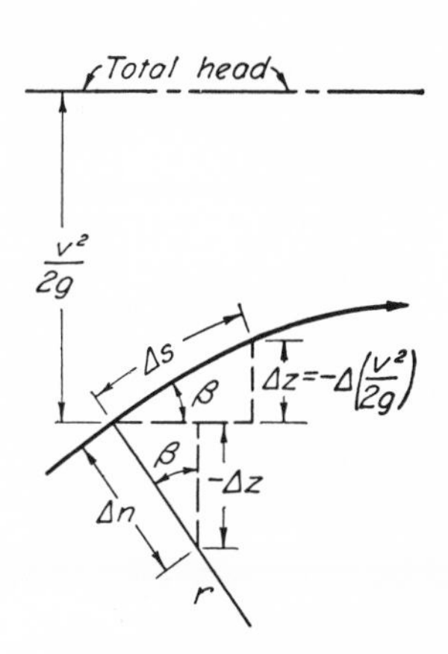

FIG. 49. Definition sketch for jet geometry.

From the equation for tangential acceleration (or from the derivative of the above expression with respect to s), the rate of change in velocity head along any stream line within the jet will be seen to vary directly with the negative sine of the angle of inclination of the filament (see Fig. 49),

$$\frac{\partial(v^2/2g)}{\partial s} = -\frac{\partial z}{\partial s} = -\sin\beta \tag{68}$$

and from the equation for normal acceleration (or from the derivative of Eq. (67) with respect to n) the ratio of the velocity head to the radius of curvature of the filament will be found to vary with the cosine of this angle:

$$\frac{v^2/2g}{r} = -\frac{1}{2}\frac{\partial z}{\partial n} = \frac{1}{2}\cos\beta \tag{69}$$

Evaluation of the velocity components at any point. Although the magnitude of the velocity vector itself governs the magnitude of the velocity head, it is convenient to resolve this vector into horizontal and vertical components; evidently,

$$v_x = v\cos\beta \qquad v_z = v\sin\beta$$

and

$$v = \sqrt{{v_x}^2 + {v_z}^2} \tag{70}$$

The acceleration vector may also be so resolved, its components obviously having the magnitudes

$$a_x = \frac{dv_x}{dt} = 0 \quad \text{and} \quad a_z = \frac{dv_z}{dt} = -g \tag{71}$$

Integration of these two equations with respect to time then yields expressions for the components of the velocity vector at any point in terms of time of travel of a particle from the reference point:

$$v_x = C_1 = (v_0)_x \tag{72}$$

$$v_z = -gt + C_2 = (v_0)_z - gt \tag{73}$$

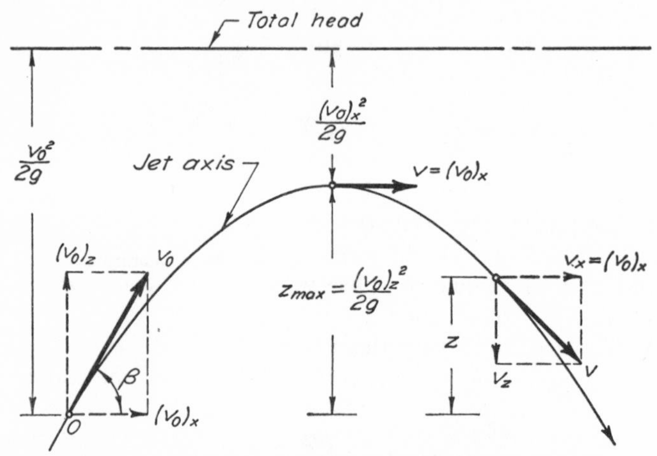

FIG. 50. Interrelationship of velocity head and jet elevation.

Since $v_x = dx/dt$ and $v_z = dz/dt$, further integration will yield expressions for the distances x and z traveled in the two directions, indicating that the trajectory of any fluid element is parabolic in form:

$$x = (v_0)_x t \tag{74}$$

$$z = (v_0)_z t - \tfrac{1}{2}gt^2 \tag{75}$$

Finally, by combining Eqs. (73) and (75) it will be found that at any instant

$$v_z^2 = (v_0)_z^2 - 2gz \tag{76}$$

Although the velocity vector is the vector sum of its two components, it will be seen from Eq. (70) that the velocity head, a scalar quantity, is simply the following algebraic sum:

$$\frac{v^2}{2g} = \frac{v_x^2}{2g} + \frac{v_z^2}{2g}$$

Although these last quantities are not heads, in the strict sense of the word, they will be seen to have a particular significance if Eqs. (67) and (76) are combined in the form

$$\frac{v^2}{2g} = \frac{(v_0)_x{}^2}{2g} + \frac{(v_0)_z{}^2}{2g} - z \tag{77}$$

Evidently, any change in velocity head due to a change in elevation will be equal to the change in $v_z{}^2/2g$, the quantity $v_x{}^2/2g$ necessarily remaining constant. It follows, therefore, that the maximum vertical elevation of the jet profile will be $z_{max} = (v_0)_z{}^2/2g$, at which point the filament will be the distance $(v_0)_x{}^2/2g$ below the line of constant total head. These relationships are shown in diagrammatic form in Fig. 50.

Example 20. Water is discharged from a 6-inch horizontal pipe flowing full at the outlet, the jet striking the ground at a horizontal distance of 10 feet and a vertical distance of 4 feet from the end of the pipe. Estimate the rate of flow.

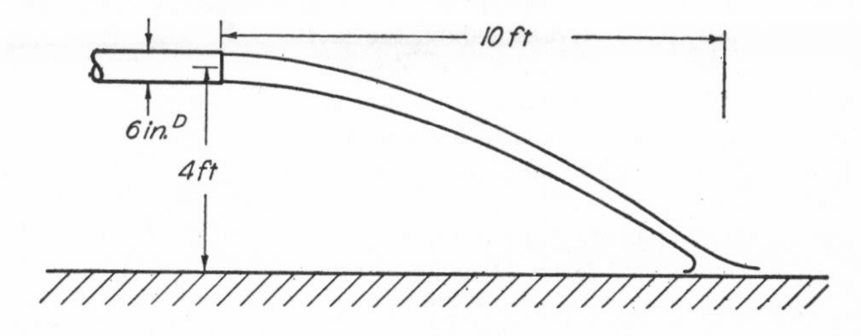

Since $(v_0)_z = 0$, it follows from Eqs. (74) and (75) that

$$x = (v_0)_x t \quad \text{and} \quad z = -\tfrac{1}{2}gt^2$$

Eliminating t,

$$v_0 = (v_0)_x = \frac{x}{\sqrt{-2z/g}} = \frac{10}{\sqrt{-2(-4)/32.2}} = 20.1 \text{ fps}$$

Hence

$$Q = Av_0 = \frac{\pi \times (\frac{1}{2})^2}{4} \times 20.1 = 3.94 \text{ cfs}$$

PROBLEMS

113. A jet of water has a velocity of 25 feet per second and is inclined at an angle of 60° to the horizontal as it leaves a nozzle. Determine (*a*) the initial radius of curvature of the jet, (*b*) the maximum elevation which the water will attain, and (*c*) the radius of curvature of the jet at its highest point.

114. If a liquid jet inclined at an angle of 30° to the horizontal is found to rise a vertical distance of 5 feet in a horizontal distance of 12 feet, what is the initial jet velocity?

115. Water leaves the bucket of a high spillway at an angle of 45° and a velocity of 40 feet per second. Determine the maximum height to which the sheet will rise, and the location of the point at which it will again reach its initial elevation.

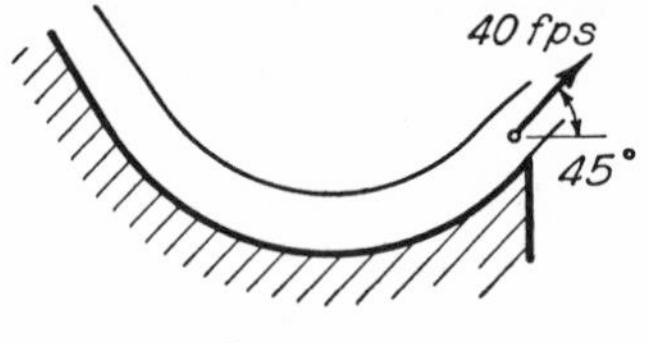

PROB. 115.

116. Liquid is discharged from a nozzle with an initial velocity of 30 feet per second. At what two nozzle inclinations will the liquid strike an object at a horizontal distance of 20 feet from the nozzle?

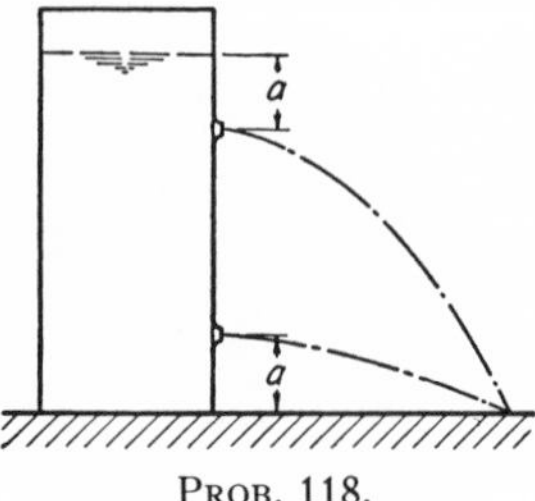

PROB. 118.

117. An orifice is to be placed in the side of a tank at such an elevation that the jet will attain a maximum horizontal distance from the tank at the level of its base. If the depth of liquid in the tank is maintained at 5 feet, what is the proper distance from the orifice to the free surface?

118. Show that the jets from two orifices in the side of a tank will intersect a plane through the base at the same distance from the tank if the head on the upper orifice is equal to the height of the lower orifice above the base.

119. If a nozzle delivers a jet of water with an efflux velocity of 50 feet per second, what is the greatest horizontal distance at nozzle level which the jet will attain as the inclination of the nozzle is varied from 0° to 90°?

120. Liquid emerges from an orifice in the bottom of a tank under a head of 3 feet. How far below the orifice will the jet diameter be reduced by 50 per cent?

121. A fireman must reach a window 85 feet above the ground with a fire stream from a nozzle having a cylindrical tip 1⅛ inches in diameter and discharging 250 gallons of water per minute. Neglecting air resistance and assuming a nozzle height of 5 feet, determine the greatest distance from the building at which the fireman can stand and still play the stream upon the window. (Note that optimum conditions do not necessarily correspond to $z_{max} = 80$ feet.)

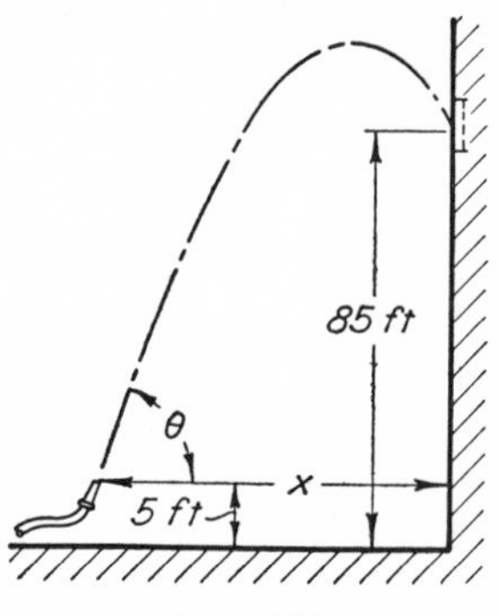

PROB. 121.

20. SIGNIFICANCE OF THE FROUDE NUMBER

Relative influence of gravity upon the flow pattern. At the end of Chapter III brief mention was made of the Euler number, a dimensionless flow parameter involving the fluid density, a velocity, and a difference in pressure intensity. For any given boundary proportions it was found that this number would have a constant magnitude—regardless of the boundary scale, the rate of flow, the density, and the absolute pressure intensity—as long as the various fluid properties other than density had no influence upon the form of the flow pattern. In the event of such influence, therefore, it was reasoned that the extent to which the Euler number would be found to deviate from this reference magnitude might well serve as a measure of the extent to which the flow was affected by each additional property.

In respect to fluid weight, it has been shown that no change in the flow pattern will result if the motion is completely defined by solid boundaries, for then the sole effect of gravitational attraction is to change the pressure intensity from point to point in direct proportion to the change in elevation. If the fluid is unconfined in any zone, however, the form of the free surface, and hence of the entire pattern of motion, will be subject to gravitational influence. That the extent of such influence is a relative matter, however, may readily be seen.

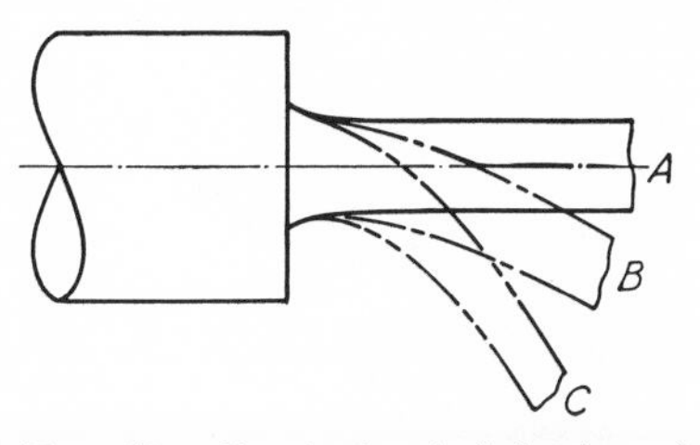

FIG. 51. Gravitational deflection of liquid jets.

Consider, for instance, the efflux of a fluid from the boundaries shown in Fig. 51. The inertia of the fluid tends to make it continue in the longitudinal direction after leaving the orifice, but the effect of gravity is to deflect it in the vertical direction; evidently, the greater the density and the velocity, the smaller the deflection in a given distance, whereas, the greater the difference in specific weight between the moving fluid and the surrounding medium, the greater the deflection will tend to be. A jet of air emerging into the atmosphere (or a submerged jet of water) would thus remain symmetrical about the longitudinal axis, regardless of how small the velocity or the density might be, as indicated by profile A in Fig. 51; that is, the gravitational effect must be nil so long as there is no difference in specific weight. But a jet of fluid of greater specific weight than that of the surrounding medium would not remain symmetrical, the asymmetry becoming the more pronounced (profiles B and C) the lower the velocity and density or the greater the relative specific weight of the moving fluid. The Euler number, which may reasonably be expected to vary with the jet profile, should therefore be a function of the several quantities (velocity, density, difference in specific weight, and a reference length) governing the relative weight influence.

Formulation of the Froude number. The only possible dimensionless combination of a velocity V, a density ρ, a difference in specific weight $\Delta\gamma$, and a length L is the quantity $\rho V^2/L\ \Delta\gamma$ or some power thereof. Just as the denominator of the Euler number was found to typify a unit force due to a pressure difference, and the numerator a unit inertial reaction to such a force, the quantity $\rho V^2/L\ \Delta\gamma$ may be considered to represent the ratio of a unit inertial reaction ($\rho V^2/L$) to a typical unit force due to gravity ($\Delta\gamma$). Evidently, the greater $\rho V^2/L$ is in comparison with $\Delta\gamma$ for a given type of flow, the smaller will be the relative effect of gravity, and vice versa. Large values of $\rho V^2/L\ \Delta\gamma$ therefore indicate small gravitational influence, and small values large gravitational influence.

The quantity $\rho V^2/L\ \Delta\gamma$, or, more conveniently, its square root, is known as the *Froude number:*

$$\mathbf{F} = \frac{V}{\sqrt{L\ \Delta\gamma/\rho}} \tag{78}$$

According to the foregoing discussion, if gravitational effects modify the flow pattern, the Euler number must be a function of this weight parameter:

$$\mathbf{E} = \phi(\mathbf{F}) \tag{79}$$

In other words, if a change in V, L, $\Delta\gamma$, or ρ causes $\mathbf{F}$ to change, $\mathbf{E}$ must then change accordingly; conversely, only if $\mathbf{F}$ is constant will $\mathbf{E}$ be constant.

Variation of E with F. Evidently, the symmetrical jet of Fig. 51 corresponds to an extremely high value of $\mathbf{F}$, the intermediate profile to a moderate value of $\mathbf{F}$, and the form showing the greatest gravitational deflection to a very low value of $\mathbf{F}$. As $\mathbf{F}$ changes, of course, $\mathbf{E}$ (i.e., C_d) must also change. Indeed, the troublesome terms of the jet and weir equations (see Section 18) which can be evaluated only by successive approximation are really forms of the Froude number. In other words, the discharge coefficient of a circular orifice, which was shown in the last chapter (Fig. 31) to depend only upon the boundary geometry in the event of zero gravitational effect, i.e.,

$$\mathbf{E} = C_d = \phi\left(\beta, \frac{d}{D}\right)$$

must now be considered to vary with the Froude number as well as the boundary geometry; that is,

$$\mathbf{E} = C_d = \phi\left(\beta, \frac{d}{D}, \mathbf{F}\right)$$

The determination of the function $\phi(\beta, d/D, \mathbf{F})$ is too complex to warrant further comment in this book, but typical results are plotted in

Fig. 52 for the conditions $\beta = 90°$ and $d/D = 0$. Indeed, the complexity of many examples of flow under gravitational influence is such that experimental studies are generally necessary to determine the corresponding relationship between the Froude number and the Euler number, or between the Froude number and the geometry of the flow pattern. Such examples, needless to say, are by no means restricted to flow through orifices and over weirs but include all phenomena in which a fluid surface is not guided by fixed boundaries.

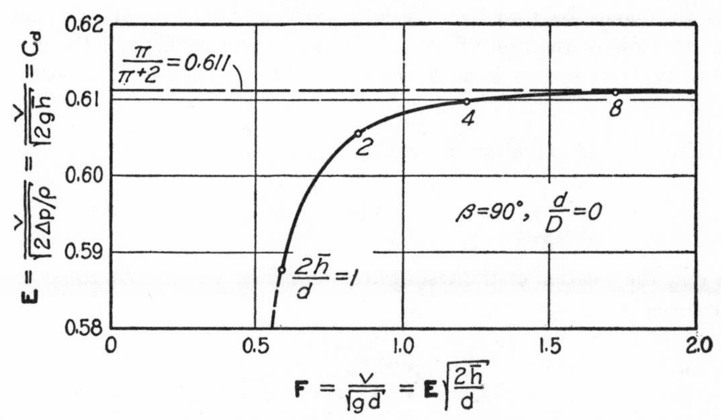

FIG. 52. Variation of the Euler number with the Froude number for a circular orifice.

Gravitational similitude. Quite apart from such study of the variation of the Euler number and the flow pattern with the Froude number, certain engineering problems frequently require, prior to construction, accurate knowledge of large-scale conditions of free-surface flow at a predetermined value of **F**. Since such problems generally involve the motion of a liquid in contact with the atmosphere, $\Delta\gamma$ becomes almost exactly equal to the liquid γ; since $\gamma/\rho = g$, the Froude number then reduces to

$$\mathbf{F} = \frac{V}{\sqrt{gL}} \tag{80}$$

If the flow in question is duplicated in miniature by model tests at an identical value of **F**, the results may then readily be transferred to prototype scale according to the ratios of the several terms in the Froude and Euler numbers. In a word, if fluid weight and fluid density are the only properties influencing two states of flow with geometrically similar boundary proportions, such states are of necessity both kinematically and dynamically similar (i.e., velocities and pressures will be similarly distributed) if the Froude numbers, and hence the

Euler numbers, are of the same magnitude. It is pertinent to note, in this connection, that the Froude number was named after an English engineer of the last century who first utilized this similarity principle in the study of the resistance of ships by means of model tests.

Example 21. Tests are to be made in a laboratory towing tank on a 1 : 25 scale model of a new ship, in order to determine the wave resistance which will be encountered by the prototype. (*a*) If the maximum speed which the prototype is expected to attain is 20 knots (nautical miles per hour), at what speed should the model be towed in the laboratory to obtain waves dynamically similar to those of the prototype? (*b*) If the wave resistance of the model is found to be ½ pound, to what resistance would this correspond at prototype scale?

(*a*) If the Froude numbers of model and prototype are to be the same,

$$\frac{V_m}{\sqrt{g_m L_m}} = \frac{V_p}{\sqrt{g_p L_p}} \quad \text{or} \quad \frac{V_m}{V_p} = \sqrt{\frac{g_m L_m}{g_p L_p}}$$

Since g at the location of the tests will probably be very nearly equal to that where the ship will operate,

$$\frac{V_m}{V_p} = \sqrt{\frac{L_m}{L_p}} = \sqrt{\frac{1}{25}} = \frac{1}{5}$$

and

$$V_m = \frac{V_p}{5} = \frac{20}{5} = 4 \text{ knots}$$

(*b*) Since the Euler numbers will then also be identical,

$$\frac{V_m}{\sqrt{2\Delta p_m/\rho_m}} = \frac{V_p}{\sqrt{2\Delta p_p/\rho_p}} \quad \text{or} \quad \frac{\Delta p_m}{\Delta p_p} = \frac{V_m^2}{V_p^2}\frac{\rho_m}{\rho_p}$$

If sea water is used in the tests, $\rho_m = \rho_p$. Since the product of Δp and an area L^2 will represent a force F,

$$\frac{F_m}{F_p} = \frac{\Delta p_m}{\Delta p_p}\frac{L_m^2}{L_p^2} = \frac{V_m^2}{V_p^2}\frac{L_m^2}{L_p^2} = \frac{L_m^3}{L_p^3}$$

Hence,

$$F_p = F_m\left(\frac{L_p}{L_m}\right)^3 = \frac{1}{2} \times \overline{25}^3 = 7810 \text{ lb}$$

PROBLEMS

122. Rotation of a cylindrical tank of water 3 feet in diameter at the constant rate of 60 revolutions per minute produces a difference in surface elevation between wall and axis of 1.38 feet. Determine, through use of the Froude criterion for similarity, the angular velocity which would produce the same relative difference in surface level in a tank of mercury 6 inches in diameter.

123. It is desired to determine through tests on a 1 : 20 scale model the wave formation produced by bridge piers when built in a river having a maximum velocity of 12 feet per second. What velocity should prevail in the model in order to produce a geometrically similar surface configuration? If the rate of flow between piers in the model is then 7 cubic feet per second, what will be the corresponding prototype flow?

124. A spillway is designed to carry a peak flow of 1200 cubic feet of water per second. What rate of flow should be established in a 1 : 30 scale model of the spillway in order to achieve similar discharge conditions?

125. Tests are to be made upon a model sea wall $\frac{1}{25}$ prototype size. If the period of the waves in the prototype is 8 seconds, what wave period should prevail in the model tests?

126. If the maximum force exerted by a wave upon the model sea wall of Problem 125 is 15 pounds, what is the magnitude of the corresponding prototype force?

127. Fresh-water tests on a 5-foot model ship yielded the accompanying curve of wave resistance versus speed. If the cruising speed of the 450-foot prototype is 15 knots, what wave resistance should it then encounter in ocean water?

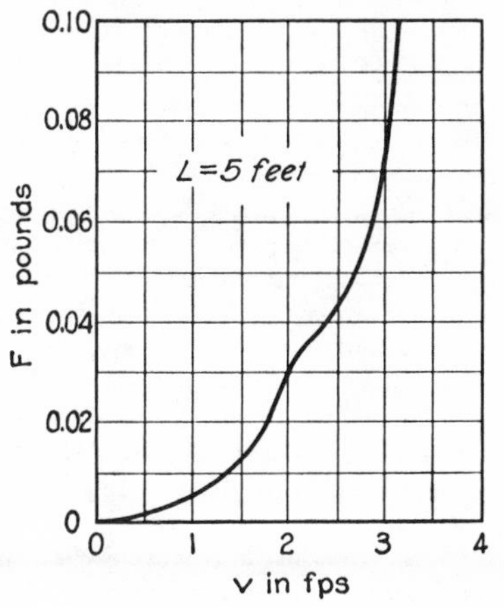

PROB. 127.

128. The curve of Problem 127 may be generalized by plotting $F/(L^2\rho V^2/2)$ against $V/\sqrt{gL}$. Prepare such a plot.

QUESTIONS FOR CLASS DISCUSSION

1. What is the difference between mass and weight? Which is measured by a beam balance and which by a spring balance?

2. State whether the following characteristics of a given body of matter can vary, and, if so, under what conditions: mass, weight, density, specific weight, specific gravity.

3. The vector f_w is invariably normal to a plane of constant elevation. To what is the vector f_p normal? To what is the vector f normal?

4. Under what circumstances is the vector f_p zero? The vector f?

5. What is meant by hydrostatic-pressure variation? When can the intensity of pressure vary hydrostatically in one direction and not in another? Under what conditions is the pressure distribution hydrostatic in all directions?

6. Distinguish between (*a*) steady and unsteady flow, (*b*) uniform and non-uniform flow, and (*c*) rotational and irrotational flow, with respect to their bearing upon the sum of velocity head, pressure head, and elevation. Which of these cases is represented by the flow net?

7. Why does the concept of fluid head have no physical significance in the motion of a gas?

8. Why can fluid weight have no influence upon the flow pattern if the fluid is fully confined by fixed boundaries?

9. Why is it necessary that the space below the nappe of a weir used for flow measurement be fully ventilated (i.e., at atmospheric pressure)? If the pressure intensity of the air below such a nappe were reduced, what would be the effect upon the nappe profile and the head on the weir for a given rate of flow?

10. Spillway profiles are generally designed according to the lower surface of a weir nappe under comparable conditions of head and height, in order to obtain atmospheric pressure at the crest and down the face of the spillway. How would the pressure distribution at the crest vary if the designed head on the spillway were exceeded?

11. If the liquid jet from a long slot is directed vertically upward, the flow net will yield the accompanying profile for the zone of maximum elevation. Discuss the distribution of velocity and pressure intensity along the line of symmetry.

12. Show that the Froude number for an orifice discharging liquid into air is propor-

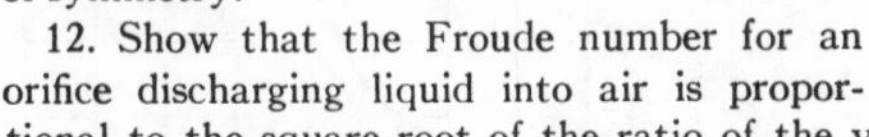

tional to the square root of the ratio of the velocity head of the jet to the orifice diameter.

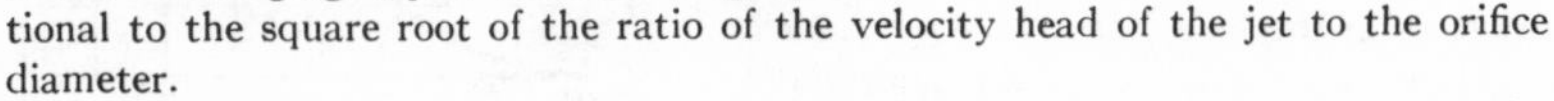

13. In studying flow under the scale model of a sluice gate, is it sufficient to make the ratio b/h the same, or must the Froude numbers also be made identical? Why?

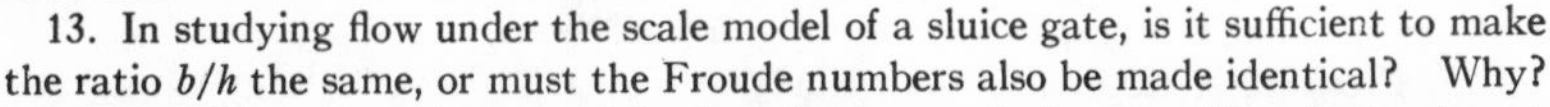

14. A spillway model is tested in the laboratory at the same Froude number at which the geometrically similar prototype structure is expected to perform during maximum discharge. If the subscripts m and p refer to model and prototype, and if the model-prototype scale is 1 : 50, show from the equalities $\mathbf{F}_m = \mathbf{F}_p$ and $\mathbf{E}_m = \mathbf{E}_p$ that the following relationships must prevail:

$$\frac{h_m}{h_p} = \frac{1}{50} \qquad \frac{V_m}{V_p} = \left(\frac{1}{50}\right)^{1/2} \qquad \frac{q_m}{q_p} = \left(\frac{1}{50}\right)^{3/2}$$

$$\frac{p_m}{p_p} = \frac{1}{50} \qquad \frac{F_m}{F_p} = \left(\frac{1}{50}\right)^{3} \qquad \frac{Q_m}{Q_p} = \left(\frac{1}{50}\right)^{5/2}$$

SELECTED REFERENCES

LEA, F. C. *Hydraulics for Engineers and Engineering Students.* 6th edition, Longmans, Green, 1938.

ROUSE, H. *Fluid Mechanics for Hydraulic Engineers.* Engineering Societies Monograph, McGraw-Hill, 1938.

FREEMAN, J. R. *Hydraulic Laboratory Practice.* American Society of Mechanical Engineers, 1929.

CHAPTER V

ONE-DIMENSIONAL METHOD OF FLOW ANALYSIS

21. PRINCIPLES OF MOMENTUM AND ENERGY

Average characteristics of the gross filament. Fluid motion of the most general type is a three-dimensional problem, in that both the velocity vector and the acceleration vector may have components in each of three coordinate directions. Under such circumstances the stream-line pattern will embody a very complex system of curves in space, the motion being subject to even an approximate analysis only if symmetrical about a longitudinal axis, as in flow around or between boundary surfaces of revolution. For this reason the illustrative material of the foregoing chapters has been restricted largely to the special case of flow in two dimensions, the comparative ease of graphical representation and analysis allowing primary emphasis to be placed upon the underlying principles of curvilinear motion.

Once such principles are fully understood, the student may safely proceed to what is known as the *one-dimensional method of analysis*, an arbitrary simplification of two- or three-dimensional conditions which permits a rapid approximate solution for even the more complex states of flow. Were fluid motion ever truly one-dimensional, it is obvious that the analysis would then be simplified considerably, for the convective acceleration would be zero at every point, and the velocity would be the same in magnitude and direction at every successive cross section. The one-dimensional method of approach, however, ignores accelerative effects in the *normal* direction only, variation in the longitudinal direction thus becoming the entire subject of study. In a word, the flow is presumed to have the characteristics of a single stream filament, and average values of velocity, pressure intensity, and elevation are considered typical of the flow as a whole at any cross section of this gross filament.

Such simplification permits the adaptation of two important relationships of hydrodynamics, the *momentum principle* and the *energy principle*, for convenient use with the continuity principle in the preliminary analysis of flow in both closed and open conduits. Oversimplification, of course, is a dangerous policy; it is therefore essential that these principles first be exactly derived from the basic equations of flow, and that the resulting errors of simplification then be carefully evaluated. For instance, the continuity equation for any flow passage, however large,

$$V_1A_1 = V_2A_2$$

is at once applicable to the one-dimensional method, and in itself involves no error. It must be recalled, however, that the integral on which it is based,

$$Q = \int v\, dA = VA$$

not only admits the possibility of velocity variation across the section but necessarily involves a section which is normal to the direction of flow at every point. These facts are of primary importance in the derivation and adaptation of the two remaining principles.

Derivation of the momentum principle. The *impulse-momentum* equation of mechanics states that the product of a force and the increment of time during which it acts (i.e., the *impulse* of the force) is equal

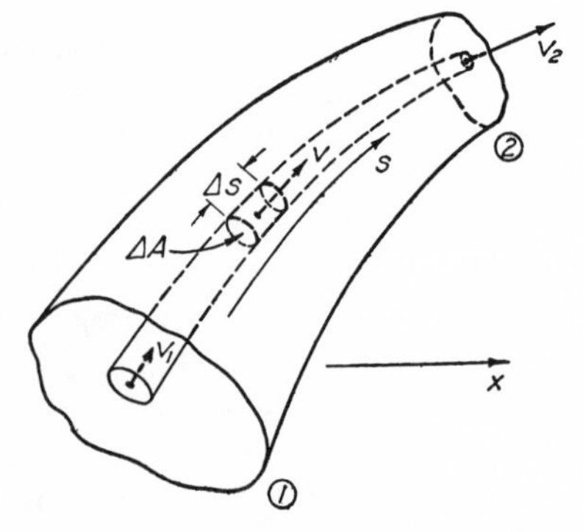

FIG. 53. Definition sketch for the momentum and energy principles.

to the resulting change in the product of the mass of the body on which the force acts and the velocity of the body (i.e., the change in the *momentum* of the body). Both impulse and momentum are necessarily vector quantities. In any direction x, therefore,

$$F_x\, dt = d(Mv)_x$$

which is evidently simply a different way of writing the Newtonian equation of acceleration, $F_x = M\, dv_x/dt$. Letting dF_x represent the differential force acting upon an incremental length of a stream filament (see Fig. 53),

$$dF_x\, dt = d(\rho\, ds\, dA\, v)_x$$

For steady flow at constant density, the right side of the equation may be transformed in the following manner:

$$d(\rho\, ds\, dA\, v)_x = \rho\, ds\, dA\, dv_x = \rho\, ds\, dA \frac{\partial v_x}{\partial s} ds$$

$$= \rho\, ds\, dA \frac{\partial v_x}{\partial s} v\, dt = \rho \frac{\partial v_x}{\partial s} ds\, dQ\, dt$$

Therefore, per unit time, the following impulse-momentum relationship must apply:

$$dF_x = \rho \frac{\partial v_x}{\partial s} ds\, dQ$$

Since dQ is constant along the filament, integration of this expression between sections 1 and 2 at once yields

$$\Delta F_x = \rho[(v_x)_2 - (v_x)_1]\, dQ$$

in which the left side represents the impulse per unit time due to all forces acting in the x direction upon the filament, and the right side the resulting rate of change in the x component of its momentum. Integration with respect to Q then yields a similar relationship for the entire zone of flow shown in Fig. 53. Since each of the terms requires further explanation, the integral is written for the present in the following general form:

$$\Sigma F_x = \rho \int [(v_x)_2 - (v_x)_1]\, dQ$$

Every fluid element composing the total body of fluid is under stress exerted by those elements with which it is in contact. Since every action involves an equal and opposite reaction, it will be seen that in the process of summation these *internal* stresses cancel one another, with the result that the summation ΣF_x involves only the x component of the weight of the fluid and of the forces exerted *externally* upon its boundary surface. In order to evaluate ΣF_x directly, it is therefore necessary to perform the process of summation either graphically or analytically from known conditions of boundary form and pressure distribution.

ΣF_x may be determined indirectly, however, through evaluation of the right side of the equation. In general, the velocity will vary across the two end sections of the zone in question, and this integration must then also proceed either graphically or analytically, from known conditions of velocity distribution. If, on the other hand, sections 1 and 2 are chosen in *uniform* regions either side of the non-uniformity under study, the momentum equation may conveniently be written in terms of the mean velocity for each section, together with a coefficient K_m embodying the effect of velocity variation across the section; that is,

$$\Sigma F_x = (K_m Q \rho V_x)_2 - (K_m Q \rho V_x)_1 \tag{81}$$

in which

$$K_m = \frac{1}{Q} \int \frac{v_x}{V_x} dQ = \frac{1}{A} \int \frac{v_x}{V_x} \frac{v}{V} dA \tag{82}$$

More will be said about the factor K_m on a later page, but it may be noted at this point that the quantity $K_m Q \rho V$ represents a rate of flow, or *flux*, of momentum past a given cross section. The momentum principle thus states that the total force acting in any direction upon a given zone of steady flow is equal to the difference in the flux of the corresponding component of fluid momentum past the two end sections.

Derivation of the energy principle. The *work-energy* equation of mechanics likewise states that the product of a force component and the distance through which it acts in the corresponding direction (i.e., the *work* done by that force component) must be equal to the change in one-half the product of the mass and the square of the corresponding component of velocity (i.e., the change in the *kinetic energy* of the mass). Thus, in any coordinate direction x

$$F_x \, dx = d\left(\frac{M v_x^2}{2}\right)$$

which is simply another way of writing the Newtonian equation $F_x = M \, dv_x/dt$. The work-energy relationship is not a vector relationship, however, despite the fact that the foregoing equation involves vector components; in other words, the total work done upon a mass is the *scalar* sum of the work done in the several coordinate directions:

$$F_s \, ds = F_x \, dx + F_y \, dy = d\left(\frac{M v_x^2}{2}\right) + d\left(\frac{M v_y^2}{2}\right) = d\left(\frac{M v^2}{2}\right)$$

Again with reference to Fig. 53, the work done by pressure gradient and weight upon the element $ds \, dA$ of a stream filament during a displacement ds will be

$$\left[-\frac{\partial}{\partial s} (p + \gamma z) \, ds \, dA \right] ds = d\left(\frac{\rho \, ds \, dA \, v^2}{2}\right) = ds \, dA \frac{\partial}{\partial s}\left(\frac{\rho v^2}{2}\right) ds$$

Introducing the identities $ds = v\,dt$ and $v\,dA = dQ$, and dividing by dt, this becomes, for steady flow,

$$-\frac{\partial}{\partial s}(p + \gamma z)\,ds\,dQ = \frac{\partial}{\partial s}\left(\frac{\rho v^2}{2}\right)ds\,dQ$$

which represents the rate at which work is done upon the element, or the *power* of the acting forces, and the corresponding rate of change in its kinetic energy. Since dQ is the same at all successive cross sections, the integral of this expression along the filament from section 1 to section 2 will be

$$(p_1 + \gamma z_1 - p_2 - \gamma z_2)\,dQ = \left(\frac{\rho v_2^{\,2}}{2} - \frac{\rho v_1^{\,2}}{2}\right)dQ$$

and its integral over the entire zone shown in Fig. 53 then becomes

$$\int (p_1 + \gamma z_1 - p_2 - \gamma z_2)\,dQ = \int \left(\frac{\rho v_2^{\,2}}{2} - \frac{\rho v_1^{\,2}}{2}\right)dQ$$

As in the momentum relationship, evaluation of either side of this equation will in general involve graphical or analytical integration based on known distributions of pressure intensity and velocity across the two end sections of the zone in question. If, however, sections 1 and 2 lie in *uniform* regions of flow either side of the non-uniformity under study, the fact that the sum $p + \gamma z$ is a constant across each section at once yields the result $(p_1 + \gamma z_1 - p_2 - \gamma z_2)Q$ for the left side of the equation. The right side may then be evaluated in terms of the mean velocity at each section, together with a coefficient K_e embodying the effect of velocity variation across the section; that is

$$(p_1 + \gamma z_1 - p_2 - \gamma z_2)Q = \left(K_e Q \frac{\rho V^2}{2}\right)_2 - \left(K_e Q \frac{\rho V^2}{2}\right)_1 \tag{83}$$

in which

$$K_e = \frac{1}{Q}\int \left(\frac{v}{V}\right)^2 dQ = \frac{1}{A}\int \left(\frac{v}{V}\right)^3 dA \tag{84}$$

Postponing, for the moment, further discussion of the factor K_e, it may nevertheless be seen that the quantity $K_e Q \rho V^2/2$ represents a rate of flow, or flux, of kinetic energy past a given section. The energy principle thus states that the rate at which work is done upon a given zone of steady flow is equal to the difference in the flux of kinetic energy past the two end sections.

If Eq. (83) is divided by Q, the result will be seen to involve a series of terms each having the dimension of energy per unit volume. A term such as $\rho V^2/2$ does, in fact, represent *kinetic energy per unit volume*, and γz may similarly be regarded as *potential energy per unit volume*. One is therefore tempted to look upon the pressure intensity p as a third form of energy, and state that the total energy per unit volume (frequently called the "specific energy") is the same at successive sections:

$$\left(K_e \frac{\rho V^2}{2} + p + \gamma z\right)_1 = \left(K_e \frac{\rho V^2}{2} + p + \gamma z\right)_2 \tag{85}$$

It must be realized, however, that p represents neither energy nor, in itself, a capacity to do work, for such work can be done only if a pressure *difference* exists at neighboring points. For example, the pressure intensity within a closed tank of liquid can be raised to an extremely high magnitude with little work (and therefore little change of energy) through an almost negligible displacement of a small piston; however, work can be done as the result of such pressure intensity only if the liquid emerges from the tank into a zone of lower pressure intensity, and then only if continued displacement of the piston (i.e., power input) maintains the pressure difference.

Simplification of principles. The foregoing principles of momentum and energy are, in themselves, quite rigorous. When used in the one-dimensional method of analysis, on the other hand, the assumption of zero normal acceleration (i.e., hydrostatic pressure distribution) at every section necessarily involves an error which varies in proportion to the degree of non-uniformity actually present at any section under study. So far as the velocity-correction coefficients K_m and K_e are concerned, their use again depends upon the accuracy required by the analysis. It will be noted, first of all, that the energy factor K_e will generally be greater than the momentum factor K_m, owing to the fact that the ratio v/V occurs to the third power in the former and only to the second power in the latter. If, for instance, the velocity distribution across a circular conduit follows the parabolic relationship, K_e will have the magnitude 2.0 and K_m the magnitude 1.33. These are probably the greatest values encountered in general flow conditions, and as the velocity variation across the section becomes less pronounced both coefficients approach 1.0 as a limit. If the velocity variation across a section is not very great, therefore, the assumption that either coefficient is equal to unity will probably not introduce errors of serious magnitude in the preliminary analyses for which the one-dimensional method is commonly used.

Unless K_e and K_m are known to be considerably greater than unity, it is usually satisfactory to adopt the following forms of the continuity, energy, and momentum relationships for steady flow:

$$Q = V_1A_1 = V_2A_2 \tag{86}$$

$$\frac{\rho V_1^2}{2} + p_1 + \gamma z_1 = \frac{\rho V_2^2}{2} + p_2 + \gamma z_2 \tag{87}$$

$$\Sigma F_x = Q\rho \Delta V_x \tag{88}$$

The continuity equation at once permits evaluation of the velocity in terms of Q and the boundary cross section. The energy equation then permits determination of the pressure change accompanying a change in cross section or elevation. Finally, the momentum equation permits solution for the total force acting upon the flow boundary at a section change.

Example 22. Fluid entering the abrupt enlargement in a circular conduit does not immediately expand but continues a short distance beyond the enlargement as a parallel jet surrounded by more or less stagnant fluid. Assuming

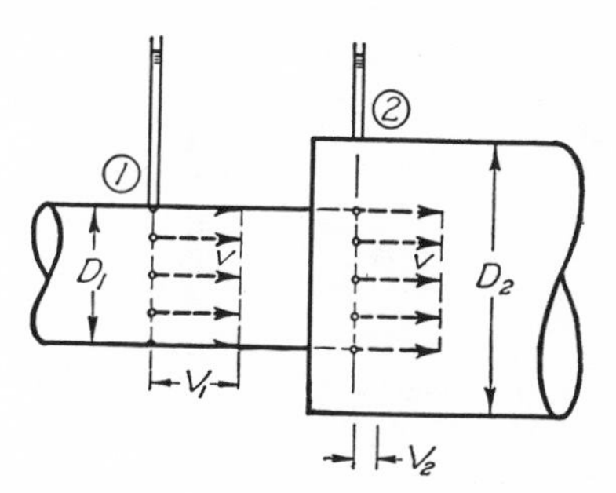

that the velocity is constant across section 1, show that the momentum and energy equations written in terms of the *mean* velocity will balance only if K_m and K_e for section 2 are properly evaluated.

At section 1, $K_m = 1.0 = K_e$.
At section 2, $V_2 = V_1D_1^2/D_2^2$, hence

$$K_m = \frac{1}{A}\int\left(\frac{v}{V}\right)^2 dA = \frac{1}{\pi D_2^2/4}\left[\left(\frac{V_1}{V_2}\right)^2 \frac{\pi D_1^2}{4}\right] = \frac{V_1^2}{V_2^2}\frac{D_1^2}{D_2^2} = \frac{D_2^2}{D_1^2}.$$

and

$$K_e = \frac{1}{A}\int\left(\frac{v}{V}\right)^3 dA = \frac{1}{\pi D_2^2/4}\left[\left(\frac{V_1}{V_2}\right)^3 \frac{\pi D_1^2}{4}\right] = \frac{V_1^3}{V_2^3}\frac{D_1^2}{D_2^2} = \frac{D_2^4}{D_1^4}$$

Since there is no fluid acceleration between sections 1 and 2,

$$\Sigma F_x = 0 \quad \text{and} \quad \Delta(p + \gamma z) = 0$$

hence

$$0 = (K_m Q \rho V)_2 - (K_m Q \rho V)_1 = Q\rho \left(\frac{D_2^2}{D_1^2} V_1 \frac{D_1^2}{D_2^2} - 1 \times V_1 \right)$$

and

$$0 = \left(K_e \frac{\rho V^2}{2} \right)_2 - \left(K_e \frac{\rho V^2}{2} \right)_1 = \frac{D_2^4}{D_1^4} \frac{\rho V_1^2 D_1^4 / D_2^4}{2} - 1 \times \frac{\rho V_1^2}{2}$$

If, on the other hand, K_m and K_e are placed equal to unity at both sections, the foregoing equations will obviously not balance:

$$0 \neq Q\rho V_1 \frac{D_1^2}{D_2^2} - Q\rho V_1$$

$$0 \neq \frac{\rho V_1^2 D_1^4 / D_2^4}{2} - \frac{\rho V_1^2}{2}$$

Example 23. A blower to be used for the demonstration of wind forces upon model airplanes is designed as shown in the accompanying sketch. If the 15-inch

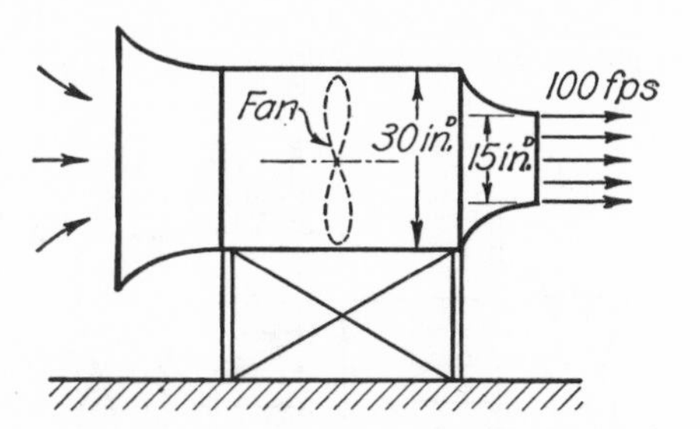

air stream is to have a maximum velocity of 100 feet per second, what must be the horsepower of the motor driving the fan, assuming an overall efficiency of 75 per cent? What will be the longitudinal thrust upon the frame supporting the conduit?

The kinetic energy per unit volume produced by the blower will be

$$\frac{\rho V^2}{2} = \frac{0.0025 \times \overline{100}^2}{2} = 12.5 \text{ ft-lb/ft}^3$$

Since no overall change in pressure intensity occurs, the required power input is

$$P = \frac{1}{\text{Eff.}} Q \frac{\rho V^2}{2} = \frac{1}{0.75} \times 100 \times \frac{\pi}{4} \times \left(\frac{15}{12} \right)^2 \times 12.5 = 2050 \text{ ft-lb/sec}$$

$$\frac{2050}{550} = 3.72 \text{ hp}$$

The force necessary to produce the corresponding acceleration of the air, which must be equal and opposite to the thrust of the air upon the conduit (and of the conduit upon the frame), is then found from Eq. (88):

$$\Sigma F = Q\rho\,\Delta V = 100 \times \frac{\pi}{4} \times \left(\frac{15}{12}\right)^2 \times 0.0025 \times (100 - 0) = 30.7 \text{ lb.}$$

Hence the thrust is -30.7 pounds (i.e., to the left).

Example 24. A nozzle at the end of a 3-inch hose produces a jet 1½ inches in diameter. Determine the longitudinal stress in the joint at the base of the nozzle when it is discharging 300 gallons of water per minute. Assume that $K_m = 1.0 = K_e$.

Since there are 7.48 gallons in 1 cubic foot,

$$Q = \frac{300}{7.48 \times 60} = 0.668 \text{ cfs}$$

$$A_1 = \frac{\pi}{4} \times \left(\frac{3}{12}\right)^2 = 0.0491 \text{ ft}^2; \quad V_1 = \frac{0.668}{0.0491} = 13.6 \text{ fps}$$

$$A_2 = \frac{\pi}{4} \times \left(\frac{1.5}{12}\right)^2 = 0.0123 \text{ ft}^2; \quad V_2 = \frac{0.668}{0.0123} = 54.3 \text{ fps}$$

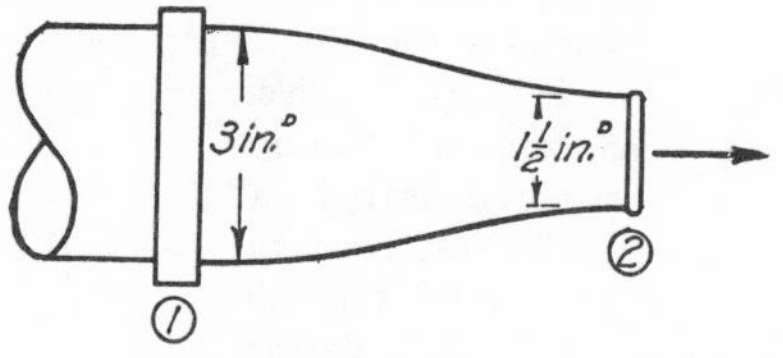

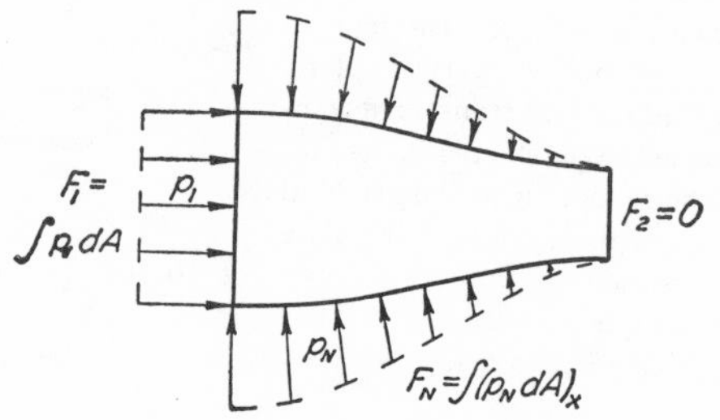

The average pressure intensity at section 1 will then be

$$p_1 = \frac{\rho}{2}(V_2^2 - V_1^2) = \frac{1.94}{2}(\overline{54.3}^2 - \overline{13.6}^2) = 2680 \text{ psf}$$

The total force producing the acceleration of the flow is

$$\Sigma F = Q\rho\,\Delta V = 0.668 \times 1.94 \times (54.3 - 13.6) = 52.8 \text{ lb}$$

Since (see free-body diagram) ΣF consists of the sum of the fluid pressure at section 1 and the longitudinal force exerted by the nozzle on the fluid,

$$\Sigma F = F_1 - F_N = p_1 A_1 - F_N$$

whence

$$F_N = p_1 A_1 - \Sigma F = 2680 \times 0.0491 - 52.8 = 78.7 \text{ lb}$$

From the Newtonian principle of action and reaction, the force exerted by the hose on the nozzle must be equal and opposite to the force exerted by the fluid on the nozzle. The joint stress is therefore 78.7 pounds tension.

PROBLEMS

(Assume that $K_e = 1.0 = K_m$ unless velocity distribution is known.)

129. The transverse velocity distribution of oil flowing in a circular pipe is represented by the parabolic equation $v = v_{\max}(1 - r^2/r_0^2)$. Determine the corresponding values of K_m and K_e.

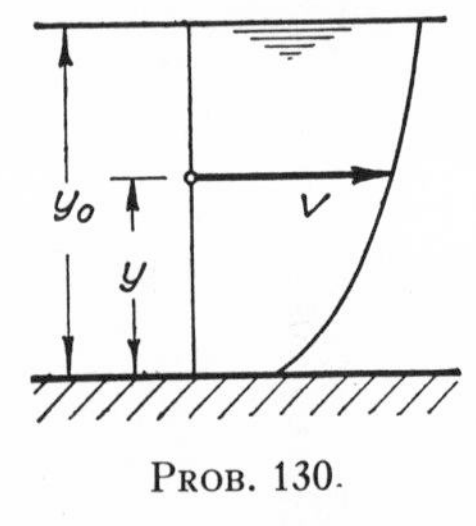

PROB. 130.

130. The velocity distribution in a very wide river 8 feet deep is found to vary from 2 feet per second at the bottom to 6 feet per second at the surface, approximately in accordance with the expression $v = 2 + 4(y/y_0)^{0.5}$. Evaluate the flux of (*a*) kinetic energy, and (*b*) momentum, per unit width of channel.

131. Determine by graphical or approximate numerical integration the values of K_e for the velocity profiles before and after the two-dimensional transition of Fig. 6.

132. If a 2-inch stream of water leaving a nozzle with a velocity of 75 feet per second and an inclination of 30° from the horizontal strikes a vertical wall at a horizontal distance of 25 feet from the nozzle, what force will it exert upon the wall?

133. A 3-inch jet of water having a velocity of 40 feet per second is deflected through an angle of 120° by a stationary curved plate. Determine the longitudinal and transverse components of force exerted upon the plate.

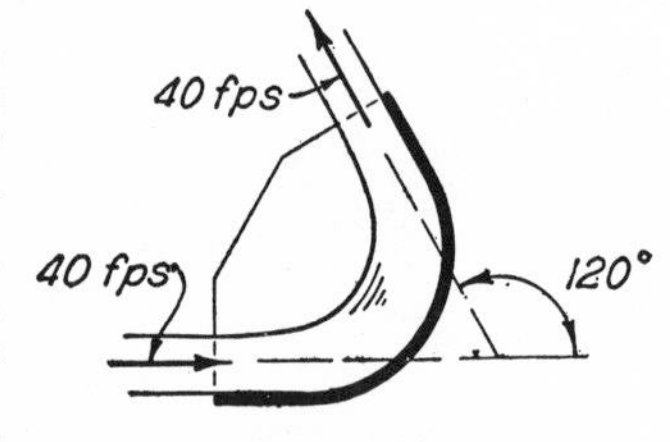

PROB. 133.

134. Determine the reduction in longitudinal force upon the plate of Problem 133 if the plate is moved in the direction of flow at a speed of 10 feet per second.

135. If the sheet of liquid discharged at the rate q_0 from a long slot strikes a plane boundary at an angle of 45°, what will be the ratio q_1/q_2 for the divided flow?

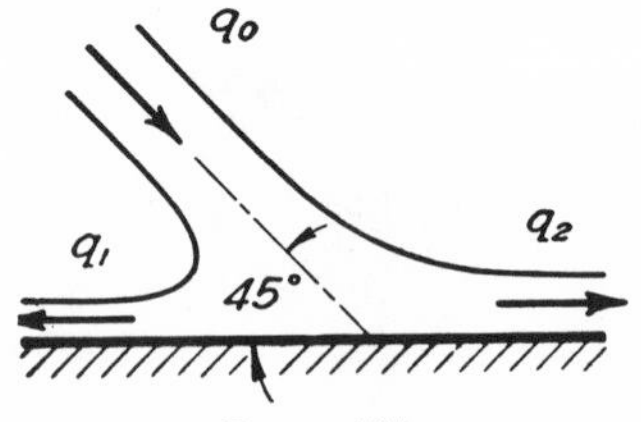

PROB. 135.

136. A water sprinkler consists of ½-inch jets at either end of a rotating arm, as shown in the accompanying sketch. What torque must be applied to the arm to hold it stationary when the velocity of the jets is 20 feet per second? If mechanical friction is ignored, what constant angular speed should the arm attain?

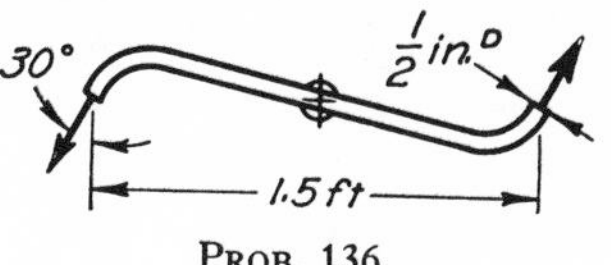

PROB. 136.

137. A large disk weighing 15 pounds is so mounted that it may move freely along a vertical axis, the plane of the disk thereby remaining horizontal. Directly below the disk is a nozzle delivering a vertical stream of water having an efflux velocity of 25 feet per second and an initial diameter of 2 inches. Assuming that the disk deflects the water horizontally, how high above the nozzle will the disk be held in equilibrium by the force of the jet?

138. An air duct changes in diameter from 5 feet to 3 feet. What rate of flow would be indicated by a head of ½ inch of water in the U-tube connected as shown?

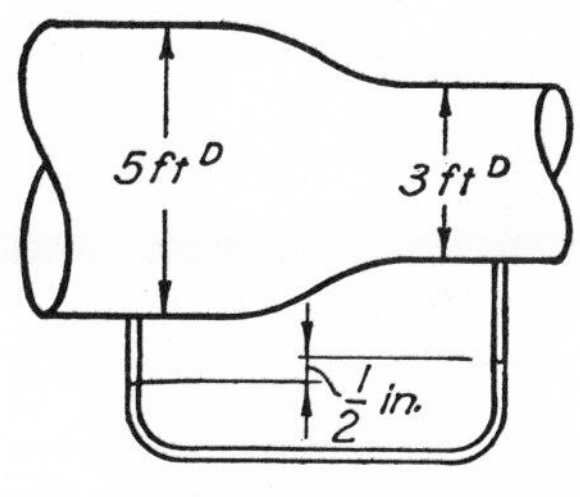

PROB. 138.

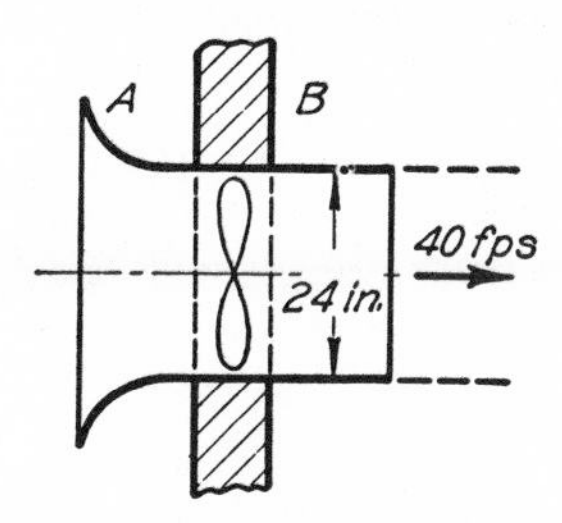

PROB. 139.

139. If a ¾-horsepower motor is required by a ventilating fan to produce a 24-inch stream of air having a velocity of 40 feet per second, what is the efficiency of the fan? Assume that the pressure intensities at A and B are the same.

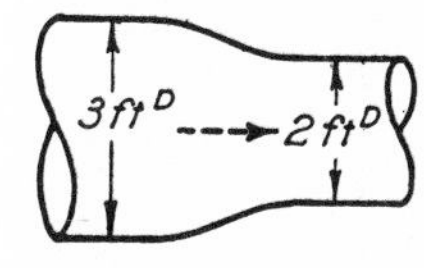

PROB. 140.

140. Water flows at the rate of 25 cubic feet per second through the pipe transition shown in the accompanying figure. If the pressure intensity at the centerline of the 3-foot section is 1.5 pounds per square inch, what will be the centerline pressure in the 2-foot section? What force will be required to produce the change in the momentum of the water as it passes through the transition?

141. A 4-inch circular orifice at the end of a 6-inch pipe yields a jet of oil 3.4 inches in diameter. What force will be exerted upon the orifice plate when the pressure intensity of the approaching flow is 2 pounds per square inch?

142. Through use of the continuity, energy, and momentum principles, show that the contraction coefficient for the so-called Borda mouthpiece or reëntrant tube is 0.5.

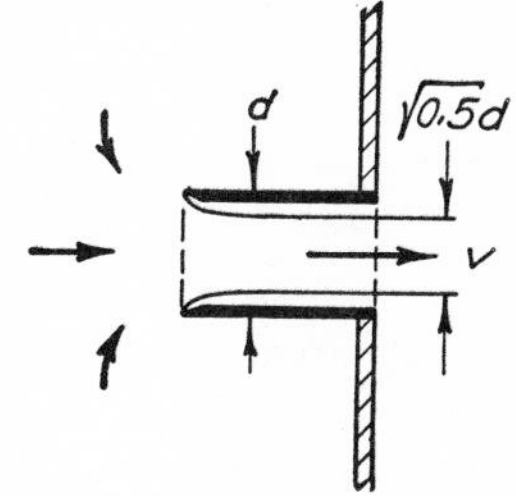

PROB. 142.

143. Determine the resultant force upon a horizontal 90° elbow in a 15-inch pipe carrying water under a mean pressure of 3 pounds per square inch if the mean velocity of flow is (a) negligible, and (b) 12 feet per second.

144. If a flow of 75 cubic feet of air per second passes through a 60° reducing elbow at the end of a pipe as shown in the figure, determine the magnitude and direction of the resultant force upon the elbow.

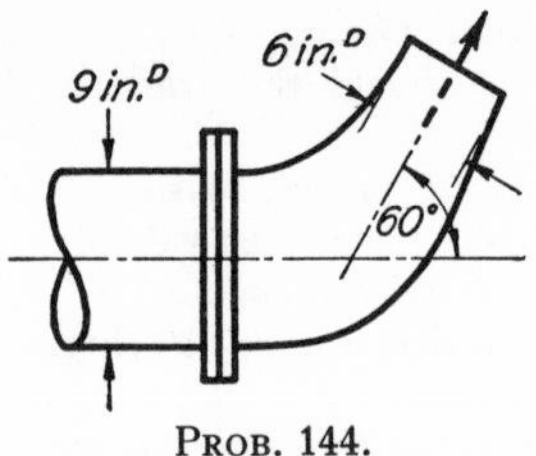

PROB. 144.

22. STEADY FLOW OF LIQUIDS IN CLOSED CONDUITS

Lines of total head and piezometric head. If all terms of Eq. (87) are divided by γ, the following form of the energy equation will result:

$$H = \frac{V_1^2}{2g} + \frac{p_1}{\gamma} + z_1 = \frac{V_2^2}{2g} + \frac{p_2}{\gamma} + z_2 \tag{89}$$

This is, of course, the Bernoulli equation for the gross filament, stating that the mean total head of the flow will be the same at all successive sections.

In dealing with liquid flow through a closed conduit, a very significant picture of relative flow conditions may be obtained by plotting

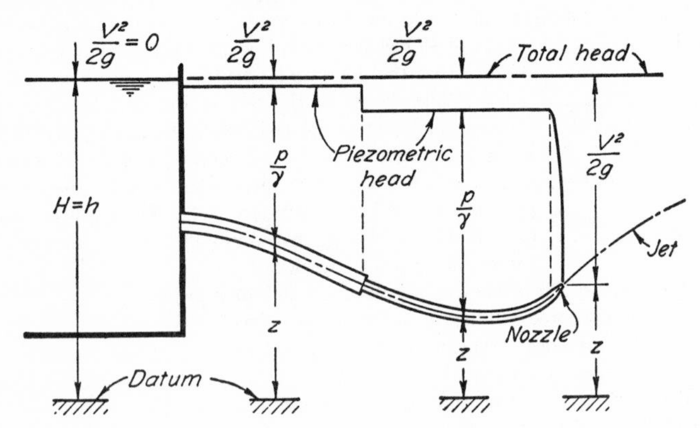

FIG. 54. Lines of total head and piezometric head for a conduit of non-uniform cross section.

to scale upon a diagram of the conduit elevation the lines of total head H and piezometric head h obtained by application of this equation to successive sections of the conduit. Such a graphical representation of flow characteristics is shown in Fig. 54 for gravity discharge from a large reservoir through a pipe terminating in a nozzle. Since the horizontal line of total head must have the same elevation as the free surface in the reservoir (a section of the "filament" at which the velocity head is essentially zero), the velocity of efflux will be governed entirely by the relative elevation of the nozzle. The rate of discharge, on the other hand, will depend not only upon the velocity of efflux but also upon the area of the jet. Knowledge of the variation in pipe diameter then permits determination of the velocity head for each cross section of the flow, subtraction of which from the total head yields the piezometric head at the corresponding section. If z denotes the centerline elevation of the conduit, the mean pressure head p/γ is evidently equal to $h - z$, the height of the line of piezometric head above the conduit axis. Though the elevation of intermediate points along the pipe axis would thus seem to affect only the pressure head, it will be recalled that reduction of the pressure intensity to a value approaching absolute zero invariably leads to cavitation and discontinuity of flow. Since zones in which the axis of the conduit rises above the line of piezometric head (as in the siphon of Fig. 55) are zones of subatmospheric pressure, it should be kept in mind that there is a physical limit to such pressure reduction.

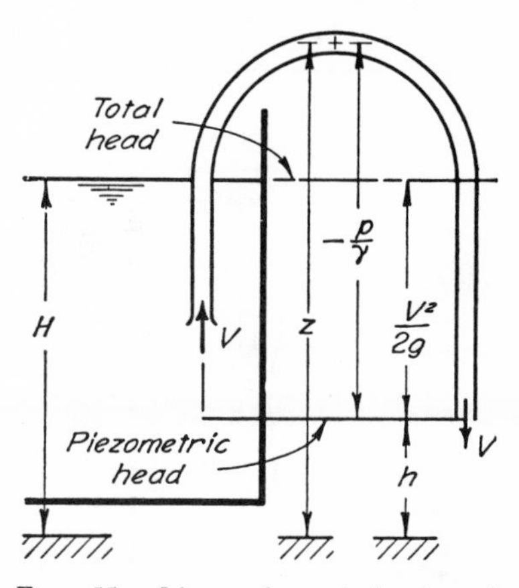

FIG. 55. Lines of total head and piezometric head for a siphon.

Change of head through hydraulic machinery. A gravity-flow system such as that shown in Fig. 54 is often used to convert the potential energy (i.e., energy of elevation) of water impounded in a reservoir into a useful form of mechanical energy by means of a water turbine driven by the free jet. Under such circumstances it becomes necessary to evaluate the available power of the flow. Since power has the dimension of energy per unit time, and since the velocity head of the jet represents energy per unit weight of fluid, it follows that the power available in a free jet is equal to the product of the volume rate of flow, the specific weight of the fluid, and the velocity head: $P = Q\gamma V^2/2g$.

On the other hand, work may be done under pressure by a pump or upon a turbine located at some intermediate point in the conduit.

Evidently, the rate at which such work is done determines the corresponding power input or output of the system, and the effect of such power input or output is to increase or decrease, respectively, the total head of the flow. Figure 56, for instance, illustrates flow conditions during the pumping of liquid from a reservoir, the increase in

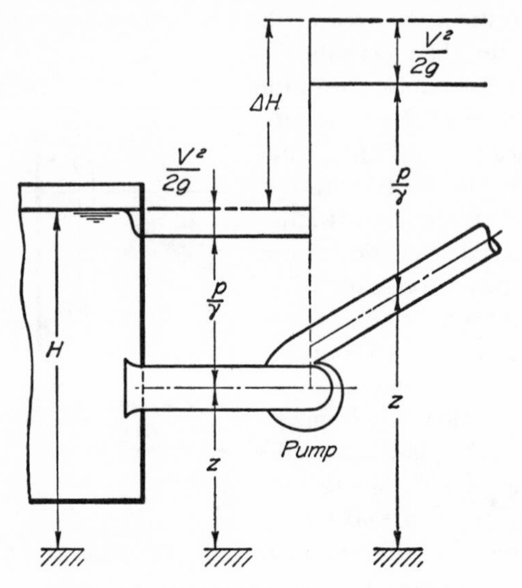

FIG. 56. Variation in head at a pump.

total head at the pump depending upon the rate of flow, the specific weight of the liquid, and the power input; that is,

$$\Delta H = \frac{P}{\gamma Q} \tag{90}$$

Assuming the change in head to occur between sections before and after the pump the Bernoulli equation then states that the total head at section 2 is equal to that at section 1 plus the intervening change:

$$\frac{V_1^2}{2g} + \frac{p_1}{\gamma} + z_1 + \Delta H = \frac{V_2^2}{2g} + \frac{p_2}{\gamma} + z_2 \tag{91}$$

The diagram of total-head and piezometric-head lines is then obtainable as before. In the case of a turbine, of course, the sign of ΔH must be negative to indicate a reduction in head, or power output by the flow.

Example 25. During a flow of 15 cubic feet per second, the gage pressure is +10 pounds per square inch in the horizontal 12-inch supply line of a water turbine, and −6 pounds per square inch at an 18-inch section of the draft tube 5 feet below. Estimate the horsepower output of the turbine under such conditions, assuming an efficiency of 85 per cent.

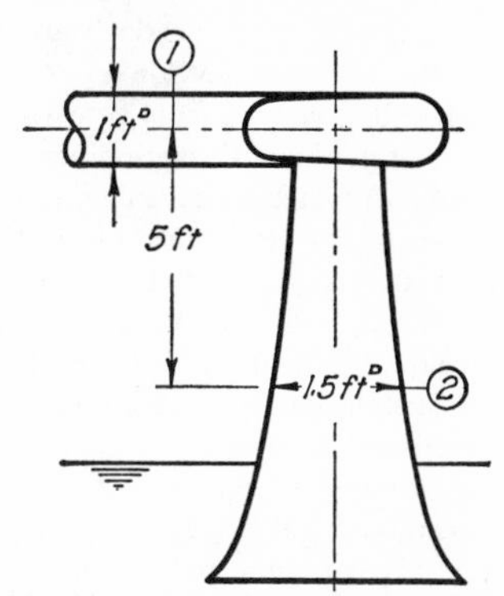

Since $Q = A_1V_1 = A_2V_2$

$$V_1 = \frac{15}{\frac{\pi}{4} \times 1^2} = 19.1 \text{ fps} \quad \text{and} \quad V_2 = \frac{15}{\frac{\pi}{4} \times \overline{1.5}^2} = 8.49 \text{ fps}$$

From Eq. (91),

$$\begin{aligned} \Delta H &= \frac{V_2^2}{2g} + \frac{p_2}{\gamma} + z_2 - \frac{V_1^2}{2g} - \frac{p_1}{\gamma} - z_1 \\ &= \frac{\overline{8.49}^2}{2 \times 32.2} + \frac{-6 \times 144}{62.4} + 0 - \frac{\overline{19.1}^2}{64.4} - \frac{10 \times 144}{62.4} - 5 \\ &= 1.12 - 13.83 + 0 - 5.66 - 23.05 - 5 \\ &= -46.4 \text{ ft.} \end{aligned}$$

The power input is, according to Eq. (90),

$$P = Q\gamma\Delta H = 15 \times 62.4 \times (-46.4) = -43{,}500 \text{ ft-lb/sec}$$

The horsepower output of the turbine is then

$$\frac{43{,}500 \times 0.85}{550} = 67.2 \text{ hp}$$

PROBLEMS

145. If the centerline of a 4-inch siphon attains a maximum elevation of 8 feet above the surface of a reservoir, and if a negative centerline pressure head of 30 feet is not to be exceeded, what is the greatest permissible rate of flow? To what outlet elevation will this rate of flow correspond?

146. From the base of an open standpipe a 4-inch pipe 50 feet in length leads to a nozzle discharging a jet 1.5 inches in diameter. If the depth of water in the standpipe is 75 feet and the nozzle outlet is 100 feet below the free surface, determine (*a*) the rate of flow, (*b*) the velocity head in the pipe and in the jet, and (*c*) the pressure head at each end of the pipe. Submit a sketch showing the lines of total head, piezometric head, and centerline elevation.

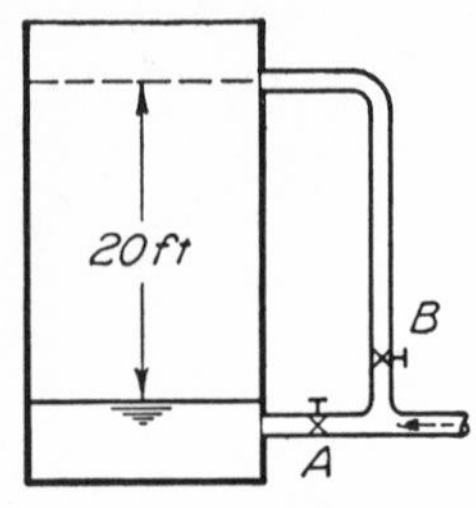

PROB. 147.

147. The water level in an open standpipe is to be raised 20 feet by pumping from a nearby reservoir. Compare the costs of pumping through pipes *A* and *B*.

148. A 6-inch pipe leading from a reservoir branches into a 4-inch pipe terminating in a 2-inch nozzle and a 3-inch pipe discharging directly into the atmosphere. If the nozzle ($C_c = 1$) is 30 feet below the reservoir surface and the 3-inch outlet is 15 feet below the reservoir surface, determine the rate of flow through each pipe. Submit a sketch showing to scale the lines of total head, piezometric head, and assumed centerline elevation.

149. When the rate of flow through the pump shown in the accompanying sketch is 3 cubic feet of water per second, a differential gage connected to piezometer taps either side of the pump indicates an 8-inch head of mercury. What horsepower is delivered by the pump to the flow?

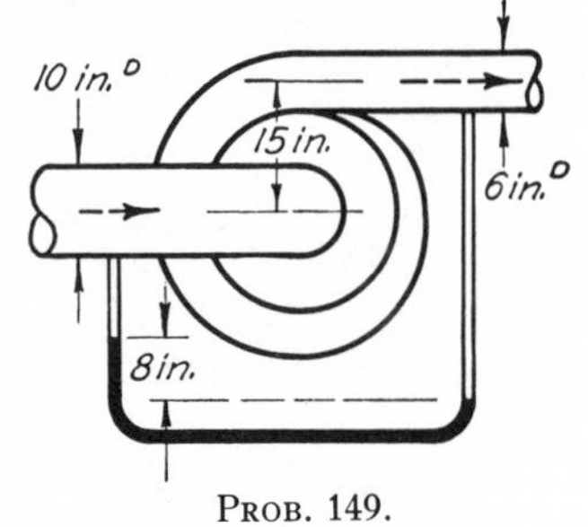

PROB. 149.

150. A blower having an efficiency of 75 per cent is to supply 500 cubic feet of air per minute to a 6-inch pipe under a pressure equivalent to 2 inches of water. If the 12-inch intake pipe draws directly from the atmosphere, what horsepower motor should be provided? What will be the pressure intensity in the intake pipe?

151. Water is pumped at the rate of 5 cubic feet per second from a reservoir under the conditions indicated. If the pump supplies energy to the flow at the rate of 15 horsepower, determine the pressure intensities at points *A* and *B*.

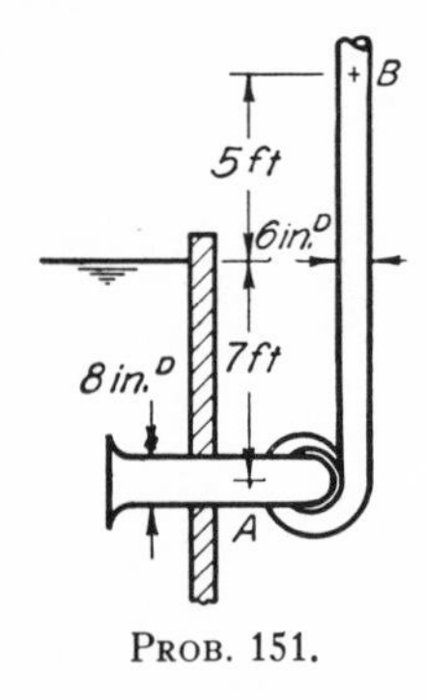

PROB. 151.

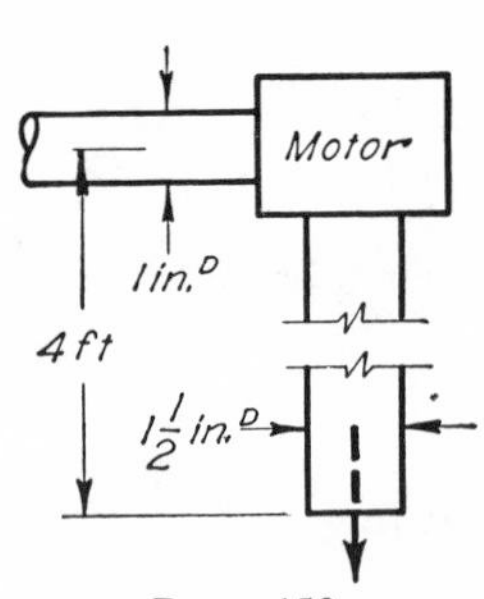

PROB. 152.

152. A small water motor discharging into the atmosphere operates under a pressure intensity of 10 pounds per square inch during a flow of 40 gallons per minute. If the motor efficiency is 65 per cent, what is its power output under these conditions?

153. A horizontal 8-inch pipe leading from a reservoir supplies water to a small turbine, and the water leaving the turbine is discharged directly into the atmosphere through a similar 8-inch pipe. If the reservoir surface lies 20 feet above the outlet pipe, and if the efflux velocity is 15 feet per second, what is the power delivered to the turbine by the flow? Sketch to scale the corresponding lines of total head, piezometric head, and centerline elevation.

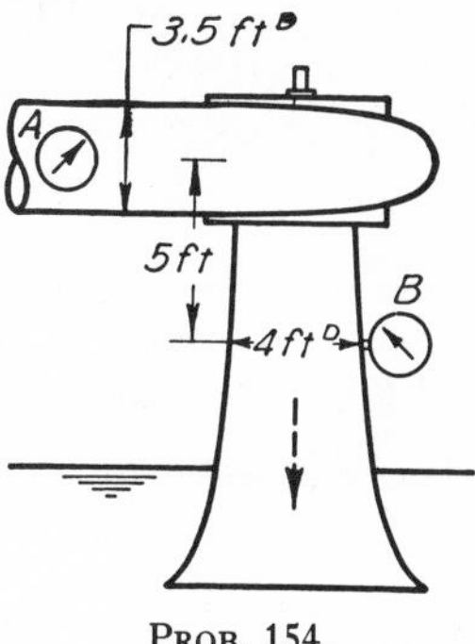

PROB. 154.

154. During a flow of 200 cubic feet per second through the turbine shown in the accompanying diagram, the pressure intensity indicated by gage A is 12 pounds per square inch. What will be the reading of gage B if the turbine is delivering 800 horsepower at 75 per cent efficiency?

23. UNSTEADY FLOW OF LIQUIDS IN CLOSED CONDUITS

Orifice discharge under falling head. The sole criterion of unsteadiness in the one-dimensional method of flow analysis is that the mean velocity at any section (and hence the rate of flow) will vary in magnitude with time. If such variation takes place, one must, strictly speaking, take into account a term for local acceleration corresponding to the quantity $\partial v_s/\partial t$ in Eq. (51) for the elementary filament. Thus, although in steady flow the total head of the gross filament is the same at all sections, in unsteady flow

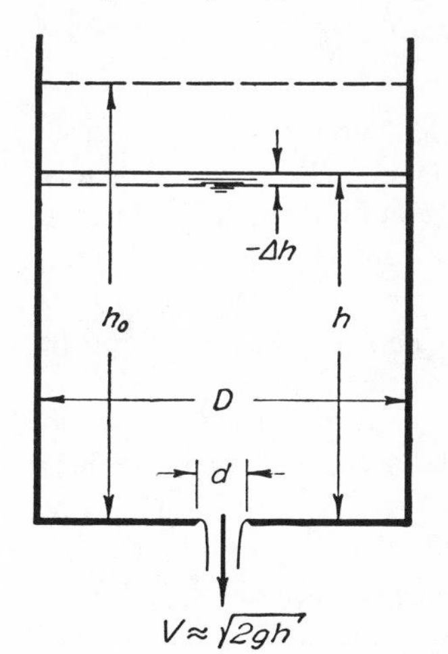

FIG. 57. Discharge from a small orifice under falling head.

$$\frac{1}{g}\frac{\partial V}{\partial t} = -\frac{\partial H}{\partial s} \tag{92}$$

In other words, if the rate of flow varies with time, the mean total head will vary with distance along the gross filament.

In some cases which are actually variable with time, to be sure, the quantity $\partial V/\partial t$ will be so small that it can be neglected without appreciable error. An illustration of such a circumstance is found in the emptying of a tank of liquid by means of an orifice, the size of the orifice cross section relative to that of the tank obviously controlling the rate at which the velocity of efflux changes with time. If, as shown in Fig. 57, the ratio d/D is very small, $\partial V/\partial t$ will also be very small. and the total head at all points will be almost exactly equal to the elevation of the free surface in the tank. The rate of efflux at any time may then be written simply as

$$Q = C_d \frac{\pi d^2}{4} \sqrt{2gh}$$

and the free surface, in accordance with the equation of continuity, will fall at the almost imperceptible rate

$$-\frac{dh}{dt} = \frac{Q}{\pi D^2/4} = C_d \frac{d^2}{D^2} \sqrt{2gh}$$

Evidently,

$$dt = -\frac{1}{C_d} \frac{D^2}{d^2} \frac{dh}{\sqrt{2gh}}$$

integration of which yields

$$t = -\frac{1}{C_d} \frac{D^2}{d^2} \frac{2}{\sqrt{2g}} \sqrt{h} + C$$

Letting $h = h_0$ when $t = 0$, evaluation of the constant of integration results in the following relationship between time and surface elevation:

$$t = \frac{1}{C_d} \frac{D^2}{d^2} \frac{2}{\sqrt{2g}} (h_0^{1/2} - h^{1/2}) \tag{93}$$

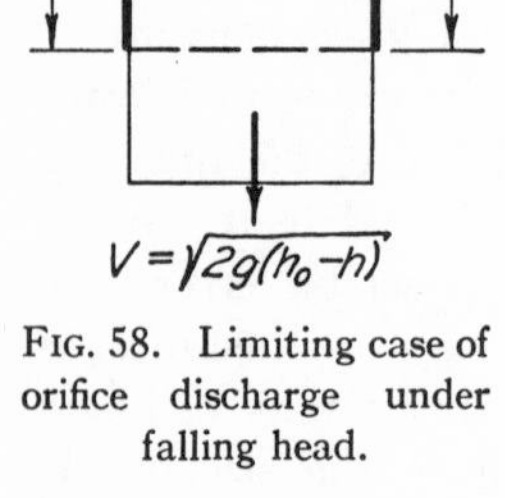

FIG. 58. Limiting case of orifice discharge under falling head.

As long as d is small in comparison with D, the foregoing treatment will yield reasonably accurate information as to the interrelationship of time, head, and rate of efflux. Consider, however, the limiting case in which $d = D$, as shown in Fig. 58. Though Eq. (93) happens to give the time required to empty such a bottomless tank (an example of free fall starting from rest), it is obvious that the highest velocity of flow $V = \sqrt{2gh_0}$ will occur just as the last drop of liquid leaves the tank, whereas according to the orifice equation $V = \sqrt{2gh}$ the last drop would leave under a zero head and hence at a zero velocity. In other words, problems in which the variation in rate of flow with time is not negligible must be analyzed in strict accordance with Eq. (92) for unsteady motion. Three representative examples of such motion will be discussed in the following paragraphs.

Pendulation of liquid in a U-tube. A very common type of variation in conduit flow with time is represented by the pendulation of a liquid column in an open U-tube (Fig. 59). Ignoring, as in the foregoing pages, viscous (or tangential) stresses, the amplitude and period of such pendulation may be evaluated as follows: Eq. (92) is first written as an integral over the length L of the column,

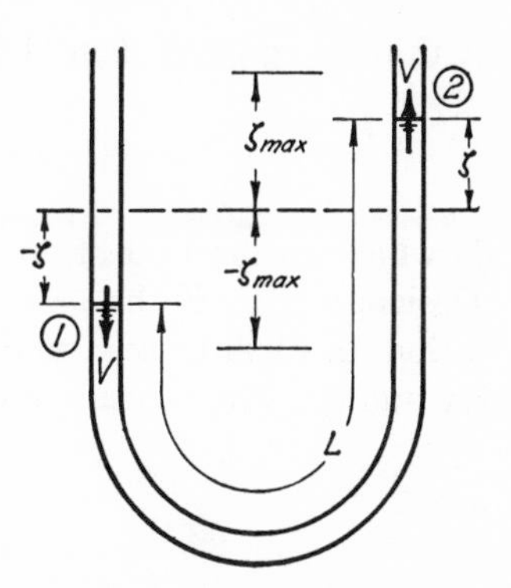

FIG. 59. Pendulation of liquid in a U-tube.

$$\frac{1}{g}\int_0^L \frac{\partial V}{\partial t}\, ds = -\int_0^L \frac{\partial H}{\partial s}\, ds$$

Since the U-tube is considered to be of uniform cross section, the quantity $\partial V/\partial t$ at the left will at any instant be the same at all sections of the tube and hence may be treated as a constant in this integration; the integral at the right will be simply the difference in total head between the two free surfaces, which may at once be expressed in terms of the instantaneous distance ζ (zeta) to either free surface from the line of static equilibrium; thus,

$$\frac{L}{g}\frac{\partial V}{\partial t} = H_1 - H_2 = \left(\frac{V^2}{2g} - \zeta\right) - \left(\frac{V^2}{2g} + \zeta\right) = -2\zeta$$

By means of the following principles of the calculus, the variable t may be eliminated for purposes of a second integration,

$$\frac{\partial V}{\partial t} = \frac{\partial V}{\partial \zeta}\frac{d\zeta}{dt} = \frac{\partial V}{\partial \zeta} V = \frac{1}{2}\frac{\partial (V^2)}{\partial \zeta}$$

Hence

$$\frac{L}{g}\frac{\partial V}{\partial t} = \frac{L}{2g}\frac{\partial (V^2)}{\partial \zeta} = -2\zeta$$

or

$$\frac{\partial (V^2)}{\partial \zeta} = -\frac{4g\zeta}{L}$$

the integral of which with respect to ζ is evidently

$$V^2 = -\frac{2g\zeta^2}{L} + C$$

Since, when $\zeta = \zeta_{max}$, $V = 0$, the constant of integration will be $C = 2g\,\zeta_{max}^2/L$, whereupon

$$V = \frac{d\zeta}{dt} = \sqrt{\frac{2g}{L}\,(\zeta_{max}^2 - \zeta^2)}$$

A third integration then yields the result

$$t = \sqrt{\frac{L}{2g}}\,\sin^{-1}\frac{\zeta}{\zeta_{max}} \tag{94}$$

in which no constant of integration need appear if $t = 0$ when $\zeta = 0$. Designating by T the *period*, or time required for a complete pendulation, this will be seen to depend only upon the acceleration of gravity and the length of the liquid column:

$$T = 2\pi\sqrt{\frac{L}{2g}} \tag{95}$$

Establishment of flow. The second example to be considered involves the gradual establishment of flow through a long pipe leading

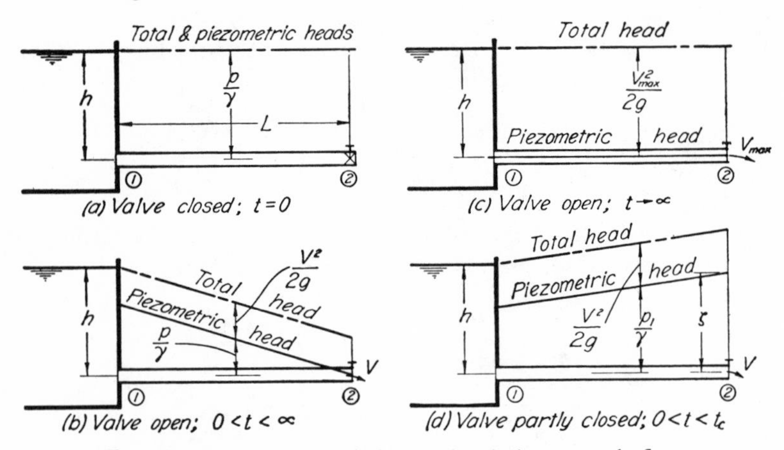

FIG. 60. Variation in head along a pipe during unsteady flow.

from a reservoir, after a valve at the end of the pipe has suddenly been opened. With reference to Fig. 60, the integral of Eq. (92) over the length L of the uniform pipe again leads to the expression

$$\frac{L}{g}\frac{\partial V}{\partial t} = H_1 - H_2$$

Letting h represent the elevation of the reservoir surface above the pipe outlet, the difference in total head at any time t after the valve has opened will be simply $h - V^2/2g$. Moreover, since V varies with time alone, the quantity $\partial V/\partial t$ may now be written as a total derivative. The expression then becomes

$$\frac{L}{g}\frac{dV}{dt} = h - \frac{V^2}{2g}$$

Evaluation of the general integral

$$\int dt = 2L \int \frac{dV}{2gh - V^2}$$

leads at once to the relationship

$$t = \frac{L}{\sqrt{2gh}} \log_e \frac{\sqrt{2gh} + V}{\sqrt{2gh} - V} + C$$

in which, since $t = 0$ when $V = 0$, the constant of integration will also be zero. Presuming that the velocity will eventually attain the value

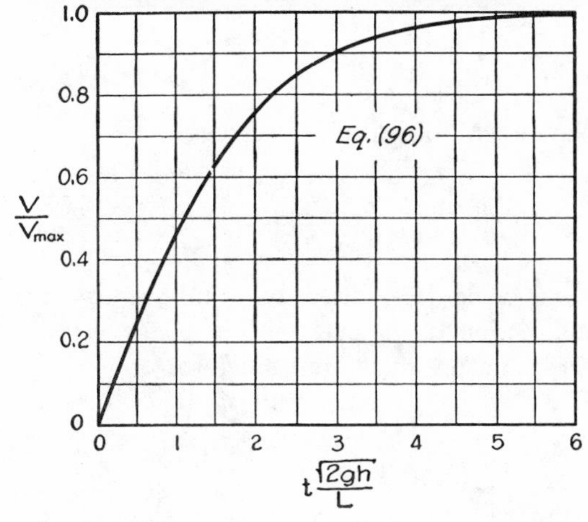

FIG. 61. Dimensionless diagram for the establishment of flow in a pipe.

$V_{\max} = \sqrt{2gh}$, the foregoing result may be written in the general dimensionless form

$$t\frac{\sqrt{2gh}}{L} = \log_e \frac{1 + V/V_{\max}}{1 - V/V_{\max}} \tag{96}$$

which is plotted in Fig. 61. Evidently, the flow will become fully established only after an infinite time. However, the interval required for V to attain, say, 99 per cent of its ultimate magnitude may readily be evaluated from the diagram; as may be seen from Eq. (96), such time will be directly proportional to the length of pipe, and inversely proportional to the ultimate velocity of flow.

Pressure rise due to valve closure. The third example to be considered is quite the reverse of flow establishment—the rapid *reduction* in rate of flow due to the closure of a valve at the end of a pipe. If, to be sure, the valve were closed in such a manner that the velocity change followed the curve of Fig. 61 from right to left, the problem would be already solved. On the other hand, appreciable departure from this form of velocity variation might lead to excessive intensities of pressure within the pipe, and hence further investigation is warranted. Assume, for example, that the valve is so operated that the velocity decreases linearly with time, varying from V_{max} to zero in the interval t_c required for complete closure; that is

$$V = V_{max}\left(1 - \frac{t}{t_c}\right)$$

whence

$$\frac{dV}{dt} = -\frac{V_{max}}{t_c}$$

As before, the integral of Eq. (92) becomes

$$\frac{L}{g}\frac{\partial V}{\partial t} = \frac{L}{g}\frac{dV}{dt} = H_1 - H_2$$

Since the flow is being decelerated, the total head must *increase* in the downstream direction, as indicated in Fig. 60*d*. Introducing the variable downstream head ζ shown in the figure,

$$\frac{L}{g}\frac{dV}{dt} = h - \zeta - \frac{V^2}{2g}$$

which becomes, upon substitution of the foregoing values for dV/dt and V,

$$-\frac{L}{g}\frac{V_{max}}{t_c} = h - \zeta - \frac{V_{max}{}^2}{2g}\left(1 - \frac{t}{t_c}\right)^2$$

or, solving for ζ,

$$\zeta = h + \frac{LV_{max}}{gt_c} - \frac{V_{max}{}^2}{2g}\left(1 - \frac{t}{t_c}\right)^2 \tag{97}$$

Evidently, at the instant the valve closure is complete the pressure head at the end of the pipe will exceed h by the amount LV_{max}/gt_c.

If the pipe is short, the velocity is low, and the time of closure t_c is long, the load on the pipe will not be excessive. The load increases as t_c decreases, however, indicating that too rapid closure of a valve at the end of a long pipe may produce sufficient stress to burst the pipe. It is therefore common practice to install an open *surge tank* upstream from the control valve, which permits very rapid valve closure without overstressing the conduit walls.

Although the foregoing examples of unsteady flow in closed conduits represent extreme simplifications of problems actually encountered, the basic method of analysis remains the same, however complex the boundary conditions may be. Pendulation of liquid in a simple U-tube, for instance, is basically characteristic of the oscillation of surface level in two interconnected reservoirs, or in a surge tank at some point in a long penstock. Such changes in conduit section, or similar section changes in problems of flow establishment or reduction, simply lead to additional terms in the foregoing analyses. In this regard, however, the fact must be mentioned that all fluid motion is opposed by internal stresses due to fluid viscosity; it is thus apparent that an oscillating column of liquid will eventually come to rest as the result of viscous action, and that viscous resistance will serve to shorten the time required for establishment of flow in a long pipe and reduce the stresses due to valve closure. The latter phenomena, moreover, are also influenced by the elastic properties of both fluid and conduit. Evaluation of the actual magnitude of such effects will be discussed in later chapters.

Example 26. A 12-inch pipe ½ mile in length ends in a 4-inch nozzle, the jet from which is used to drive a turbine. If the ultimate velocity head of the

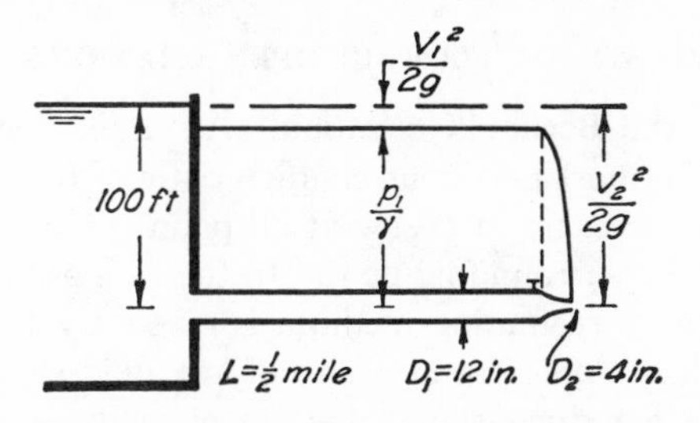

jet is 100 feet, estimate the time required for the turbine wheel to reach 95 per cent of full speed after the control valve has been opened.

Since

$$\frac{L}{g}\frac{dV}{dt} = h_o - \frac{V_2^2}{2g} = h_o - \frac{V_1^2}{2g}\frac{A_1^2}{A_2^2}$$

$$\int dt = 2L\frac{A_2^2}{A_1^2}\int\frac{dV}{2gh_0(A_2^2/A_1^2) - V^2}, \quad t = \frac{L}{\sqrt{2gh_0}}\frac{A_2}{A_1}\log_e\frac{1 + V/V_{max}}{1 - V/V_{max}}$$

From Fig. 61, when $V/V_{max} = 0.95$, $t\sqrt{2gh}/L = 3.7$, whence

$$t = \frac{L}{\sqrt{2gh_0}}\frac{3.7}{A_1/A_2} = \frac{0.5 \times 5280}{\sqrt{2 \times 32.2 \times 100}} \times \frac{3.7}{9} = 13.6 \text{ seconds}$$

PROBLEMS

155. Estimate the time required to empty a cylindrical tank 5 feet in diameter through a 2-inch sharp-edged circular orifice in the bottom, if the tank is originally filled with water to a depth of 4 feet.

156. Leakage in the bottom of a storage tank for oil is found to cause the surface level to decrease from 10 feet to 9 feet 8 inches in 24 hours. What further drop in level would occur during an additional 6 days?

157. After a heavy rain the water in a reservoir is observed to waste over a relief spillway under a 2-foot head, the surface level thereafter falling 3 inches in 8 hours. Assuming the continued decrease in head to follow the same functional relationship, estimate the additional time required for the head to be reduced to a magnitude of 1 foot. (Note that $Q \sim h^{3/2} \sim -dh/dt$.)

158. As a substitute for a surveyor's level in field construction, a long hose is filled with water and the required elevations are measured from the free surfaces at the upturned ends of the hose. To avoid error, an estimate must be made of the period of oscillation of the water column. What would be the period if the hose were 1000 feet long?

159. A turbine is supplied by a 10-inch pipe leading 800 feet from a reservoir. If the rate of flow during normal operation is 6 cubic feet of water per second, estimate the minimum time required for the turbine to reach 95 per cent of capacity from complete shut-off.

160. Assuming that the turbine of Problem 159 is closed down in such a manner as to produce a linear rate of velocity change, compute the increase in line pressure which would result from closure in (*a*) 10 seconds, and (*b*) 100 seconds.

24. STEADY FLOW IN OPEN CHANNELS

Adaptation of the Bernoulli equation. As emphasized in the foregoing chapter, the most pertinent characteristic of free-surface flow is the fact that the pressure intensity at all points of the free surface is equal to that of the surrounding fluid. In the case of liquid flow in an open channel, the surrounding medium is evidently the atmosphere, relative to which the pressure intensity of any fluid is generally measured. Since the one-dimensional method of analysis assumes essentially uniform motion, it becomes apparent that the plane of piezometric head in such analysis of open-channel flow should coincide with the free surface. Therefore, with reference to Fig. 62, the Bernoulli equation

$$\left(\frac{V^2}{2g} + h\right)_1 = \left(\frac{V^2}{2g} + h\right)_2$$

may for present purposes be written in terms of the velocity head, the depth of flow, and the elevation of the channel floor:

$$\left(\frac{V^2}{2g} + y + z_0\right)_1 = \left(\frac{V^2}{2g} + y + z_0\right)_2$$

If, in addition, the velocity terms are transformed through use of the relationship $Q = VA$, in which A is expressible in terms of y, the Bernoulli and continuity relationships may be combined in the following significant form:

$$\frac{Q^2}{2gA_1^{\,2}} + y_1 = \frac{Q^2}{2gA_2^{\,2}} + y_2 + \Delta z_0 \qquad (98)$$

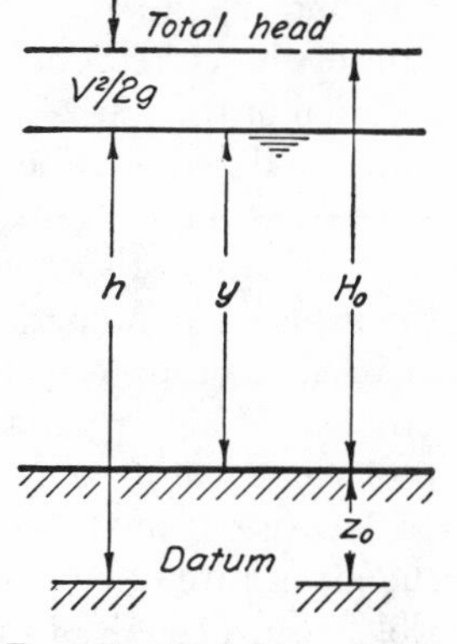

FIG. 62. Definition sketch for open-channel flow.

Equation (98) will be found sufficient for the analysis of any type of open-channel transition discussed in the remainder of this section. Indeed, the channel may have any form of cross section whatever, and the change in boundary elevation may likewise be accompanied by a change in cross-sectional form, provided only that the sections are

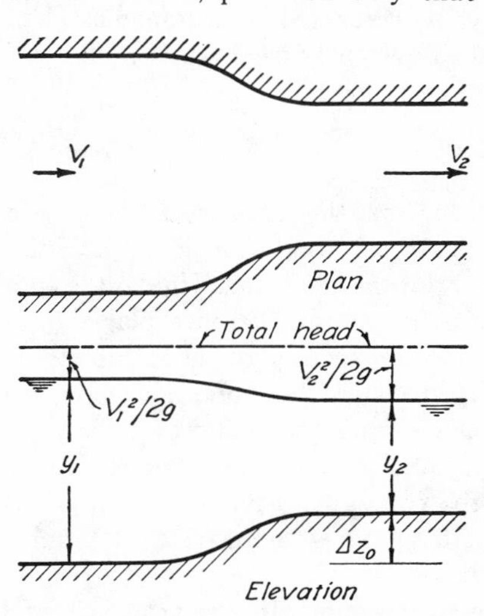

FIG. 63. Change in surface elevation at a channel contraction.

sensibly uniform before and after the transition. As a typical example, consider the reduction in flow section shown in plan and longitudinal profile in Fig. 63. If the depths y_1 and y_2 are known, Eq. (98) will permit evaluation of the corresponding rate of flow; or, if the rate of flow and the initial depth are known, the equation will permit evaluation of the corresponding change in surface elevation. Although calculation of the rate of flow is quite straightforward, it should be noted that the determination of either depth will involve the trial-and-error solution of a cubic equation.

Significance of specific head. At the same time that Eq. (98) yields the most ready numerical solution of a specific open-channel problem, in itself it yields very little descriptive information as to the general problem of depth variation. Indeed, not until the solution has been performed will it be known whether the depth in Fig. 63 will increase or decrease as fluid passes the transition at the given rate, or whether the given rate of flow at the given depth will even be physically possible. Such conditions may, however, be foretold by investigating the interrelationship of the several terms in the general expression (see Fig. 62)

$$H_0 = \frac{Q^2}{2gA^2} + y \tag{99}$$

Owing to the fact that the variation of A with y depends upon the form of channel cross section, the following discussion will, for the sake of simplicity, be restricted to rectangular channels of constant width; Eq. (99) then evidently becomes

$$H_0 = \frac{q^2}{2gy^2} + y \tag{100}$$

With reference to Fig. 62, the quantity $H_0 = H - z_0$ will be seen to represent the height of the total-head line above the channel floor, a quantity known as the *specific head*. Unlike the total head H, which is measured from a horizontal reference plane, the specific head H_0 will evidently vary from section to section if the elevation z_0 of the floor changes. The depth y, therefore, must obviously vary with H_0, even though q remains constant; q, however, may also be regarded as a variable if conditions for different rates of flow through the same transition are to be compared. Since there are evidently two independent variables, it becomes necessary to determine the effect upon y of each one in turn.

The specific-head diagram. If q is arbitrarily held constant, Eq. (100) will permit the evaluation of y for any magnitude of H_0, or vice versa, successive computations of this sort yielding a specific-head diagram such as that shown in Fig. 64. For any specific head above a certain minimum value, two *alternate depths* are seen to be possible;

that is, the given rate of flow may take place at a small depth and high velocity or at a large depth and low velocity, with identical values of the specific head. Below the minimum value of H_0, on the other hand, the given rate of flow cannot occur. The magnitude of y_c, the *critical*

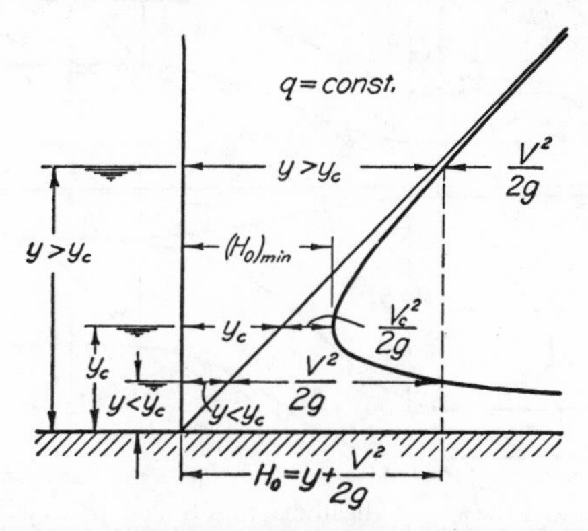

FIG. 64. Specific-head diagram, showing alternate and critical depths.

depth at which the minimum value of H_0 occurs, may be determined by setting equal to zero the derivative of H_0 with respect to y,

$$\frac{dH_0}{dy} = 0 = -\frac{q^2}{gy^3} + 1$$

whence, in terms of the constant rate of flow q, or of the corresponding *critical velocity* $V_c = q/y_c$,

$$y_c = \sqrt[3]{\frac{q^2}{g}} = 2\frac{V_c^2}{2g} \tag{101}$$

Further inspection of the specific-head diagram will disclose the following pertinent fact: If the initial depth of flow is greater than the critical, a decrease in H_0 will result in a decrease in depth; but, if the initial depth of flow is smaller than the critical, a decrease in H_0 will result in an increase in depth. Thus, with reference to Fig. 65*a*, two totally different surface profiles (depending upon whether $y_1 > y_c$ or $y_1 < y_c$) can be produced by a rise in channel boundary, even though the rate of flow and the specific head are the same for both. If the boundary rise (i.e., the decrease in H_0) is sufficiently great, moreover, it should be apparent that the depth at section 2 will approach the critical, as indicated in Fig. 65*b*. Since this represents the minimum specific head at which the rate of flow is physically possible, it follows that further rise in the lower boundary would of necessity cause the

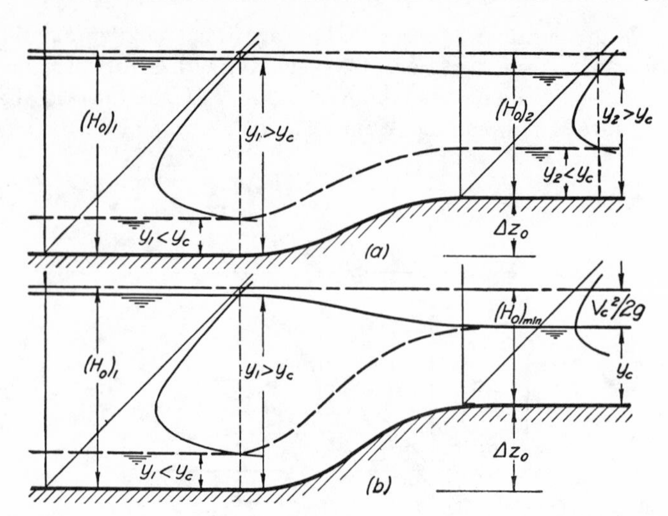

FIG. 65. Application of the specific-head diagram to flow past a rise in channel floor.

depth y_1 (and hence the total head) to increase sufficiently for the given flow to be reestablished.

The discharge diagram. If, in turn, the specific head is to be treated as a constant, Eq. (100) should be solved for q:

$$q = \sqrt{2gy^2(H_0 - y)} \tag{102}$$

A plot of the interrelationship between q and y for any given value of the specific head, in accordance with Eq. (102), will then yield the *discharge diagram* shown in Fig. 66. As in the case of the specific-head diagram for constant q, it will be seen that two alternate depths are again possible, but now only so long as q does not exceed a certain maximum value beyond which flow is physically impossible at the given specific head. If the critical depth is defined as that corresponding to maximum discharge for a particular specific head, y_c may be evaluated by setting equal to zero the derivative of q with respect to y; thus,

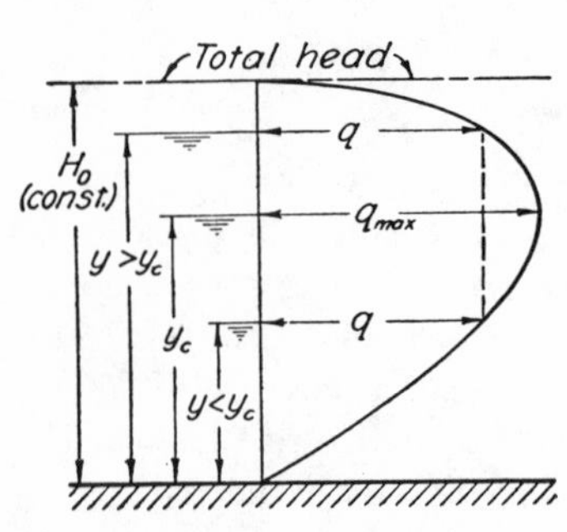

FIG. 66. Discharge diagram, showing alternate and critical depths.

$$\frac{dq}{dy} = 0 = \frac{2g(2yH_0 - 3y^2)}{2\sqrt{2gy^2(H_0 - y)}}$$

whence,

$$y_c = \frac{2}{3}H_0 = 2\frac{V_c^2}{2g} \tag{103}$$

Since the discharge diagram permits study of the relationship between depth and rate of flow under conditions of constant specific head, it is directly applicable to the analysis of discharge through a channel leading from a reservoir of constant surface elevation. Assuming that such a channel terminates abruptly in a free fall, and that

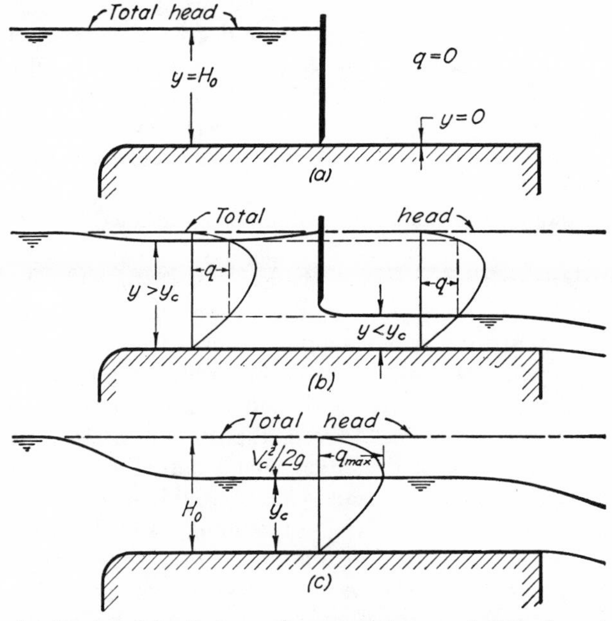

FIG. 67. Application of the discharge diagram to the control of flow by a sluice gate.

a sluice gate is located some distance from either end, it will be seen that the gate is the only means of regulating the flow. If the gate is fully closed (see Fig. 67*a*), q will obviously have a magnitude of zero, and the depths $y = H_0$ upstream and $y = 0$ downstream from the gate will correspond to the limits of the discharge diagram. If the gate opening is greater than zero but less than y_c, alternate depths will be established on the upstream and downstream sides of the gate (Fig. 67*b*) corresponding to the intermediate rate of flow shown on Fig. 66. But since the maximum rate of flow for the given specific head will occur when $y = y_c$, it is apparent that raising the gate beyond the height y_c will have no further influence upon the discharge, which will then continue at the critical depth as shown in Fig. 67*c*.

The reader must be reminded, at this point, that application of the one-dimensional method of flow analysis will yield accurate quantitative results only in proportion to the extent to which the underlying assumptions are fulfilled. The assumption of hydrostatic conditions will obviously preclude the possibility of obtaining therewith more than a rough approximation of actual conditions in zones of appreciable curvature. In Fig. 65, for instance, the form of the free surface immediately above the sloping boundary is merely a qualitative indication of the surface profile; and in Fig. 67 it is evident that the simplified equations give no clue to the degree of jet contraction as the flow emerges from under the sluice, or to the surface curvature in any of the several zones of acceleration. Indeed, only the mathematical principles represented by the flow net permit a general analysis of two-dimensional conditions involving appreciable curvature of the stream lines.

Critical depth and the Froude number. Although the foregoing examples of free-surface flow were characterized by boundary and depth variations in the vertical plane alone, comparable changes in depth would result were the boundary variation entirely horizontal, i.e., a rapid narrowing or widening of an otherwise uniform channel. There would again be two alternate depths of flow for any discharge or specific head, the critical depth and critical velocity having the same significance as before. Indeed, the one-dimensional method of analysis for any form of transition would differ from the foregoing only to the extent that the mean velocity would have to be expressed as the ratio of the total discharge to the total area of the cross section, as in Eq. (98). It must be noted, however, that horizontal variations in boundary form are likely, in flows at velocities above the critical, to produce local disturbances leading to diagonal waves at the free surface. Complexities of this nature are obviously beyond the scope of this book. On the other hand, attention should be called to the fact that the Froude criterion of dynamic similarity for such gravitational phenomena provides a ready means of transferring experimental results to homologous boundary forms of any size, regardless of the complexity of the surface configuration, in terms of the following one-dimensional flow characteristics.

In constant discharge, it will be recalled, the critical depth was found to indicate flow with minimum specific head. With constant specific head, on the other hand, the critical depth was found to yield the maximum rate of discharge. Either case, however, was characterized by the same ratio of velocity head and depth; that is, from Eqs. (101) and (103), $(V_c^2/2g)/y_c = \frac{1}{2}$. If a Froude number for liquid motion of this nature is written in terms of the mean velocity, the depth of flow, and the acceleration of gravity, $\mathbf{F} = V/\sqrt{gy}$, it will be seen that a numerical value of unity corresponds to critical conditions, regardless of whether the discharge is a maximum or the specific

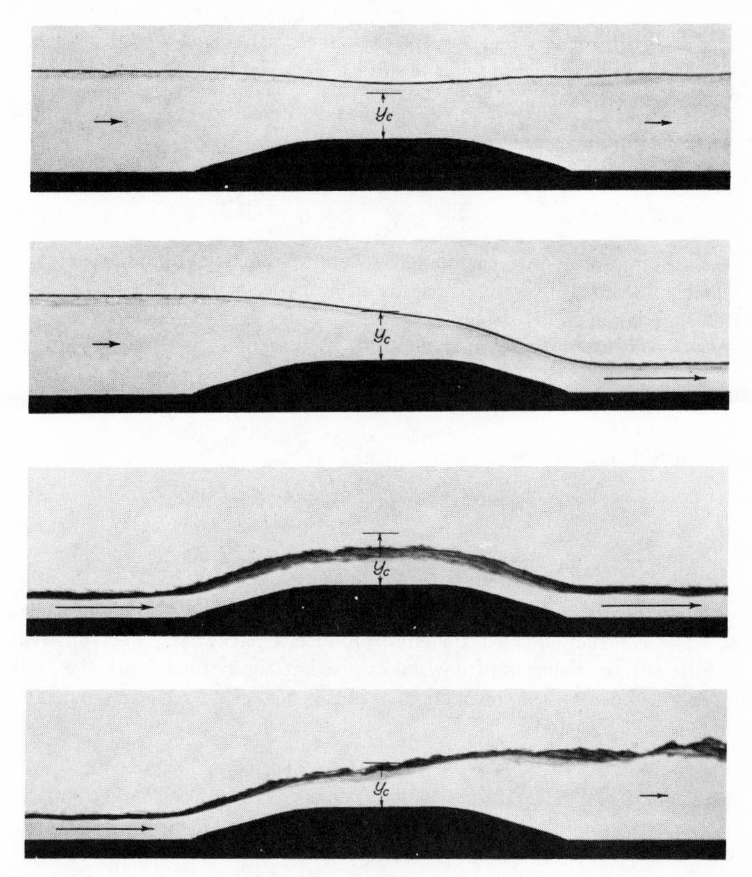

PLATE IX. Changes in flow regime produced by a local increase in boundary elevation.

head a minimum. Evidently, this particular form of $\mathbf{F}$ is directly dependent upon the ratio of velocity head and depth. It therefore follows that Froude numbers less than unity indicate flow at depths greater than the critical depth and at velocities less than the critical velocity; and Froude numbers greater than unity indicate flow at depths less than the critical depth and at velocities greater than the critical velocity. For any given form of boundary transition, the configuration of the free surface should then be a unique function of the Froude number of the approaching flow, a fact of great importance in checking the design of hydraulic structures by scale models.

Example 27. Water flows at a velocity of 8 feet per second and a depth of 6 feet in an open channel of rectangular cross section. If the channel width is reduced from 10 feet to 7 feet and the bottom elevation is increased 1 foot at a given section, to what extent will the surface elevation be affected by the boundary contraction?

The Froude number of the approaching flow is

$$\mathbf{F}_1 = \frac{V_1}{\sqrt{gy_1}} = \frac{8}{\sqrt{32.2 \times 6}} = 0.575$$

Since $\mathbf{F}_1$ is less than unity, the boundary contraction should produce a drop in surface elevation *if* this does not require a lower specific head at the contracted section than is physically possible.

Since $Q = VA$ and $q = Q/b$, in the contracted section

$$q_2 = \frac{V_1 A_1}{b_2} = \frac{10 \times 6 \times 8}{7} = 68.6 \text{ cfs/ft}$$

Hence, for this rate of flow,

$$y_c = \sqrt[3]{\frac{q^2}{g}} = \sqrt[3]{\frac{68.6^2}{32.2}} = 5.26 \text{ ft}$$

and

$$(H_0)_{\min} = \tfrac{3}{2} y_c = \tfrac{3}{2} \times 5.26 = 7.89 \text{ ft}$$

But this evidently requires a greater total head than that of the approaching flow. The contraction must therefore produce a backwater effect upstream, the entire surface rising until flow takes place at the critical depth (i.e., minimum specific head) at the contraction, as in Fig. 65*b*. Under such circumstances $y_2 = y_c = 5.26$ feet and

$$\frac{Q^2}{2gA_1^2} + y_1 = (H_0)_{\min} + \Delta z_0$$

$$\frac{\overline{480}^2}{2 \times 32.2 \times (10y_1)^2} + y_1 = 7.89 + 1$$

Solution by successive approximation yields $y_1 = 8.38$ feet, indicating that the contraction produces a 2.38-foot increase in upstream depth.

PROBLEMS

161. At what two depths could a flow of 400 cubic feet of water per second at a specific head of 7 feet be carried by a rectangular channel 9 feet wide? Evaluate the Froude number for each depth.

162. For purposes of discharge measurement, the width of a rectangular channel is reduced from 9 feet to 7 feet and the floor is raised 1 foot in elevation at a given section.

When the depth of the approaching flow is 6 feet, what rate of flow would be indicated by a 5-inch drop in surface elevation at the contracted section?

163. A rectangular channel 10 feet wide carries 300 cubic feet of water per second at a depth of 5 feet. If the width is to be reduced to 7 feet beyond a given section, what change in bottom elevation will produce a zero change in surface elevation?

164. The design of a three-dimensional channel transition is checked by means of tests on a 1:20 scale model, from which it is found that the flow will overtop the model structure when the discharge exceeds 0.85 cubic foot per second. To what capacity of the prototype structure does this rate of flow correspond?

165. What rate of discharge per unit width will yield depths of 7 feet and 1 foot, respectively, upstream and downstream from a sluice gate? What is the corresponding height of gate opening?

166. If a rectangular channel 8 feet wide carries a flow of 250 cubic feet per second at a depth of 5 feet, what change in surface elevation will be produced by a local rise in floor level of ½ foot?

167. What is the maximum increase in floor elevation in Problem 166 which will not affect the upstream depth?

168. A sluice gate under a 10-foot head (Fig. 48) discharges a sheet of water 2 feet deep. Compute by means of the momentum principle the horizontal force per unit width exerted by the flow upon the gate.

25. OPEN-CHANNEL WAVES AND SURGES

Celerity of the elementary gravity wave. In the foregoing section, methods were given for an approximate analysis of the disturbance of uniform flow by a stationary boundary transition in which the resulting pattern of motion remained constant with time. The sluice gate, for instance, was considered fixed in some given position, and conditions of steady flow for various positions were compared. A change in gate opening with time, however, would no longer yield a pattern of steady motion, for the velocity and depth would vary not only in the neighborhood of the gate but also over a great distance in both the upstream and downstream directions. Such phenomena evidently represent examples of distinctly unsteady motion, differing from their closed-conduit counterparts in the fact that not only the rate of flow but also the area of the flow section must be considered variable with time. Only if the surface configuration is displaced without changing form can such problems readily be analyzed; the procedure then reverts to the one-dimensional method for steady flow, through proper translation of the point of observation.

Consider, for example, the two-dimensional propagation of a very low *wave*, or surface undulation, through a channel of liquid otherwise at rest, as indicated in Fig. 68*a*. Designating by c the *celerity* of the wave (i.e., its rate of movement relative to the liquid as a whole), it follows that the picture of unsteady motion may be transformed into a steady one by translating the point of observation toward the left at the velocity $v = c$; in the eyes of the observer, at each section there is added vectorially the relative velocity $V = c$ toward the right, with the result shown in Fig. 68b. Since this transformed picture has the characteristics of steady motion, the total head may now be considered constant; differentiation of Eq. (100) will then yield the expression

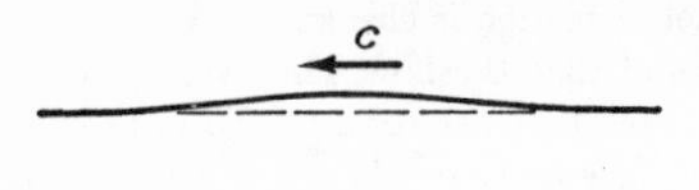

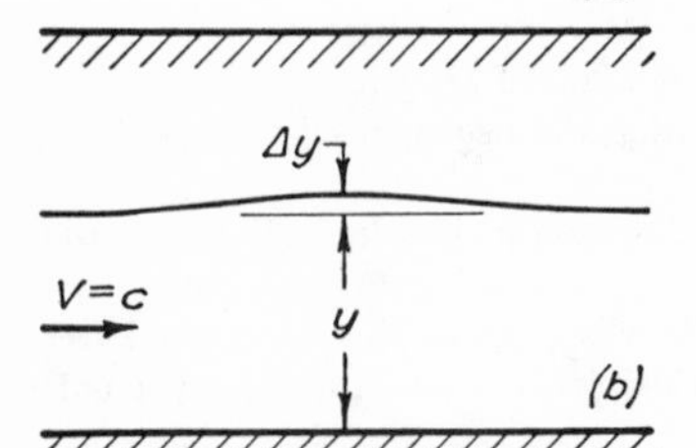

FIG. 68. Definition sketch for the analysis of wave propagation.

$$0 = \frac{V\,dV}{g} + dy$$

and upon combining therewith the continuity relationship $Vy = \text{constant}$ written in the differential form

$$V\,dy + y\,dV = 0$$

it will be seen that

$$V = c = \sqrt{gy} \tag{104}$$

Evidently, a very low wave will be propagated through an open channel at a rate depending only upon the gravitational acceleration and the liquid depth. (Lest the reader assume that Eq. (104) applies as well to waves in the ocean, it must be noted that the corresponding celerity of waves having a length λ (lambda) which is small in comparison with the depth y is of the form $c = \sqrt{g\lambda/2\pi}$ and hence independent of the depth.)

So long as the liquid is otherwise at rest, the actual velocity of a wave will be identical with its celerity. But if the liquid itself is in motion, the velocity v_w of the wave with respect to a stationary observer will depend upon both the velocity V of the liquid and the celerity c of the wave, since c is measured *with respect to the liquid.* That is,

$$v_w = V \pm c \tag{105}$$

wherein c is considered positive when its direction is the same as that of V, and negative when its direction is the opposite. It now becomes apparent that, if the uniform flow has a velocity greater than the celerity of a small wave, such a disturbance cannot travel in the upstream direction with respect to a stationary observer. As a matter of fact, if the Froude number is written in the form $\mathbf{F} = V/\sqrt{gy}$, it will be seen to indicate the ratio of the actual velocity of flow to the celerity of a very small wave. Froude numbers smaller than unity therefore correspond to flows at velocities less than the elementary wave celerity, and Froude numbers greater than unity to flows at velocities greater than the elementary wave celerity; only if the Froude number is less than unity can a small wave travel upstream.

Waves and surges of finite amplitude. Needless to say, not all waves are of small amplitude. Depth variations of appreciable magnitude, however, involve a certain amount of curvilinear motion, which cannot be analyzed by means of the present one-dimensional method. Suffice it to say that the celerity of a solitary wave of finite size and stable form has been shown both analytically and experimentally to exceed that of the elementary wave by an amount depending upon its relative height; that is, as a first approximation

$$c \approx \sqrt{gy}\left(1+\frac{3}{2}\frac{\Delta y}{y}\right)^{1/2} \tag{106}$$

It must be noted, moreover, that a wave of this type becomes unstable as the ratio $\Delta y/y$ approaches unity, the crest developing a sharp peak and finally breaking; beyond (or even near) this limit, Eq. (106) obviously has no physical significance.

A different type of unsteady motion, known as a *surge*, is produced by a rapid increase in the rate (and hence the depth) of flow. So long

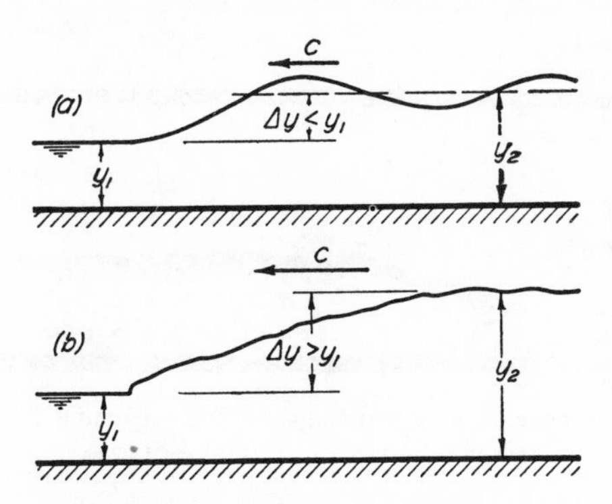

FIG. 69. Profiles of the undular and the breaking surge.

as $\Delta y/y$ is less than unity (Fig. 69*a*), the surge will take the form of a series of undulations about the greater depth. If its relative height exceeds unity, however, breaking of the first wave will produce an abrupt discontinuity of the liquid surface (Fig. 69*b*). Either type of surge will nevertheless attain a stable form, and the celerity of propagation will again depend upon the elementary wave celerity and the relative change in elevation. Such conditions are shown schematically in Fig. 70*a*, in which the surge is assumed to travel to the left with the celerity c in an originally quiet liquid. Although this unsteady picture may again be transformed into a steady one by adding the velocity $V = c$ to the right at every section (Fig. 70*b*), the existence of the surface discontinuity at the wave front makes it inadvisable to proceed on the assumption of constant total head. The momentum and continuity equations, however, must still be satisfied. Thus, the only external forces acting upon the free body of liquid between sections 1 and 2 are those resulting from hydrostatic distribution of pressure over the vertical sections; equating the total force in the direction of flow to the

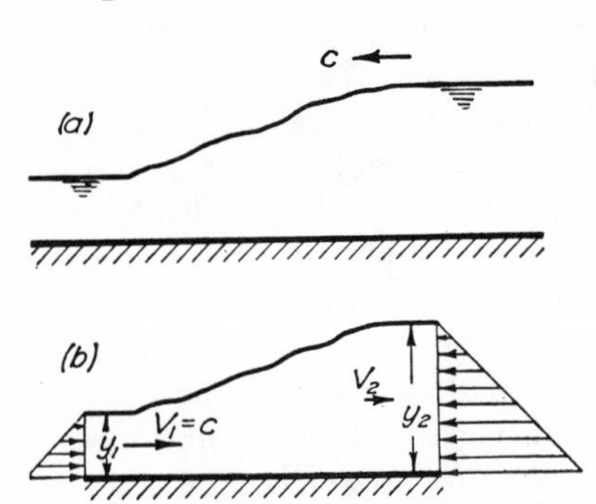

FIG. 70. Definition sketch for the analysis of surge propagation.

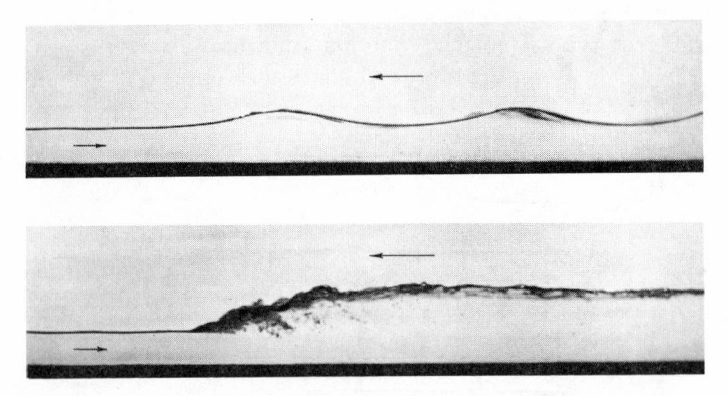

PLATE X. Surges traveling upstream against the oncoming flow—a phenomenon often encountered in tidal estuaries.

product of the rate of discharge, the density, and the velocity change (refer to Eq. 88), it will be seen that

$$\frac{\gamma y_1^2}{2} - \frac{\gamma y_2^2}{2} = q\rho(V_2 - V_1)$$

Introduction of the continuity relationship

$$q = V_1 y_1 = V_2 y_2$$

finally yields for the surge celerity the expression

$$V_1 = c = \sqrt{gy_1}\left[\frac{1}{2}\frac{y_2}{y_1}\left(\frac{y_2}{y_1} + 1\right)\right]^{1/2} \tag{107}$$

Characteristics of the hydraulic jump. If a surge were formed, for instance, by the partial closure of a sluice gate, the disturbance would travel upstream with the celerity c relative to the oncoming flow, but with the velocity $v_w = V - c$ relative to the sluice or to a stationary observer. Evidently, were the velocity of the oncoming flow exactly equal to the celerity of the surge, the disturbance would remain fixed in position relative to the observer—a phenomenon known as the *hydraulic jump* (see Fig. 71). Since V_1 is then equal to c, Eq. (107) may be rewritten in the form

$$\frac{V_1}{\sqrt{gy_1}} = \mathbf{F}_1 = \left[\frac{1}{2}\frac{y_2}{y_1}\left(\frac{y_2}{y_1} + 1\right)\right]^{1/2} \tag{108}$$

This will serve as an illustration of a previous statement that the Froude number is the primary criterion for similarity of such

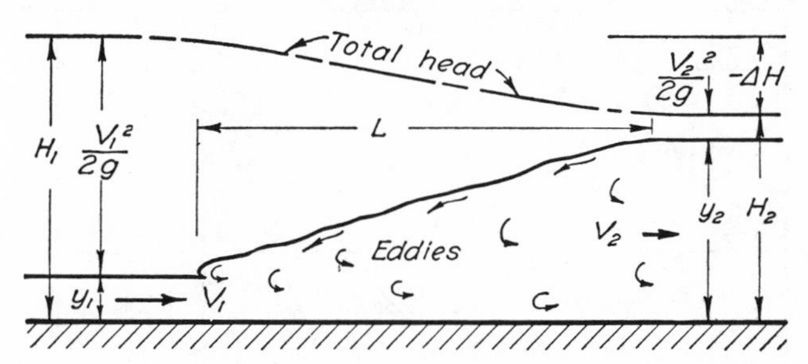

FIG. 71. Characteristics of the standing surge or hydraulic jump.

gravitational phenomena; not only may the ratio of the initial and final depths of the hydraulic jump be expressed, from Eq. (108), as a function of $\mathbf{F}_1$ alone,

$$\frac{y_2}{y_1} = \frac{1}{2}\left(\sqrt{1 + 8\mathbf{F}_1^2} - 1\right) \tag{109}$$

but experimental measurements of the length L of the transition (see Figs. 71 and 72) show that its ratio to either y_1 or y_2 is also a unique function of the Froude number.

Had it been assumed in the foregoing development that the total head was the same at sections 1 and 2, a result at variance with Eq. (107) would have been obtained. This is due to the fact that the abrupt discontinuity at the head of the surge involves a continuous transformation of kinetic energy of translation into kinetic energy of rotation (i.e., turbulence), a process which will be discussed in greater detail at a later point. Through use of the Bernoulli relationship and

PLATE XI. Silhouette of the hydraulic jump ($\mathbf{F} \approx 7$); air bubbles entrained at the breaking front indicate the violence of the eddies as the flow expands.

Eq. (108) it is therefore possible to evaluate the change ΔH in total head produced thereby, the ratio of this change to either y_1 or y_2 again being a function of only the Froude number, as indicated in Fig. 72.

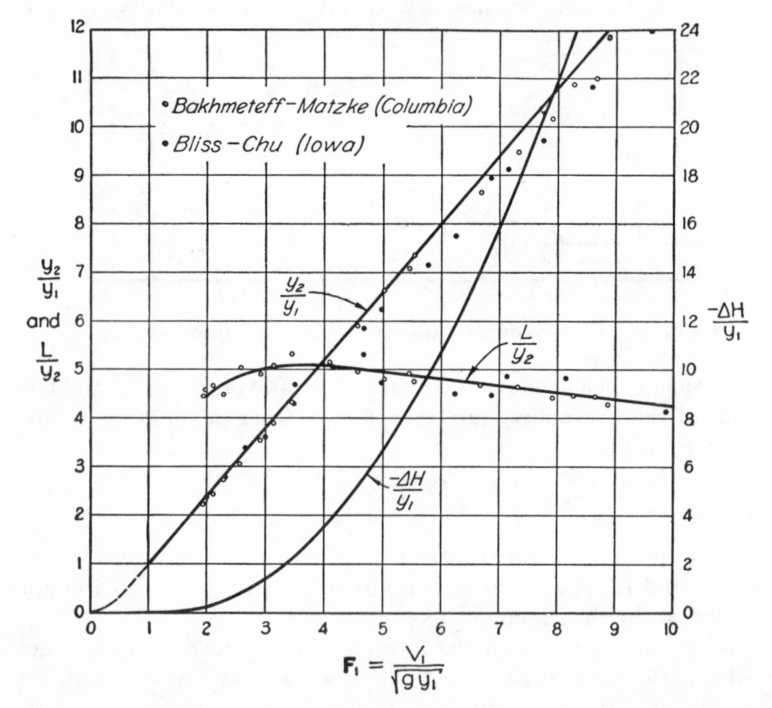

FIG. 72. Hydraulic-jump characteristics as dimensionless functions of the Froude number.

Noteworthy is the fact that ΔH may represent more than 50 per cent of the initial specific head at high values of $\mathbf{F}$; this results in frequent use of the hydraulic jump to reduce the energy of flow at the foot of chutes and spillways.

As in Chapter IV, the analysis of free-surface flow in the present chapter has been restricted to the case of a liquid in contact with the atmosphere. The extension of this analysis to the relative motion of fluids having more nearly the same specific weight is unfortunately beyond the scope of this book. Nevertheless, the reader might well be reminded at this point that nearly all open-channel phenomena which have been mentioned in these pages are closely duplicated in the atmosphere, in reservoirs and lakes, and in the ocean, by the relative movement of fluid masses differing only slightly in specific weight as the result of temperature variation or of the presence of dust or silt in suspension. That only the time scale differs considerably may be seen by writing the Froude number $V/\sqrt{gy} = V/\sqrt{y\gamma/\rho}$ in the more general form $V/\sqrt{y\,\Delta\gamma/\rho}$, which, as noted in Section 20, simply indicates that the gravitational influence is reduced in relative magnitude. Indeed, just as the simpler form of the Froude number has been used as a similitude parameter in open-channel studies, the more general form may well serve as a guide for obtaining similarity of all weight effects in the experimental study of atmospheric and oceanic gravity currents.

Example 28. The depth and velocity of flow in a rectangular channel are 4 feet and 6 feet per second, respectively. If the rate of inflow at the upstream end of the channel is abruptly increased to the extent that the depth is doubled in magnitude, determine the absolute velocity of the resulting surge.

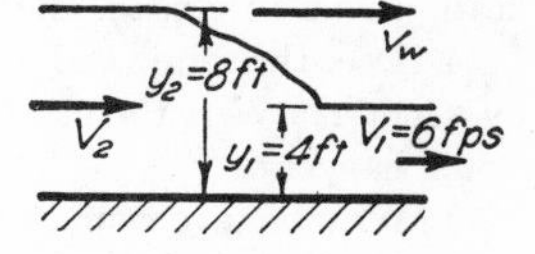

Conditions of steady flow are first simulated by adding the velocity v_w to the left. Then, according to Eq. (107),

$$v_w - V_1 = \sqrt{gy_1}\left[\frac{1}{2}\frac{y_2}{y_1}\left(\frac{y_2}{y_1}+1\right)\right]^{1/2}$$

and

$$v_w = 6 + \sqrt{32.2 \times 4}\,[\tfrac{1}{2} \times \tfrac{8}{4} \times (\tfrac{8}{4} + 1)]^{1/2} = 6 + 19.6 = 25.6 \text{ fps}$$

PROBLEMS

169. Assuming water to flow in a wide channel at a mean velocity of 9 feet per second and a depth of 2 feet, determine the absolute velocities of waves of very small amplitude moving (*a*) with the flow and (*b*) against the flow, and the absolute velocities of waves of 1-foot amplitude moving (*c*) with the flow and (*d*) against the flow.

170. Gaillard Cut in the Panama Canal extends approximately 8¾ miles from the Pedro Miguel Lock to Gatun Lake, the mean depth in the cut being 41 feet. Estimate the time required for the wave produced by filling the lock to reach the lake.

171. A river flowing toward the sea at a velocity of 3 miles per hour and a depth of 7 feet encounters an incoming tide which increases the depth to 15 feet. This gives rise to a "bore" or traveling surge which moves rapidly up the river. Estimate the speed of the bore in this instance. What speed and direction will the flow have after the bore has passed?

172. The depth and velocity of flow in a rectangular channel are 3 feet and 5 feet per second, respectively. If the rate of inflow at the upstream end is abruptly doubled, what will be the height and absolute velocity of the resulting surge?

173. The depth and velocity of flow in a rectangular channel are 3 feet and 5 feet per second, respectively. If a gate at the downstream end of the channel is abruptly closed, what will be the height and absolute velocity of the resulting surge?

174. A hydraulic jump forming below a sluice gate produces a change in depth from 2 feet to 5 feet. What are (*a*) the head on the gate, (*b*) the rate of flow per unit width, and (*c*) the change in total head within the jump?

175. Water flowing down a spillway at the rate of 100 cubic feet per second per foot width leaves the apron horizontally with a mean velocity of 40 feet per second. Determine (*a*) the depth of tailwater necessary to form a hydraulic jump, (*b*) the approximate length of the jump, and (*c*) the total head of the flow beyond the jump.

QUESTIONS FOR CLASS DISCUSSION

1. What simplifying assumptions are involved in the one-dimensional method of flow analysis? Describe states of motion in which each of these assumptions will lead to appreciable error.

2. Cite examples or problems in which the one-dimensional method has, in effect, already been used in Chapters II–IV.

3. Show the relationship of the quantities $V^2/2g$, p/γ, and z to the lines of total head, piezometric head, and elevation.

4. Does the total head increase or decrease as fluid passes through a pump driven by a motor? Were the motor to stop before a valve on the line could be closed, what would happen?

5. What change would result in the analysis of the establishment of flow in a long pipe if the control valve were located (*a*) at the midpoint and (*b*) at the upstream end of the pipe?

6. When the valve on a drinking fountain is opened, the jet at once rises higher than its final level. Why does this not agree with the general equation for flow establishment?

7. What is the difference between the total head and the specific head in open-channel flow?

8. If the Froude number characterizing flow in an open channel is less than unity, will (*a*) an increase in channel width, and (*b*) an increase in bottom elevation, cause the surface elevation to increase or decrease?

9. Define the critical depth in four different ways.

10. An overflow dam is to be built in a channel carrying a given uniform flow. Sketch the change in the lines of total head and piezometric head which would be produced.

11. Why is it desirable to have a hydraulic jump form at the toe of a spillway? How may this be attained?

12. Discuss the restriction of Eq. (86) to conditions of steady flow, using as illustration a surge moving through otherwise quiet water.

13. Under what circumstances can a wave or surge move upstream against the oncoming flow?

14. Why is the Froude number a sufficient criterion for the characteristics of a standing wave or surge, but not for those of a moving wave or surge?

15. Distinguish between the type of wave encountered on a beach and that discussed in this chapter.

16. An equation of the type $c = \sqrt{gy}$ refers to waves of very great length in comparison to the depth y, whereas an equation of the type $c = \sqrt{g\lambda/2\pi}$ refers to waves having a length λ which is smaller than twice the depth. In which general category are (*a*) tides, and (*b*) storm waves, on the ocean?

17. In what manner could the movement of a dust cloud in the atmosphere be studied by means of liquids in a laboratory flume?

SELECTED REFERENCES

Gibson, A. H. *Hydraulics and Its Applications.* 4th edition, Constable, 1930.

Rouse, H. *Fluid Mechanics for Hydraulic Engineers.* Engineering Societies Monograph, McGraw-Hill, 1938.

Keulegan, G. H., and Patterson, G. W. "Mathematical Theory of Irrotational Translation Waves." *Journal of Research of the National Bureau of Standards*, vol. 24, 1940.

CHAPTER VI

EFFECTS OF VISCOSITY ON FLUID MOTION

26. VISCOUS RESISTANCE TO FLUID DEFORMATION

Viscosity characteristics. Attention was called in Chapter II to the fact that fluid motion generally involves not only translation but also rotation and deformation of any elementary portion of the fluid mass. No further attention was paid to the process of deformation, but a special case of motion was investigated in which it was assumed that rotation did not occur. It so happens that the mathematical concept of irrotational flow corresponds to conditions in which the influence of fluid viscosity is quite negligible. If, therefore, viscosity plays a leading role in any state of motion, not only can the flow no longer be considered irrotational, but the process of fluid deformation becomes of primary importance.

Like the deformation of elastic solids, the deformation of a viscous fluid gives rise to a system of normal and tangential stresses at all points within the zone of deformation, the normal stresses being measured (as before) in terms of pressure intensity and the tangential stresses in terms of *intensity of shear.* The resistance of solid matter to shear depends upon its shearing modulus of elasticity; the corresponding fluid property is the *dynamic* (or so-called "absolute") *viscosity.* Here the analogy reaches an end, however, for an elastic solid will deform until a state of equilibrium between the external forces and the internal resistance to shear is reached, whereas a viscous fluid will continue to deform so long as the external forces are applied. That is, in a fluid it is the *rate* of deformation, rather than the magnitude, which is the essential criterion for force equilibrium. The intensity of fluid shear, in other words, is directly proportional to the rate of fluid deformation, the constant of proportionality representing the dynamic viscosity of the fluid.

If, in addition to weight and pressure, one takes into account the shearing stress exerted upon a fluid element, it is possible to extend the Eulerian equations of acceleration to motion in which viscosity is no longer a negligible factor. The resulting expressions, known as the Navier-Stokes equations, are of great significance in the advanced study of fluid motion, but their complexity discourages their application in an elementary study of this nature. It will be necessary, therefore, to restrict the following analysis to steady, uniform flow, treating the more complex phenomena qualitatively rather than from a rigorous mathematical point of view.

Viscosity measurement. Probably the simplest application of the proportionality between shearing stress and rate of deformation is found in a method frequently used to measure the viscosity of liquids. As shown schematically in Fig. 73, a thin film of the liquid to be tested is enclosed between two coaxial cylinders. The inner cylinder is suspended by a torsion wire or spring, the torque-deflection ratio of which is known from calibration. The outer cylinder may be rotated independently at any desired speed. For any given liquid at constant temperature, data from a series of runs at different rotational speeds will follow a straight line passing through the origin of a torque-speed diagram, as illustrated in Fig. 74. The linear nature of such a plot indicates that the viscosity of a true fluid is independent of the rate of shear. The slope of the line, on the other hand, will vary directly with the viscosity, curve A, for instance, corresponding to a liquid having a viscosity approximately three times as great as that of curve B. Thus, through comparative tests with a liquid of known viscosity (such as water), the viscosity of any liquid at any desired temperature may rapidly be determined.

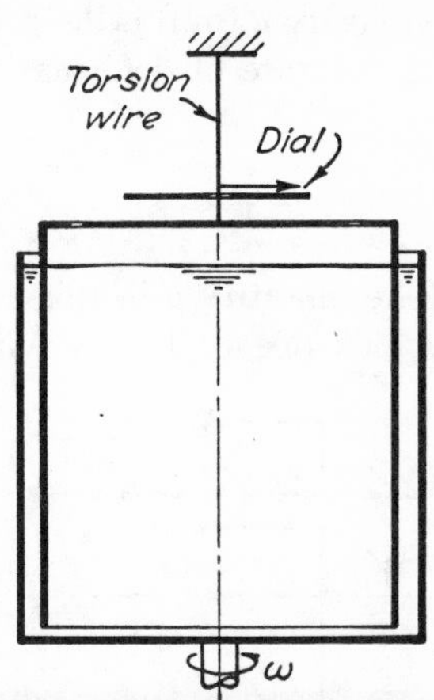

FIG. 73. Apparatus for the measurement of liquid viscosity.

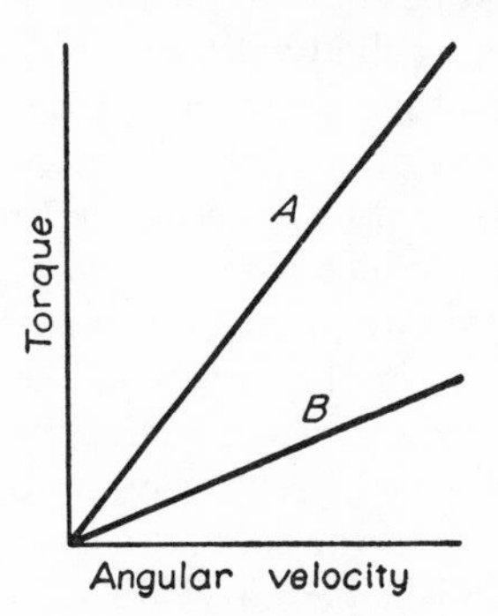

FIG. 74. Torque-speed diagram for viscous liquids.

In order to use a *viscosimeter* of this nature for direct, rather than relative, measurement, it is necessary that the spacing of the cylinders be very small in comparison with either radius. Under such circumstances, the flow pattern approaches that represented by Fig. 75, in which fluid enclosed by two parallel boundaries is deformed by the steady motion of one boundary past the other. The rate of deformation per unit fluid volume at any point may be expressed in terms of the velocity gradient dv/dy in the transverse direction, because (refer to figure) this measures the rate of change of the corner angle α of a fluid element. Since the dynamic viscosity μ (mu) is defined as the ratio of the intensity of shear τ (tau) to the rate of deformation, it follows that

$$\tau = \mu \frac{dv}{dy} \tag{110}$$

The quantity μ in this equation must be regarded as a fluid property which does not vary with the state of motion. Evidently, in order that τ will not have the physically impossible magnitude of infinity at any point within the fluid, the velocity must vary continuously across the flow section; that is, there can be no slipping (abrupt changes in velocity) between neighboring zones of flow. The same condition must hold true at the boundaries themselves, the velocity of a viscous fluid at the plane of contact with a solid boundary being identical with that of the boundary itself. In other words, the velocity of the fluid in contact with the stationary boundary of Fig. 75 must be zero, and the fluid velocity at the moving boundary must be equal to the boundary velocity v_0. The velocity of the intervening fluid will then vary linearly between these two limits, and the resulting constancy of the gradient dv/dy will indicate a corresponding constancy of the intensity of shear throughout the zone of motion. The total force necessary to move the upper boundary will be equal to the product of the intensity of shear and the boundary area; similarly, the rate at which work is done by this force will be equal to the product of the force and the boundary velocity. If, now, the upper boundary is considered to represent the inner surface of a rotating cylinder of relatively large diameter, measurement of the rotational speed and the torque exerted by the enclosed fluid upon the inner cylinder will permit evaluation of the fluid viscosity.

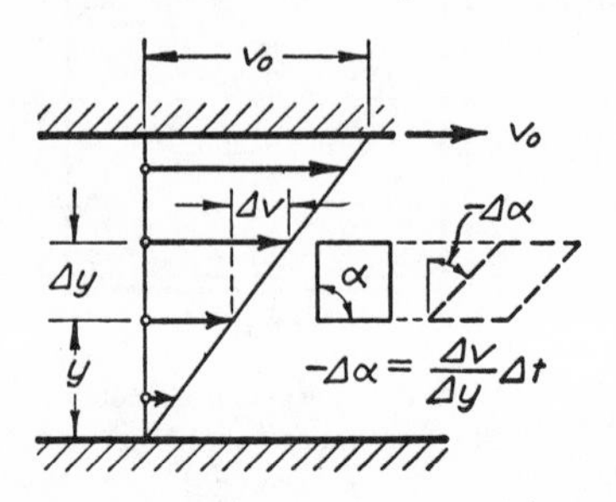

FIG. 75. Deformation of fluid by a moving boundary.

Example 29. If the cylinders shown schematically in Fig. 73 are 10 inches high, and 6.02 and 6.00 inches in diameter, respectively, what is the viscosity of a liquid which produces a torque of 0.75 pound-foot upon the inner cylinder when the outer one rotates at the rate of 90 revolutions per minute?

The tangential velocity of the outer cylinder will be

$$v = r\omega = \frac{6.02}{2 \times 12} \times \frac{2 \times \pi \times 90}{60} = 2.37 \text{ fps}$$

Hence the rate of deformation is

$$\frac{dv}{dy} = \frac{v}{\Delta r} = \frac{2.37}{(3.01 - 3.00)/12} = 2.85 \times 10^3 \text{ fps/ft}$$

The intensity of shear is equal to the torque divided by the radius and the area of either cylindrical surface:

$$\tau = \frac{0.75}{\frac{3}{12} \times \pi \times \frac{6}{12} \times \frac{10}{12}} = 2.29 \text{ psf}$$

Therefore, from Eq. (110),

$$\mu = \frac{\tau}{dv/dy} = \frac{2.29}{2.85 \times 10^3} = 8.04 \times 10^{-4} \text{ lb-sec/ft}^2$$

PROBLEMS

176. What would the radial spacing of the cylinders of Example 29 have to be in order to yield the same torque-speed ratio with water at 60° Fahrenheit?

177. Two plane boundaries lie a uniform distance of $\frac{3}{16}$ inch apart, the space between being filled with a liquid having a viscosity of 10^{-3} pound-second per square foot. What force would be required to move edgewise through the liquid a $\frac{1}{8}$-inch plate 15 inches square at a velocity of 6 inches per second, assuming the same clearance to exist on both sides of the plate?

178. Lateral stability of a long 6-inch shaft is obtained by means of a 10-inch stationary bearing having an internal diameter of 6.008 inches. If the space between bearing and shaft is filled with a lubricant having a viscosity of 5×10^{-3} pound-second per square foot, what power will be required to overcome viscous resistance when the shaft is rotated at a constant rate of 180 revolutions per minute?

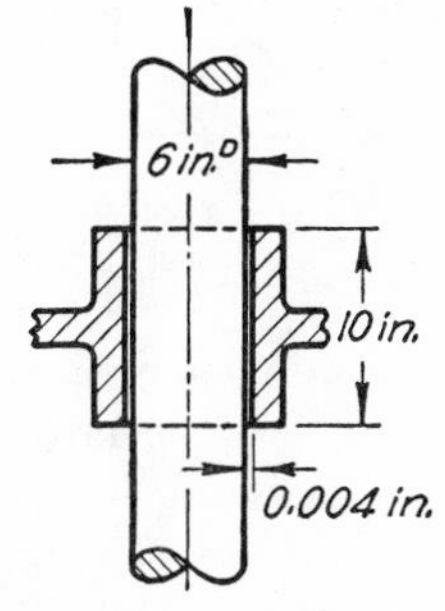

PROB. 178.

179. A sleeve 5 inches long encases a vertical rod 1 inch in diameter with a radial clearance of 0.001 inch. If, when immersed in oil having a viscosity of 3×10^{-3} pound-second per square foot, the effective weight of the sleeve is 2 pounds, how rapidly will it slide down the rod?

180. An alternative form of the rotating viscosimeter consists of a disk pivoted above a stationary boundary, the liquid (or gas) to be tested filling the very small space between the parallel surfaces; means must then be provided for measuring the driving torque and rotational speed. Determine, by integration of Eq. (110) over the lower surface of the disk, the torque-speed ratio which would be obtained for a liquid having a viscosity of 3×10^{-3} pound-second per square foot with an 8-inch disk and a boundary spacing of 0.05 inch.

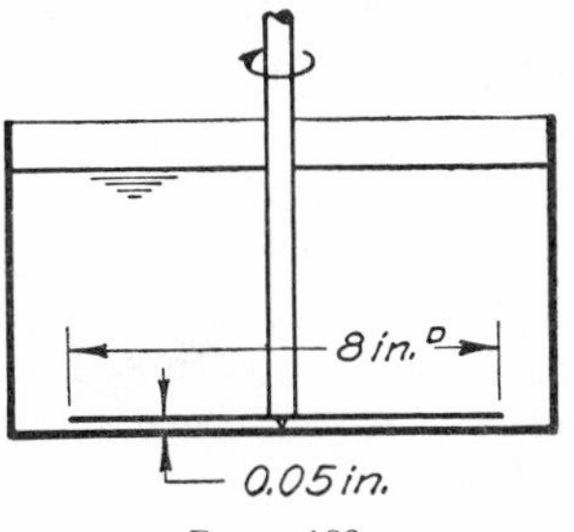

PROB. 180.

181. If the inner cylinder of a viscosimeter such as that shown in Fig. 73 is inverted, the added shear between the two bases will play a decided role in the resulting torque. Estimate, by integration of Eq. (110) over the base area, the percentage additional torque which would exist under the conditions of Example 29 if the bottom spacing were 0.01 inch.

27. PRESSURE GRADIENT IN STEADY, UNIFORM, VISCOUS FLOW

Interdependence of shear and pressure gradients. In the foregoing chapters it was found that the intensity of pressure must generally be expected to vary from point to point in a given flow (*a*) if the fluid accelerates in the corresponding direction, or (*b*) if the fluid weight has a component in that direction. It now will be shown that a pressure gradient must generally exist (*c*) if relative motion within the fluid gives rise to a state of viscous shear which varies from point to point.

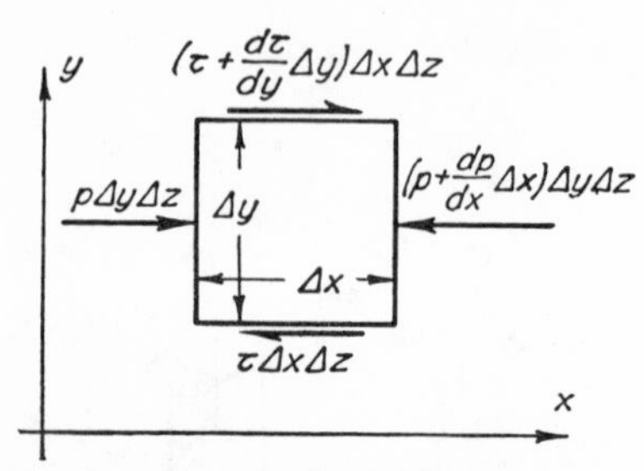

FIG. 76. Definition sketch for fluid pressure and shear.

Assume, for instance, a free body of fluid having the form of an elementary cube of volume Δx Δy Δz, of which Fig. 76 shows a cross section in the xy plane. For the condition of two-dimensional steady flow parallel to the x axis, there will be no shear upon the two parallel faces Δx Δy, the only forces acting in the direction of motion x being the normal and tangential stresses indicated in the illustration. The summation of these horizontal stresses, which must necessarily be equal to zero because of the absence of acceleration, yields the result

$$\Sigma F_x = -\left(\frac{dp}{dx}\,\Delta x\right)\Delta y\,\Delta z + \left(\frac{d\tau}{dy}\,\Delta y\right)\Delta x\,\Delta z = 0$$

Dividing by the volume $\Delta x\ \Delta y\ \Delta z$ and letting this approach zero, it follows that at any point

$$\frac{dp}{dx} = \frac{d\tau}{dy} \tag{111}$$

Evidently, therefore, a pressure gradient in steady, uniform flow must depend not only upon the existence of viscous shear, but also upon the variation of the intensity of shear across the flow. Since the flow under consideration is parallel (i.e., since p varies hydrostatically over all normal cross sections), it follows that the pressure gradient must be the same at all points along every stream line. For the particular conditions of Fig. 75, the pressure gradient has the constant magnitude of zero, since the magnitude of τ is everywhere the same. It is not zero, however, for flow between stationary boundaries, as will be seen from the following analysis.

Flow between parallel boundaries. With reference to the parallel, stationary boundaries shown in Fig. 77, it is apparent that the absence of both relative motion and acceleration in the normal direction must yield a constant magnitude of the gradient dp/dx (and hence of the gradient $d\tau/dy$) in all zones so long as the flow continues at a steady rate in the x direction. Integration of Eq. (111) with respect to y then results in the expression

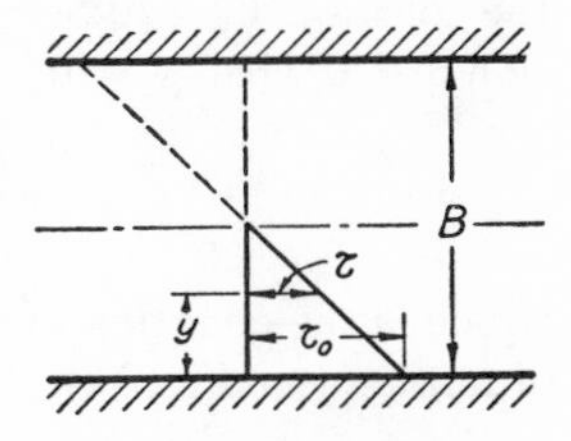

FIG. 77. Distribution of shear between stationary parallel boundaries.

$$\tau = \frac{dp}{dx}y + C_1$$

Evidently, the intensity of shear will vary linearly with distance from the boundary, for only then can $d\tau/dy$ be the same at all points. Owing to conditions of symmetry, however, it is also evident that the intensity of shear must have the same magnitude at either boundary, from which it follows (see Fig. 77) that the intensity of shear midway between the two boundaries must be zero. In other words, at the point $y = B/2$,

$$\tau = 0 = \frac{dp}{dx}\frac{B}{2} + C_1$$

Use of this result to eliminate the constant of integration from the initial equation for τ then yields the general relationship

$$\tau = -\frac{dp}{dx}\left(\frac{B}{2} - y\right) \tag{112}$$

In passing, it may be noted that the intensity of boundary shear will be directly proportional to the pressure gradient and the boundary spacing; that is, when $y = 0$, Eq. (112) becomes

$$\tau_0 = -\frac{dp}{dx}\frac{B}{2} \tag{113}$$

Since, from Eq. (110), $\tau = \mu\, dv/dy$, Eq. (112) may be written in terms of the velocity gradient:

$$\frac{dv}{dy} = -\frac{1}{\mu}\frac{dp}{dx}\left(\frac{B}{2} - y\right)$$

Integration with respect to y then yields the expression

$$v = -\frac{1}{\mu}\frac{dp}{dx}\left(\frac{By}{2} - \frac{y^2}{2}\right) + C_2$$

the constant $C_2 = 0$ being evaluated by means of the boundary condition $v = 0$ when $y = 0$. Thus,

$$v = -\frac{1}{2\mu}\frac{dp}{dx}(By - y^2) \tag{114}$$

which is the equation of a parabola having its vertex at the centerline of motion (see Fig. 78). At the point $y = B/2$ the velocity is therefore

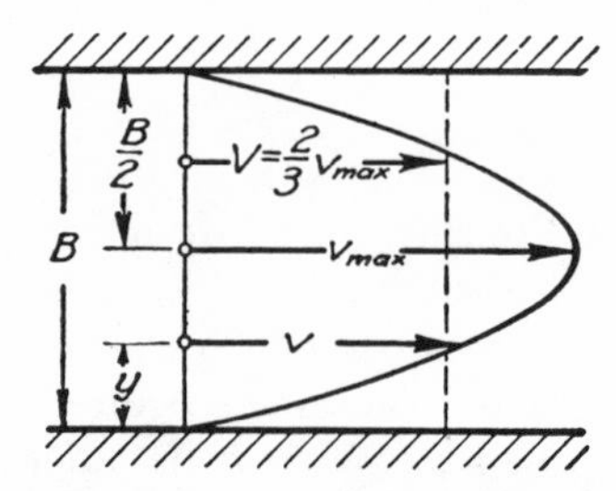

FIG. 78. Velocity distribution between stationary parallel boundaries.

a maximum; from the geometry of the parabola [or as might be determined through application of Eqs. (7) and (9)], it follows that the mean velocity of flow is two-thirds the maximum, whence

$$V = \tfrac{2}{3}\, v_{\max} = -\frac{2}{3}\frac{1}{2\mu}\frac{dp}{dx}\left(\frac{B^2}{2} - \frac{B^2}{4}\right) = -\frac{dp}{dx}\frac{B^2}{12\mu}$$

Finally, solving for $-dp/dx$,

$$-\frac{dp}{dx} = \frac{12\mu V}{B^2} \tag{115}$$

Since the flow is uniform, integration of Eq. (115) with respect to x will yield the pressure drop $p_1 - p_2$ corresponding to any length $x_2 - x_1 = L$ of the parallel boundaries:

$$p_1 - p_2 = \frac{12\mu VL}{B^2} \tag{116}$$

Equations of Poiseuille and Stokes. For flow through a horizontal circular tube, the counterpart of Eq. (112) may be obtained by expressing the summation of horizontal forces acting upon the cylindrical element of fluid shown in Fig. 79:

$$\pi r^2 p - \pi r^2 \left(p + \frac{dp}{dx} \Delta x\right) - 2\pi r \, \Delta x \, \tau = 0$$

whence

$$\tau = -\frac{dp}{dx}\frac{r}{2} \tag{117}$$

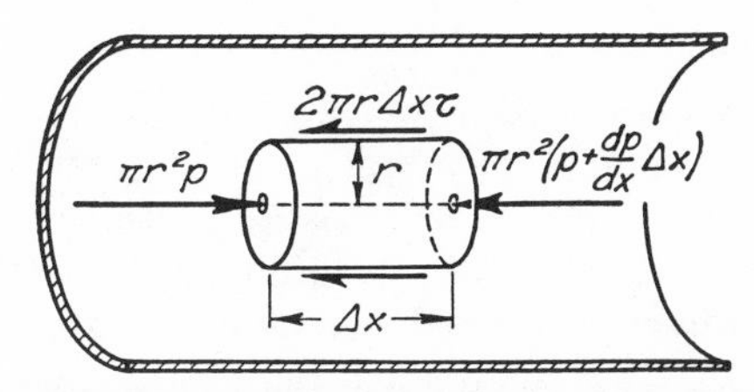

FIG. 79. Pressure and shear in flow through a circular tube.

Evidently τ is zero at the centerline, while at the boundary [the counterpart of Eq. (113)],

$$\tau_0 = -\frac{dp}{dx}\frac{r_0}{2} \tag{118}$$

Upon introduction of Eq. (110), Eq. (117) becomes, since $dy = -dr$,

$$\frac{dv}{dy} = -\frac{dv}{dr} = -\frac{dp}{dx}\frac{r}{2\mu}$$

and integration with respect to r then yields

$$v = \frac{dp}{dx}\frac{r^2}{4\mu} + C$$

Since $v = 0$ when $r = r_0$, it will be found that $C = -(dp/dx)r_0^2/4\mu$, whereupon

$$v = -\frac{1}{4\mu}\frac{dp}{dx}(r_0^2 - r^2) \tag{119}$$

which is again the equation of a parabola (Fig. 80). From the geometry of a paraboloid of revolution [or as may be shown by applying Eqs. (6) and (8)],

$$V = \tfrac{1}{2}v_{\max} = -\frac{dp}{dx}\frac{r_0^2}{8\mu}$$

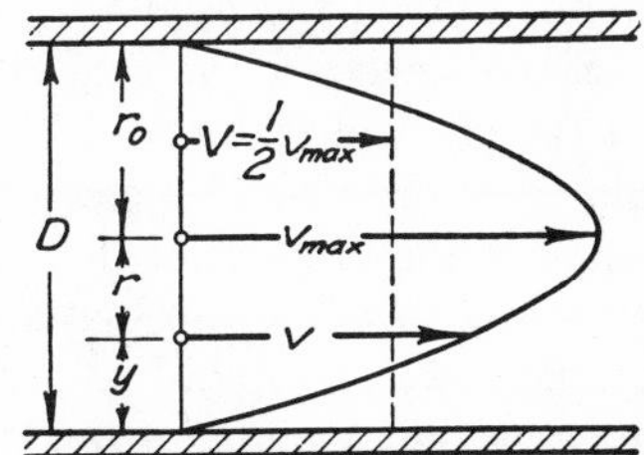

FIG. 80. Velocity distribution in flow through a circular tube.

Finally, introducing the diameter $D = 2r_0$ and solving for $-dp/dx$,

$$-\frac{dp}{dx} = \frac{32\mu V}{D^2} \tag{120}$$

Since the conduit is of uniform diameter, integration of Eq. (120) with respect to x will yield the pressure drop $p_1 - p_2$ corresponding to any tube length $x_2 - x_1 = L$:

$$p_1 - p_2 = \frac{32\mu VL}{D^2} \tag{121}$$

This is commonly known as the *equation of Poiseuille.*

Equations (116) and (121) are generally typical of flow in which viscosity is the sole fluid characteristic influencing the pattern of motion, the rate of decrease of pressure intensity in the longitudinal direction being directly proportional to the mean velocity and the dynamic viscosity. It is to be noted that, in the absence of acceleration, the fluid density does not enter the problem. Equations such as these form the basis of analysis in problems of lubrication and the flow of all fluids of high viscosity. Closely related to such flow between boundaries is flow around a solid body, likewise under the restriction that viscosity is the only fluid property affecting the pattern of motion (i.e., that accelerative effects are so small as to be negligible). Although its derivation is beyond the scope of this book, the expression for the longitudinal force F exerted by a slowly moving viscous fluid upon a small sphere, known as the *equation of Stokes*, is of interest at this point:

$$F = 3\pi D\mu v_0 \tag{122}$$

This equation finds particular application to the fall of relatively small bodies through fluids of relatively high viscosity, F then representing the effective weight of the body in question (i.e., its actual weight minus that of the fluid which it displaces).

Since power, or the rate of doing work, is dimensionally equivalent to the product of force and velocity, it is evident that the power required to move an immersed body through a fluid at a constant rate is equal to the product of the velocity and the resistance to motion (i.e., $P = vF$). In flow through a conduit, on the other hand, the pressure gradient will be found to represent the average work done upon an average unit volume of fluid per unit distance traveled. The rate at which work must be done (i.e., the power required) to maintain viscous flow through a horizontal conduit must then be equal to the product of the pressure gradient, the length of conduit, and the rate of flow; that is,

$$P = -\frac{dp}{dx} LQ = Q(p_1 - p_2) \tag{123}$$

Example 30. What horsepower is required per mile of line to overcome the viscous resistance to the flow of crude oil ($\mu = 4 \times 10^{-3}$ lb-sec/ft²) through a horizontal 3-inch pipe at the rate of 5000 gallons per hour?

Converting gallons per hour to cubic feet per second,

$$Q = \frac{5000}{7.48 \times 60 \times 60} = 0.186 \text{ cfs}$$

The mean velocity is then

$$V = \frac{Q}{A} = \frac{0.186}{\frac{\pi}{4} \times \left(\frac{1}{4}\right)^2} = 3.78 \text{ fps}$$

From Eq. (121)

$$p_1 - p_2 = \frac{32\mu VL}{D^2} = \frac{32 \times 4 \times 10^{-3} \times 3.78 \times 5280}{0.25^2} = 40{,}900 \text{ psf}$$

and from Eq. (123)

$$P = Q(p_1 - p_2) = 0.186 \times 40{,}900 = 7610 \text{ ft-lb/sec}$$

$$\frac{7610}{550} = 13.8 \text{ hp}$$

PROBLEMS

182. Crude oil is pumped at the rate of 10 gallons per minute through a horizontal 2-inch pipe, with a drop in pressure intensity of ½ pound per square inch per 100 feet of line. What is the viscosity of the oil?

183. Determine the velocity of fall through air ($\mu = 3.7 \times 10^{-7}$ pound-second per square foot) of mist droplets having a diameter of 10^{-4} foot.

184. The riveted plates of a storage tank for oil are lapped 3 inches at all joints. If a joint 30 feet below the oil surface has been sprung to such an extent that a clearance of 0.01 inch exists between the lapped plates, what loss of oil in cubic feet per day per foot of joint will occur? Assume the oil to have a viscosity of 5×10^{-4} pound-second per square foot and a specific weight of 55 pounds per cubic foot.

185. In the injection molding of a certain plastic, the heated compound is forced under a pressure of 100 pounds per square inch through a tube ¼ inch in diameter and 10 inches in length. If the mean velocity of flow is then 15 inches per second, what is the viscosity of the compound? What longitudinal force will be exerted upon the tube?

186. An oil having a viscosity of 2×10^{-3} pound-second per square foot and a specific gravity of 0.9 flows through a horizontal 1-inch pipe with a pressure drop of 0.5 pound per square inch per foot of pipe. Determine (*a*) the rate of flow in pounds per minute, (*b*) the intensity of shear at the pipe wall, (*c*) the total drag per 100 feet of pipe, and (*d*) the power required per 100 feet to maintain the flow.

187. During a heavy rain, silt-laden water flows into a reservoir having a maximum depth of 25 feet. If the finest particles of silt have a nominal diameter of 3×10^{-5} foot and a specific gravity of 2.65, estimate the time required for all the silt to settle to the bottom (assume $\mu = 2 \times 10^{-5}$ pound-second per square foot).

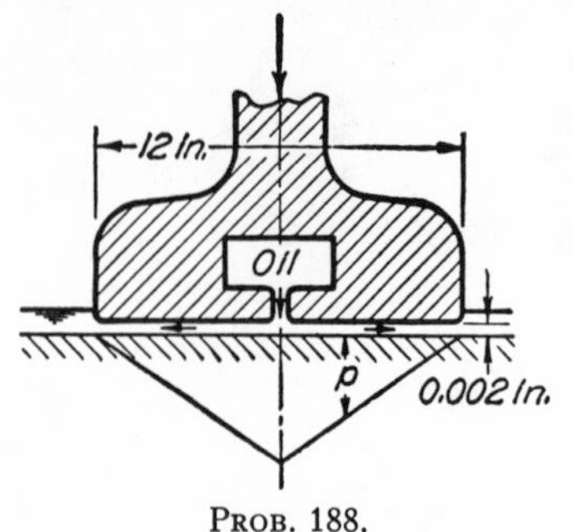

PROB. 188.

188. Certain bearing surfaces (such as those of the Palomar telescope) are lubricated by means of oil introduced under pressure at the midpoint of the bearing. Assuming two-dimensional flow from an injection slot as shown in the sketch, what rate of feed in gallons per hour per foot of length would be necessary to preserve a clearance of 0.002 inch between bearing surfaces supporting a load of 2500 pounds per square foot if the oil had a viscosity of 6×10^{-3} pound-second per square foot?

28. DISSIPATION OF ENERGY THROUGH VISCOUS SHEAR

Generation of heat in a moving fluid. The work done upon a fluid element by unbalanced pressures of the surrounding fluid was found, in the foregoing chapters, to produce a change in either the kinetic or the potential energy of the element. Since movement in the horizontal direction can produce no change in potential energy, and since the kinetic energy is constant in steady, uniform motion, one can only conclude that work done in maintaining viscous flow must result in a continuous *generation of heat*, much as in the case of solid friction. That is, the deformation of a fluid element during flow will give rise to viscous stresses within the element which tend to resist such deformation, and the work done by the external forces in deforming the element must therefore be equal to the heat produced by the internal viscous shear.

In the examples of steady, uniform flow just discussed, the forces which do work upon each fluid element may be traced ultimately to an external agency such as hydraulic machinery, or, in gravity flow, to the fluid weight itself, either of which may be considered a source of mechanical energy. On the other hand, if any fluid is once in a state of motion, work will be done within the fluid regardless of whether or not external agencies continue to act upon the fluid as a whole; in other words, generation of heat may also take place at the expense of the kinetic energy of the flow, for the velocity of such motion will gradually approach zero as the kinetic energy is transformed into heat. It is convenient, therefore, to regard the process of viscous resistance to fluid deformation as a general means whereby mechanical energy is converted into thermal energy.

A measure of the rate of such energy transformation may be obtained, for the particular condition of steady, parallel flow, through reference to Fig. 76. Although the summation of the longitudinal forces upon the element is necessarily equal to zero, there remains the couple due to shearing stresses in opposite directions along the horizontal faces of the moving element. The rate at which work is done by this couple must be equal to the product of its moment and the angular velocity $\Delta v/\Delta y$ of a line normal to the two faces. Thus, as a first approximation,

$$\Delta P \approx (\tau \, \Delta x \, \Delta z) \, \Delta y \, \frac{\Delta v}{\Delta y}$$

Dividing by the volume of the element and letting this volume approach zero, the rate at which work is done per unit volume at any point (which must be equal to the rate of dissipation of energy per unit volume at that point) will be simply the product of the intensity of shear and the velocity gradient; since $\tau = \mu \, dv/dy$, this quantity will have the alternative forms

$$\lim_{\Delta x \, \Delta y \, \Delta z \to 0} \frac{\Delta P}{\Delta x \, \Delta y \, \Delta z} = \tau \frac{dv}{dy} = \frac{\tau^2}{\mu} = \mu \left(\frac{dv}{dy}\right)^2 \tag{124}$$

A plot of this function for parallel flow between stationary boundaries is shown in Fig. 81, from which it is quite apparent that the zone of greatest generation of heat is in the neighborhood of the boundary, where the shear and the velocity gradient are highest, while at the centerline the rate of dissipation is zero. Evidently, integration of this quantity over the cross section of flow must yield the total rate of energy dissipation per unit distance in the direction of motion. For the particular case of horizontal flow through a circular tube (see page 157),

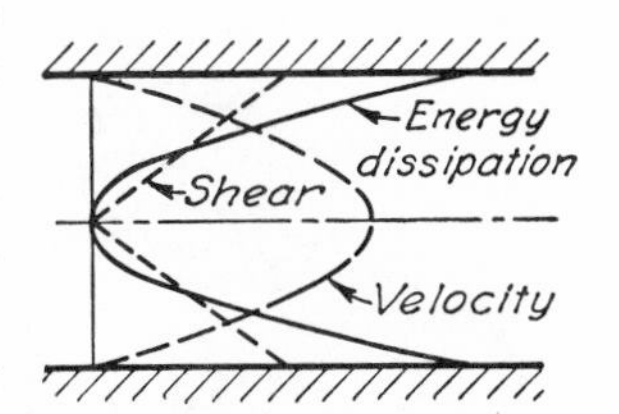

FIG. 81. Dissipation of energy in viscous flow.

$$\mu \left(\frac{dv}{dy}\right)^2 = \frac{r^2}{4\mu}\left(-\frac{dp}{dx}\right)^2$$

the integral then having the form:

$$\int_0^{r_0} \frac{r^2}{4\mu}\left(-\frac{dp}{dx}\right)^2 2\pi r \, dr = \frac{\pi r_0^4}{8\mu}\left(-\frac{dp}{dx}\right)^2$$

$$= \frac{\pi r_0^4}{8\mu} \frac{32\mu V}{D^2}\left(-\frac{dp}{dx}\right) = Q\left(-\frac{dp}{dx}\right)$$

From Eq. (123) it will be apparent that this expression is identical with that for the power per unit length necessary to maintain the flow, from which it follows that in steady, uniform, horizontal flow the product $Q(p_1 - p_2)$ represents, as indeed it must, not only the average rate at which work is done, but also the average rate at which heat is generated in a given length of horizontal conduit. In a word, the energy utilized in maintaining such flow is completely transformed into heat.

Variation of the piezometric head. If the conduit through which a viscous fluid moves is not horizontal, it may still be considered that the rate of change in pressure due to viscous shear will be of the form

$$-\left(\frac{dp}{ds}\right)_{\text{visc}} = \frac{32\mu V}{D^2}$$

There will, however, be an additional variation in p due to the gravitational (i.e., hydrostatic) effect discussed in Chapter IV:

$$-\left(\frac{dp}{ds}\right)_{\text{grav}} = \gamma \frac{dz}{ds}$$

The net rate of change will simply be the sum of the two,

$$-\frac{dp}{ds} = \frac{32\mu V}{D^2} + \gamma \frac{dz}{ds}$$

or

$$-\frac{d(p + \gamma z)}{ds} = -\frac{\gamma\, dh}{ds} = \frac{32\mu V}{D^2} \tag{125}$$

Evidently, it is the gradient of the quantity $p + \gamma z = \gamma h$ which is the criterion for energy dissipation if the flow is not horizontal. That is, with reference to Fig. 82, it will be seen that the piezometric head must be constant over any normal section, but must decrease in the direction of flow as the result of viscous resistance to deformation. Integration of Eq. (125) over the distance L and division by γ then yields for steady, uniform flow the expression

$$h_1 - h_2 = \frac{32\mu VL}{\gamma D^2} \tag{126}$$

and the power equation becomes simply

$$P = Q\gamma(h_1 - h_2) \tag{127}$$

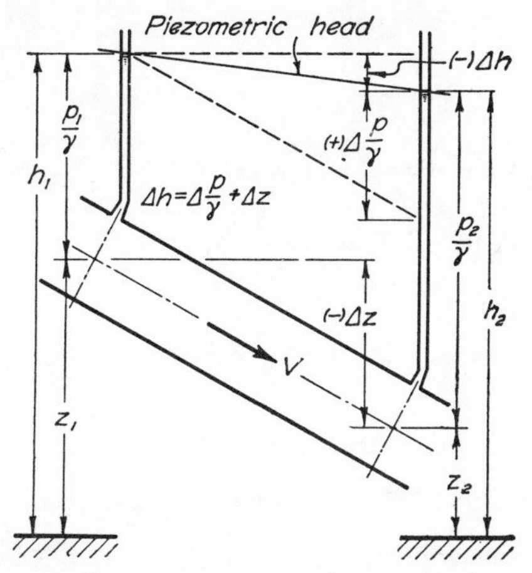

FIG. 82. Variation in piezometric head due to viscous shear.

In this connection it is interesting to note the application of the corresponding expression for flow between parallel plates [see Eq. (115)] to steady, uniform flow with a free surface. As shown in Fig. 83, the slope of the free surface and of the plane boundary must correspond to the gradient of the piezometric head h, the depth of flow y being equal to one-half the spacing B of the boundaries shown in Fig. 78. It therefore follows that

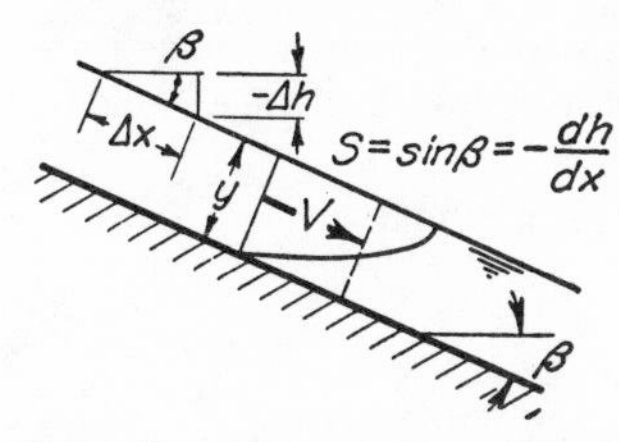

FIG. 83. Viscous flow with a free surface.

$$S = \frac{3\mu V}{\gamma y^2} \tag{128}$$

Example 31. Glycerine is circulated, for test purposes, through 80 feet of 1-inch tubing by means of a small pump set into the line. During flow, pressure gage A on the discharge side of the pump reads 25 pounds per square inch, while gage B, at the midpoint of the line and 15 feet above gage A, reads 2 pounds per square inch. (a) If the viscosity is 10^{-2} pound-second per square foot and the specific gravity is 1.25, determine the rate of flow. (b) If the line were completely insulated, what would be the rate of rise in temperature due to viscous resistance (specific heat = 0.576 British thermal unit per pound per degree)?

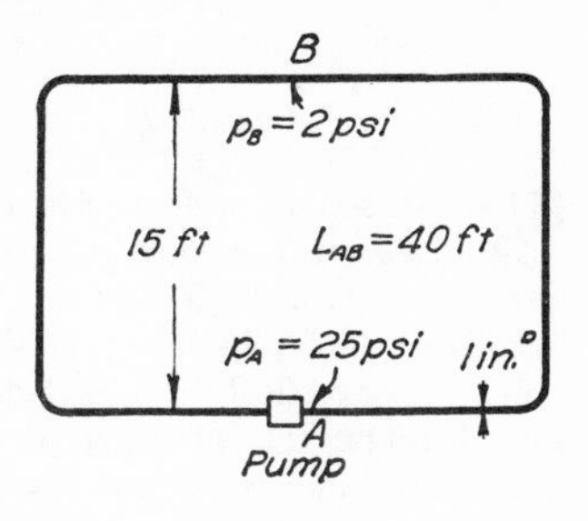

(a) According to Eq. (126)

$$h_1 - h_2 = \frac{p_1}{\gamma} + z_1 - \frac{p_2}{\gamma} - z_2 = \frac{32\mu VL}{\gamma D^2}$$

or, multiplying by γ and solving for the velocity,

$$V = \frac{D^2}{32\mu L}[p_1 - p_2 + \gamma(z_1 - z_2)]$$

Upon direct substitution of the given values,

$$V = \frac{(\frac{1}{12})^2}{32 \times 10^{-2} \times \frac{80}{2}}[(25 - 2) \times 144 + 62.4 \times 1.25 \times (0 - 15)] = 1.16 \text{ fps}$$

Therefore

$$Q = VA = 1.16 \times \frac{\pi}{4} \times \left(\frac{1}{12}\right)^2 = 0.00633 \text{ cfs}$$

or, since there are 7.48 gallons in 1 cubic foot,

$$Q = 0.00633 \times 7.48 \times 60 = 2.85 \text{ gpm}$$

(b) In order to evaluate the rate of temperature rise, it is first necessary to obtain a measure of the energy dissipation per second per pound of fluid. Since h has the dimension of energy per unit weight,

$$\frac{h_1 - h_2}{L} V = \frac{32\mu V^2}{\gamma D^2} = \frac{32 \times 10^{-2} \times \overline{1.16}^2}{62.4 \times 1.25 \times (\frac{1}{12})^2} = 0.793 \frac{\text{ft-lb}}{\text{lb-sec}}$$

Since 1 Btu = 778 ft-lb, and since 0.576 Btu will produce a rise in temperature of 1° F in 1 pound of glycerine, the resulting rate of temperature change would be

$$\frac{0.793}{778 \times 0.576} \times 60 \times 60 = 6.37° \text{ F per hour}$$

PROBLEMS

189. Crude oil is pumped at the rate of 0.5 cubic foot per second through a 4-inch pipe sloping as shown. The specific gravity of the oil is 0.92, and the dynamic viscosity 2.5×10^{-3} pound-second per square foot. By what amount will the gage pressure at point B differ from that at point A?

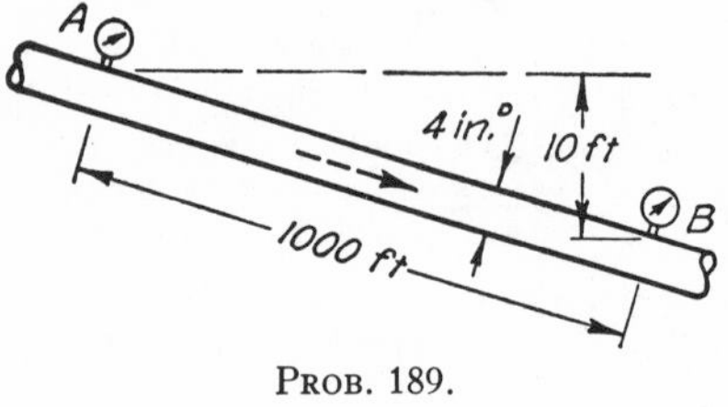

PROB. 189.

190. In a certain manufacturing process a liquid having a viscosity of 4×10^{-4} pound-second per square foot and a specific gravity of 1.2 flows down a vertical sheet of glass. If the maximum velocity of flow must not exceed ½ foot per second, what will be the limiting thickness of the liquid stream?

191. Evaluate the power required to pump 5 gallons of glycerine per minute at 100° Fahrenheit through 500 feet of 1-inch pipe discharging freely into an open tank 15 feet above the level of the supply reservoir.

192. Two vertical cylindrical containers 6 inches in diameter are connected by a 20-inch rubber tube having a ¼-inch bore. If the containers are partly filled with water at 50° Fahrenheit and one is then raised 2 inches, at what initial rate will the

surface level change in either container? (Note that the head decreases at twice the rate at which either surface level varies.)

193. The tank with the slender outlet tube shown in the accompanying sketch is used to measure the viscosity of liquids. What viscosity would be indicated by an efflux of 1 cubic inch of liquid in 30 seconds under the given head if the specific gravity of the liquid is 0.9?

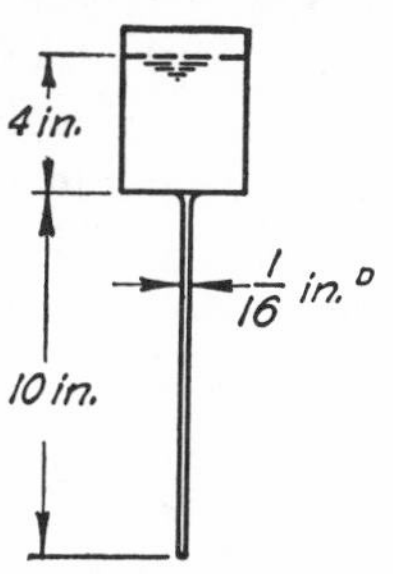

PROB. 193.

194. If the viscosimeter of Example 29 were completely insulated, what length of time would be required to raise the temperature 1° Fahrenheit (assume a specific weight of 55 pounds per cubic foot and a specific heat of 0.5 British thermal unit per pound per degree)?

195. At a given point in a lubricated bearing the unit shear within the oil film is 2 pounds per square foot. If the viscosity of the oil is 5×10^{-4} pound-second per square foot, the density is 1.85 slugs per cubic foot, and the specific heat is 0.5 British thermal unit per pound per degree Fahrenheit, determine (*a*) the local velocity gradient and (*b*) the rate of temperature rise which would result if there were no loss of heat.

196. If the viscosimeter reservoir of Problem 193 were 3 inches in diameter and filled with water at 70° Fahrenheit to the level indicated, what time would be required for the level to drop 2 inches?

197. What length of time would be required for the difference in surface level in Problem 192 to decrease to (*a*) 1 inch, and (*b*) to zero?

29. SIGNIFICANCE OF THE REYNOLDS NUMBER

Variation of the Euler number in non-uniform viscous flow. As was shown in Section 27, the steady motion of a viscous fluid between parallel boundaries must be accompanied by a pressure gradient which is directly proportional to the dynamic viscosity and the mean velocity of flow and inversely proportional to the square of a length characterizing the boundary spacing. It is noteworthy that the fluid density does not appear in such a relationship, owing to the complete absence of accelerative effects. In non-uniform flow, on the other hand, density cannot in general be ignored. With reference to the convergent boundaries of Fig. 84, for instance, one would expect a drop in pressure intensity from point to point in the mean direction of flow, as the result not only of viscous shear but also of the increase in kinetic energy as the section narrowed.

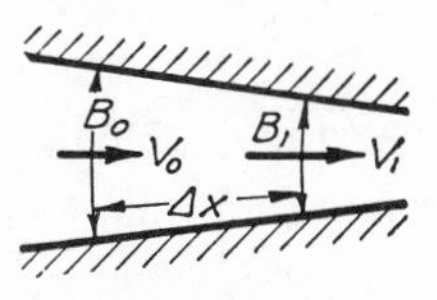

FIG. 84. Definition sketch for accelerated viscous flow.

Although the analysis of viscous and accelerative effects combined is not a simple matter, certain qualitative conclusions may at once be reached. If the fluid viscosity is high and the velocity or density is low, it is probable that the influence of viscous resistance upon the pressure gradient will by far outweigh that of the inertial resistance to acceleration; conversely, if the viscosity is low and either the velocity or the density is high, the pressure gradient necessary to produce the mass acceleration will probably be considerably greater than that necessary to maintain the flow against the viscous resistance. Evidently, the principles developed in Chapter III correspond to the limiting case in which viscous effects are negligible in comparison with inertial effects, under which circumstances (see Fig. 84)

$$p_0 - p_1 = \frac{\rho V_1^2}{2} - \frac{\rho V_0^2}{2}$$

The other extreme may be characterized by the expression

$$p_0 - p_1 = C\frac{\mu V_0 L}{B_0^2} = C'\frac{\mu V_0}{B_0}$$

in which C and C' are numerical factors depending only upon the boundary geometry [compare with Eqs. (116) and (121)]; in this case, inertial effects are negligible in comparison with those due to viscosity.

Between these two limits, needless to say, lies a vast range of flow conditions for even the specific boundary form under discussion. However, as a first approximation such conditions may be described by an equation of the Bernoulli type which combines the two foregoing relationships:

$$p_0 - p_1 = \frac{\rho V_1^2}{2} - \frac{\rho V_0^2}{2} + C'\frac{\mu V_0}{B_0}$$

Letting Δp represent the difference $p_1 - p_0$ and dividing each term by $\rho V_0^2/2$, one obtains the dimensionless expression

$$\frac{\Delta p}{\rho V_0^2/2} = 1 - \left(\frac{V_1}{V_0}\right)^2 - \frac{C'\mu V_0/B}{\rho V_0^2/2}$$

The term at the left, when written in the form $V_0/\sqrt{2\Delta p/\rho}$, will be seen at once to correspond to the Euler number introduced in Chapter III. The ratio V_1/V_0, evidently, is again a function only of the boundary geometry. The last term of the equation, on the other hand, is obviously variable between wide limits, for the denominator is a measure of the inertial characteristics of the non-uniform motion and the numerator is a measure of the viscous characteristics, either of which may range from nearly zero well toward infinity. It follows, therefore, that the Euler number is not a constant in viscous flow, since for any given boundary geometry it will vary with the ratio $(\mu V_0/B_0)/(\rho V_0^2/2)$

—or simply $VB\rho/\mu$. In other words, there exists the general functional relationship

$$\frac{V}{\sqrt{2\Delta p/\rho}} = \varphi\left(\frac{VB\rho}{\mu}\right)$$

The Reynolds number as a flow parameter; similitude. A dimensionless quantity of the type $VB\rho/\mu$ is commonly known as the *Reynolds number* **R**, a parameter which characterizes the relative importance of viscous action, much as the Froude number was shown to characterize the relative importance of gravitational action, in steady, non-uniform flow. Since the viscosity and the density appear in the Reynolds number as a ratio, it is convenient to treat this ratio of fluid properties as a property in itself. In the event of gravitational action, it will be recalled, the ratio of specific weight to density was replaced by the factor $g = \gamma/\rho$, of which the dimension [length/time^2] is purely kinematic. The ratio μ/ρ will likewise be found to be kinematic in nature, since its dimension is [length^2/time]; it is therefore called the *kinematic viscosity* ν (nu). Since

$$\nu = \frac{\mu}{\rho} \tag{129}$$

the Reynolds number, like the Froude number, will be seen to involve only a length, a velocity, and a fluid property:

$$\mathbf{R} = \frac{VL\rho}{\mu} = \frac{VL}{\nu} \tag{130}$$

The larger the Reynolds number, evidently, the less important will be the influence of viscosity upon the flow pattern, an infinite value of **R** corresponding to flow in which the viscous resistance to deformation plays no role whatever in comparison with the inertial resistance to acceleration. The smaller the Reynolds number, on the other hand, the more important the role of viscosity, a value of **R** approaching zero corresponding to flow in which inertial effects are negligible in comparison. Since the Euler number, for any given boundary geometry, is then a function of the Reynolds number, the relationship

$$\mathbf{E} = \varphi(\mathbf{R}) \tag{131}$$

should provide a general guide in the evaluation, whether analytical or experimental, of the characteristics of viscous flow under specific boundary conditions.

Efflux from an orifice, already used to illustrate the significance of the Euler and Froude numbers, likewise provides an example for the present discussion. Since the coefficient of discharge C_d is then equal to the Euler number, C_d may be expected to vary with **R**. Although the form of variation cannot be evaluated analytically, experiments have resulted in the functional relationship plotted in Fig. 85 for a circular orifice in the side of a large tank. As is to be expected, the greatest deviation from the value 0.611 of Table I is found at the low end of the abscissa scale, where the relative influence of viscosity is a

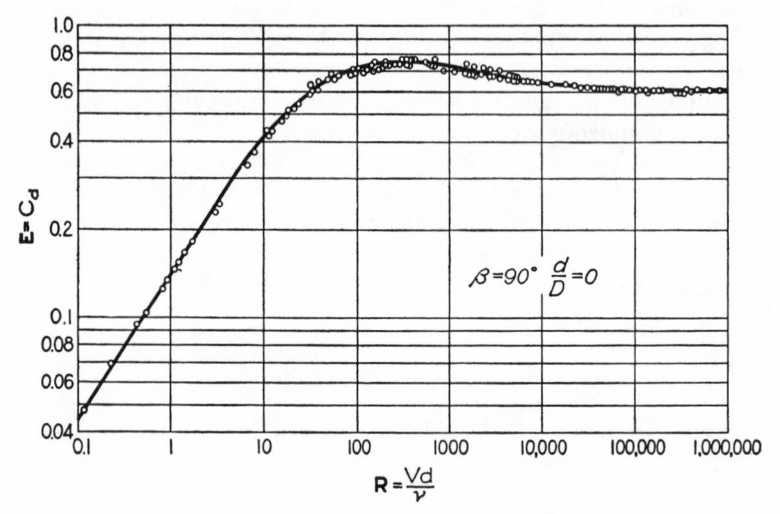

FIG. 85. The Euler number as a function of the Reynolds number for a circular orifice, after Lea.

maximum. While the curve approaches the horizontal with increasing values of **R**, it should be noted that the limiting magnitude of C_d for $\mathbf{R} = \infty$ is somewhat below 0.611, owing to the loss of energy due to viscous deformation (that is, the velocity along the solid boundary must invariably be zero, however great the Reynolds number may become).

For any particular boundary conditions, a specific value of **R** is thus seen to denote a specific Euler number and hence a specific dynamic pattern of flow, regardless of the absolute magnitudes of the linear scale, velocity, density, and viscosity. That is, so far as viscous action is concerned, two states of motion may be said to be dynamically similar if the boundaries are geometrically similar and if the Reynolds numbers (and therefore the Euler numbers) have the same magnitude. This principle of similitude is of considerable practical significance in the interpretation of measurements made on scale models of flow structures in which viscous and inertial effects are of primary importance.

Example 32. The drag of harbor currents upon a submerged mine is studied, for convenience, in a wind tunnel, at a scale of 1:3. What wind velocity should be used to simulate a 5-mile-per-hour tidal current? To what prototype resistance would a model resistance of 3 pounds correspond?

In order that the two states of flow may be dynamically similar, the following equality must obtain:

$$\frac{V_m L_m}{\nu_m} = \frac{V_p L_p}{\nu_p}$$

Assuming that, for sea water, $\nu_p = 1.4 \times 10^{-5}$ square foot per second and, for air, $\nu_m = 1.5 \times 10^{-4}$ square foot per second,

$$\frac{V_m}{V_p} = \frac{L_p}{L_m}\frac{\nu_m}{\nu_p} = \frac{3}{1} \times \frac{1.5 \times 10^{-4}}{1.4 \times 10^{-5}} = 32.2$$

whence

$$V_m = 32.2 V_p = 32.2 \times 5 = 161 \text{ mph}$$

Since the Euler numbers must also be equal,

$$\frac{V_m}{\sqrt{2\Delta p_m/\rho_m}} = \frac{V_p}{\sqrt{2\Delta p_p/\rho_p}}$$

Noting that homologous forces are proportional to $\Delta p\, L^2$,

$$\frac{F_p}{F_m} = \frac{\Delta p_p}{\Delta p_m}\frac{L_p^2}{L_m^2} = \frac{V_p^2}{V_m^2}\frac{\rho_p}{\rho_m}\frac{L_p^2}{L_m^2}$$

Assuming that $\rho_p = 1.99$ slugs per cubic foot and $\rho_m = 0.0025$ slug per cubic foot,

$$F_p = F_m\left(\frac{V_p}{V_m}\right)^2 \frac{\rho_p}{\rho_m}\left(\frac{L_p}{L_m}\right)^2 = 3 \times \left(\frac{5}{161}\right)^2 \times \frac{1.99}{0.0025} \times \left(\frac{3}{1}\right)^2 = 20.8 \text{ lb}$$

PROBLEMS

198. A new form of flow meter when tested in the laboratory yields a differential pressure of 9 pounds per square inch for a flow of 3 cubic feet of water per second through a 6-inch pipe at 60° Fahrenheit. At what rate should water at the same temperature flow through a geometrically similar meter in a 24-inch pipe to produce dynamically similar conditions? What differential pressure should then be measured?

199. At what rate should air at 100° Fahrenheit flow through the 6-inch pipe meter of Problem 198 if dynamic similarity of the flow pattern is to obtain? Evaluate the corresponding differential pressure. (Assume $\rho = 0.0023$ slug per cubic foot.)

200. Wind effects upon a barrage balloon are determined by means of a 1:15 scale model in a wind tunnel. What air velocity in the tunnel would correctly simulate a 20-mile-per-hour wind against the prototype? To what prototype drag would a 150-pound model drag correspond?

201. Tests are to be made upon a 1:30 scale model of a Great Lakes freighter. If the cruising speed of the prototype is 12 knots, what velocity should be given the

model in a towing tank according to the similarity criterion of (*a*) Froude, and (*b*) Reynolds?

202. Because of the relative ease in making internal velocity traverses, the flow pattern in a proposed water turbine is studied at a 1:10 scale reduction using air instead of water. If the operating temperature in both model and prototype is assumed to be 60°, what flow ratio must prevail to attain dynamically similar conditions? What will be the conversion ratio for (*a*) velocities, and (*b*) changes in pressure intensity?

203. Power requirements of a small submarine cruising submerged at 8 knots are studied under dynamically similar conditions in a high-density wind tunnel at 1:10 scale. What prototype horsepower would be indicated by a model drag of 65 pounds when the tunnel pressure is 5 atmospheres and the temperature 100°?

30. INSTABILITY OF VISCOUS FLOW

Disturbance of laminar motion. In deriving relationships for the mean rate of energy dissipation due to viscous shear, conditions were restricted to the elementary case of steady, uniform flow, under which circumstances the stream lines not only remain parallel to one another but also indicate velocities which do not change with time. As mentioned in the foregoing section, similar relationships will hold for shear in steady, non-uniform motion, the stream lines and velocities then varying in form from one boundary to the next but remaining unchanged with time. As a matter of fact, much the same considerations would apply to the unsteady, non-uniform flow of a viscous fluid, provided only that the individual stream lines were governed in general form by the geometry of the solid boundaries. The term *laminar* is applied to flow of this nature, because the stream lines, or, rather, stream "surfaces," appear to divide the entire region of flow into an orderly series of fluid laminae or layers conforming generally to the boundary configuration.

The existence—whatever the cause may be—of local disturbances in the flow of a viscous fluid is evidently not compatible with this concept of laminar motion. Under certain conditions, to be sure, any local fluctuations in velocity will gradually be damped by the additional viscous stresses which are produced, the flow then eventually becoming truly laminar in nature. Under other conditions, on the contrary, the viscous stresses involved may not be sufficient to quell such fluctuations before the disturbance has spread throughout the entire zone of motion. The flow is then said to be *turbulent* and may no longer be treated according to the foregoing equations for laminar motion. Since minor disturbances are always present in a moving fluid, and since turbulence is one of the most important phenomena of fluid motion, the question whether a given state of flow is inherently stable or unstable must invariably be answered.

Development of a general instability parameter. The inertial tendency of any fluid toward instability may be illustrated schematically by assuming the velocity distribution shown in Fig. 86*a* to exist in a fluid of finite density but zero viscosity. If, for any reason, the stream line at the velocity discontinuity begins to deviate from a straight line (i.e., if a disturbance occurs), as indicated in Fig. 86*b*, according to the flow net the local velocity will decrease as the neighboring stream lines diverge and increase as they converge. As the velocity decreases, however, the pressure intensity must increase, and vice versa, with the result that the pressure differences across the discontinuity will tend to increase the displacement of the stream lines at all points in the direction of the initial deviation. As the process continues, the zone of discontinuity will eventually develop into a series of vortices of finite size, as shown by Figs. 86*c* and 86*d*.

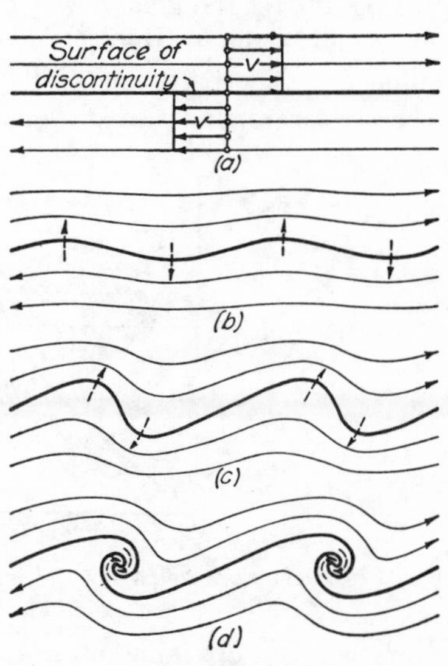

FIG. 86. Instability of flow at a surface of discontinuity.

If the viscosity of the fluid under consideration is not zero, the following modification of this schematic picture is to be expected. An abrupt discontinuity of the velocity distribution will not be possible, since this would require an infinite local intensity of shear; the steeper the local velocity gradient, however, the more nearly the conditions of Fig. 86 are approached. Similarly, the higher the fluid density, the more pronounced will be the tendency to produce eddies if an initial disturbance is present. The higher the viscosity, on the other hand, the more will the internal shear tend to oppose further differences in velocity. The presence of a solid boundary will likewise be a stabilizing influence, since the nearer the initial disturbance is to the boundary the smaller must be the lateral displacement.

With reference to Fig. 87, showing the velocity distribution of a viscous fluid near a plane boundary, it will be seen that the velocity gradient, density, viscosity, and wall distance may be combined to yield a stability parameter χ (chi) having the following dimensionless form:

$$\chi = \frac{y^2 \rho \, dv/dy}{\mu} \tag{132}$$

According to the foregoing discussion, the greater the numerator of this parameter the greater will be the tendency toward local instability, whereas the greater the denominator the greater the stabilizing influence. The parameter will necessarily be zero at the boundary, where y is zero, and also at a great distance from the boundary, where

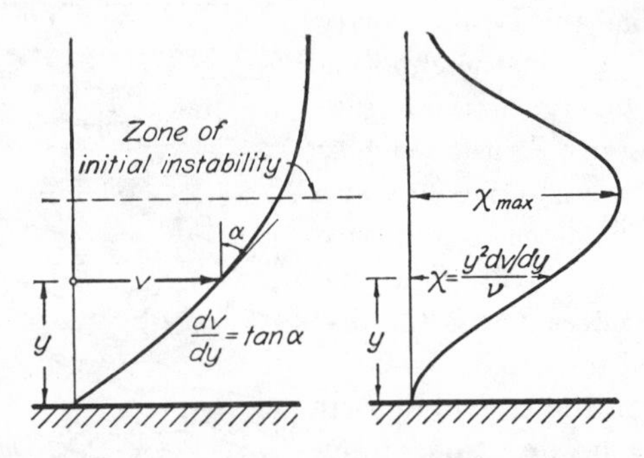

FIG. 87. Definition sketch for the instability of laminar flow near a boundary.

dv/dy is zero. At some intermediate distance, however, it will attain a maximum value, as shown in the figure. This then is the zone at which eddies may be expected to occur if the flow is disturbed in any way—provided that the magnitude of χ exceeds a certain critical value. Experiments for various boundary conditions indicate that this critical value has a magnitude of about 500. In general, therefore, if the parameter χ does not approach 500 at any point in a given flow, such flow is inherently stable; but if this critical value is much exceeded at any point, eddies will develop once a disturbance of sufficient magnitude occurs.

Reynolds' upper and lower critical limits. Since, for any given boundary geometry, the Reynolds number characterizes the flow as a whole, one would also expect a particular value of **R** to mark the limit of stability for a particular boundary form. Indeed, this was first shown to be true, for flow through a circular tube, by Osborne Reynolds, an English scientist of the last century for whom the parameter **R** was named. In Fig. 88 is shown schematically a modification of Reynolds' equipment intended for demonstration purposes. A glass tube with rounded inlet leads from a large tank and terminates in a valve by which the rate of flow can be controlled. At the inlet is a small jet permitting the introduction of a fine stream of dye. If the tank is filled with water and allowed to stand for several hours, and if the control valve is then opened slightly, injection of the dye will

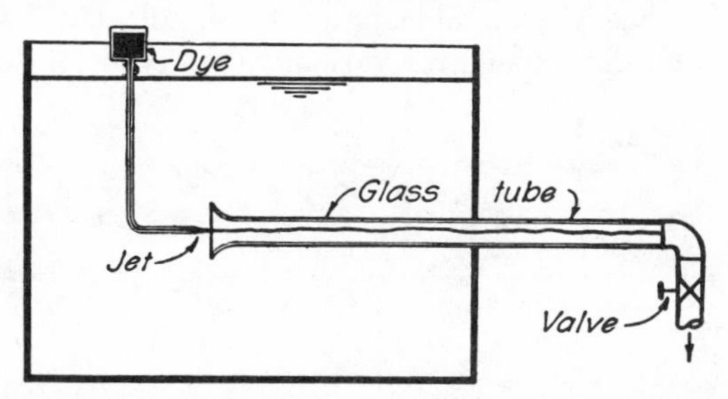

FIG. 88. Reynolds apparatus for demonstrating instability of flow through a tube.

yield a colored filament which passes through the tube so steadily that it scarcely seems to be in motion. The flow will continue in this stable state as the valve is opened farther, until eventually a velocity is reached at which the thread of dye begins to waver (Fig. 89). Continued increase in the rate of flow will make the fluctuations more intense, the dye at last rapidly diffusing over the entire cross section of the tube.

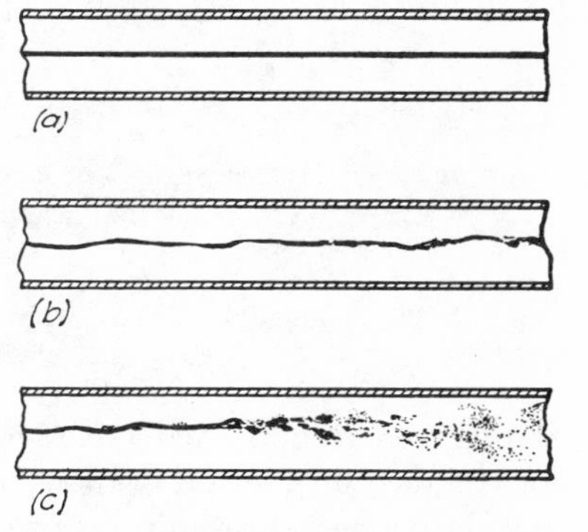

FIG. 89. Behavior of the dye filament with increasing Reynolds number.

Reynolds found that instability of flow would occur at a magnitude of the parameter $\mathbf{R} = VD/\nu$ of about 12,000, regardless of the tube diameter. More recently, however, it has been proved that the magnitude of the Reynolds number at which laminar flow becomes unstable depends largely upon the nature of the disturbances present in the flow. For instance, critical values of $\mathbf{R}$ over 40,000 have since been reached with Reynolds' original equipment, largely as the result of increasing the stilling time, improving the rounding of the inlet, and eliminating all possible vibration. The so-called *upper critical* Reynolds number—the maximum value of $\mathbf{R}$ at which laminar flow is physically possible—would therefore appear to be indeterminate, the disturbances necessary to produce instability simply decreasing in magnitude as $\mathbf{R}$ becomes larger.

There exists, however, for any given boundary geometry a *lower critical* value of **R** below which disturbances of any magnitude are eventually damped by viscous action. For flow through circular tubes, the lower critical Reynolds number is approximately $VD/\nu = 2000$, a value which has considerable practical significance. Comparable limits have been determined for various other types of flow, whether

PLATE XII. Photographs of the transition from laminar flow to turbulent flow in a glass tube.

through or around fixed boundaries of different forms, and whether for confined or free-surface flow, although these limits differ considerably in numerical magnitude, owing in part to the effect of boundary form upon the flow stability and in part to the flow dimension chosen as the length parameter. Invariably, however, flow which is characterized by a Reynolds number less than the lower critical is stable to disturbances of any magnitude, and hence is inherently laminar in nature; indeed, it is only this lower critical value which is of practical importance, since a moving fluid is seldom so free from local disturbances that laminar flow will persist at higher values of **R**.

Example 33. Certain operations in the manufacture of chemicals involve the flow of liquid in a thin, smooth sheet down an inclined plate of glass. If the Reynolds number exceeds the critical value of 500, turbulence may develop; moreover, if the Froude number exceeds 2, surface waves may form. Determine the maximum permissible rate of flow per unit width, and the corresponding slope of the glass plate, for a liquid having a viscosity of 6×10^{-5} pound-second per square foot and a specific gravity of 1.2.

Evidently, the following conditions limit both the depth and the mean velocity, the product of which determines the rate of flow:

$$\frac{Vy}{\nu} = 500 \quad \text{and} \quad \frac{V}{\sqrt{gy}} = 2$$

Upon eliminating V and solving for y,

$$\frac{500\nu}{y} = 2\sqrt{gy}, \quad y = \sqrt[3]{\frac{250{,}000\nu^2}{4g}}$$

For the given fluid characteristics,

$$y = \sqrt[3]{\frac{250{,}000 \times \left(\dfrac{6 \times 10^{-5}}{1.94 \times 1.2}\right)^2}{4 \times 32.2}} = 0.0109 \text{ ft}$$

and

$$V = 2\sqrt{gy} = 2\sqrt{32.2 \times 0.0109} = 1.18 \text{ fps}$$

Therefore,

$$q = Vy = 1.18 \times 0.0109 = 0.0129 \text{ cfs/ft}$$

and, from Eq. (128),

$$S = \frac{3\mu V}{\gamma y^2} = \frac{3 \times 6 \times 10^{-5} \times 1.18}{62.4 \times 1.2 \times 0.0109^2} = 0.024 = \sin^1 \alpha$$

$$\alpha = 1° \, 22'$$

PROBLEMS

204. Submit on logarithmic paper a plot of pipe diameter against the lower critical velocity for air and water at 60° Fahrenheit.

205. The value $\mathbf{R} = 1.0$ is sometimes taken as the approximate limit of validity of the Stokes equation, inertial effects thereafter becoming appreciable. Determine the maximum size of mist droplets which will still follow the Stokes relationship in air at 60° Fahrenheit.

206. From the equation for the velocity distribution in laminar flow through a tube, derive an expression for χ_{max} in terms of V, D, and ν. To what value of this parameter does the critical value $\mathbf{R} = 2000$ correspond?

207. Although blood is not a true fluid, in that its viscosity varies somewhat with rate of shear, it may be assumed for present purposes to be approximately 5 times as viscous as water. If the mean velocity of the blood stream through a ¼-inch artery is about 6 inches per second, what fraction of the lower critical velocity does this represent?

208. What is the greatest permissible rotational speed (i.e., that at which $\chi_{max} = 500$) of the viscosimeter of Example 29 with oil of the given viscosity, assuming its specific gravity to be 0.9?

209. What is the greatest depth of oil for which $\mu = 10^{-4}$ pound-second per square foot and $\rho = 1.7$ slugs per cubic foot which could exist in the reservoir of Problem 193 without exceeding the lower critical velocity of flow?

31. CHARACTERISTICS OF FLUID TURBULENCE

Nature of the mixing process; energy dissipation. Once the flow of a viscous fluid has become unstable in any region, the resulting eddies will spread rapidly over the entire flow section, thereby producing a complex pattern of motion which varies continuously with time. This phenomenon is known as *fluid turbulence.* Since the stream lines are then not only hopelessly intertwined but changing in form from instant to instant, turbulent flow is evidently the complete antithesis of laminar flow, in which neighboring "layers" of fluid remain distinct from one another except for the minor effect of molecular diffusion. Indeed, fluid turbulence is often likened to molecular motion on a greatly enlarged scale, the microscopic intermixing of molecules finding a counterpart in the macroscopic intermixing (i.e., eddy motion) of finite masses of fluid.

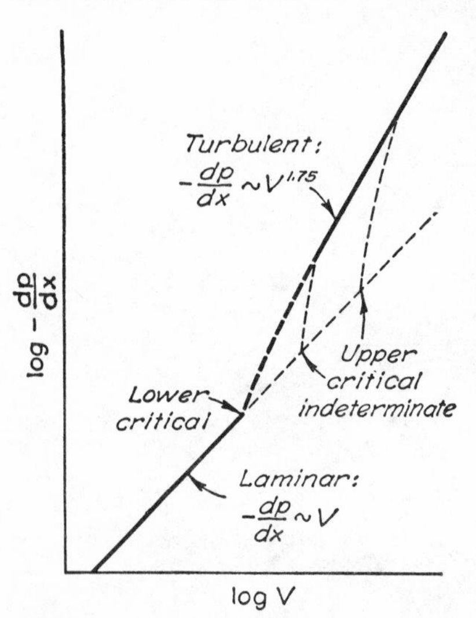

FIG. 90. Comparison of the rates of energy dissipation in laminar and turbulent flow.

Not only does turbulence involve a continuous interchange of fluid between neighboring zones of flow, as indicated by the rapid diffusion of color in the Reynolds apparatus, but at the same time the rate of energy dissipation is greatly increased, as may be seen from the abrupt change in the relationship between pressure gradient and mean velocity plotted in Fig. 90. That these effects are closely related will be evident from the following reasoning. Just as coloring matter is carried from one part of the flow to another by the mixing process of turbulence, fluid with high kinetic energy is carried from the central region toward the boundaries, while fluid with low kinetic energy is carried from the boundary region toward the center, with the result that the average velocity distribution for turbulent flow becomes far more uniform than that for laminar flow, as indicated in Fig. 91. In fact, far from being a parabola, the distribution curve is practically logarithmic in form. At the boundaries, therefore, where the eddy motion must reduce to a minimum, the velocity gradient is much higher than in wholly laminar motion, whence the intensity of boundary shear $\tau_0 = \mu(dv/dy)_{(y=0)}$ must be increased by a corresponding amount. It follows from a relationship such as Eq. (118) that the pressure gradient, a measure of the energy dissipation, will be far greater in turbulent flow.

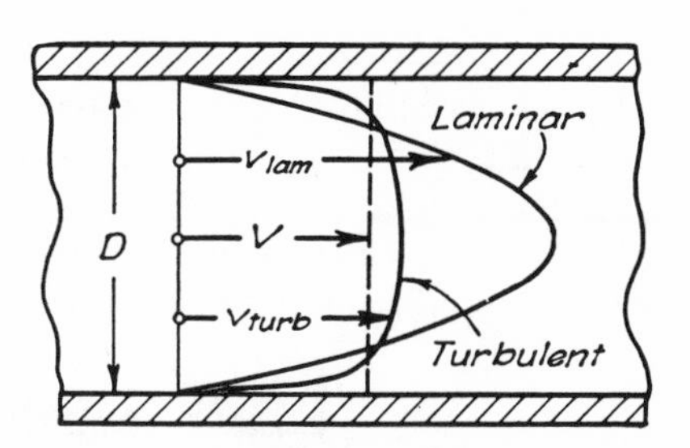

FIG. 91. Comparison of the mean velocity-distribution curves in laminar and turbulent flow.

Nevertheless, if the variation of the mean intensity of shear across the section is to remain linear, as it must in uniform flow, it is evident that τ will be increased at all points by an amount proportional to the increase in τ_0. How this is effected by the turbulence will be clear if one notes that the distribution curve of Fig. 91 represents the *temporal mean* velocity $\bar{v}$ in the longitudinal direction, whereas at any instant the complex pattern of the eddy motion results in deviations from the temporal mean value in every direction at all points; such instantaneous deviations in the longitudinal direction are indicated schematically in Fig. 92. In other words, one may visualize turbulent flow as a haphazard and ever-changing system of eddies superposed on the mean motion of the fluid, the viscous stresses within each individual eddy resulting in a rate of energy dissipation which is far in excess of that due to the mean flow alone.

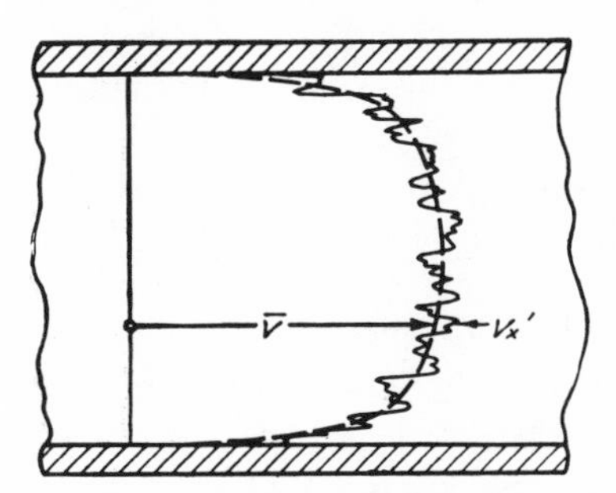

FIG. 92. Instantaneous distribution of velocity in turbulent flow.

Significance of the eddy viscosity. Since the pattern of the turbulence varies continuously with time, knowledge of the instantaneous distribution of velocity is useful only if a continuous record with time (see Fig. 93) is available for statistical analysis. For this reason, use is sometimes made of the rough similarity between molecular motion and turbulence by writing the mean intensity of shear due to turbulence [compare with Eq. (110)] in the following form:

$$\bar{\tau} = \eta \frac{d\bar{v}}{dy} \tag{133}$$

The bar over the velocity now represents the temporal mean value at any point, as plotted in Figs. 92 and 93, while η (eta) is the so-called *eddy viscosity*, analogous to the molecular viscosity μ. This method eliminates the necessity of considering the instantaneous viscous stresses within the eddies, through expressing the average effect of

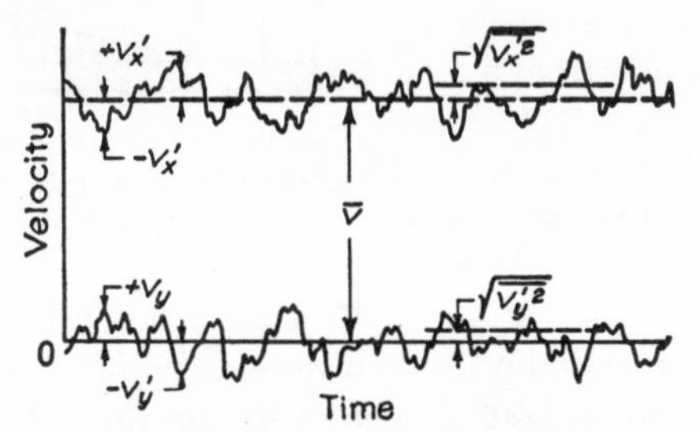

FIG. 93. Local variation of the velocity with time.

these stresses in terms of the mean velocity gradient and a parameter characteristic of the turbulence.

In the kinetic theory of gases it has been shown that the viscosity μ of a gas depends upon the density of the gas, the mean free path of the molecules, and the mean molecular velocity. If one continues the analogy between such microscopic molecular motion and the macroscopic eddy motion of fluid turbulence, then the eddy viscosity should depend upon the density of the fluid, the mean size l of the eddies, and some characteristic velocity of the eddy motion—the most appropriate form of the last parameter being the root-mean-square $\sqrt{\overline{v'^2}}$ of the instantaneous velocity deviations from the mean velocity of the primary motion (see Fig. 93). Since μ is a fluid property, its magnitude is the same at all points of a given flow (provided that the temperature is constant). The parameter η, however, is a fluid characteristic only in so far as it includes the fluid density; otherwise it is a characteristic of motion and may reasonably be expected to vary from point to point in the flow. In fact, by dividing η by ρ there is obtained a factor which is characteristic of the flow alone:

$$\frac{\eta}{\rho} = \epsilon \tag{134}$$

The factor ϵ (epsilon) is evidently a *kinematic eddy viscosity* comparable to ν, the kinematic molecular viscosity. This parameter should then depend only upon the eddy size l and the eddy velocity $\sqrt{\overline{v'^2}}$.

Although the measurement of turbulence characteristics is a very tedious statistical process, the results of such measurements are extremely significant. These are shown schematically in Fig. 94 for flow through a pipe of circular cross section. It will be noted that the average eddy size varies from zero at the boundary to a maximum at the centerline, the velocity of fluctuation on the other hand being a maximum near the boundary and decreasing toward the center; the product $\epsilon = l\sqrt{\overline{v'^2}}$ therefore becomes a maximum approximately midway between centerline and wall. Evidently, the mixing effect is most pronounced in this region, approaching a low value in the central zone but attaining a minimum of zero at the wall. This knowledge is particularly important—not only in the analysis of energy dissipation, but also in problems of heat transfer and the suspension of finely divided material (dust or silt) by the process of turbulent convection. Indeed, since ϵ provides a direct measure for the mixing process, it is frequently called the *diffusion coefficient*.

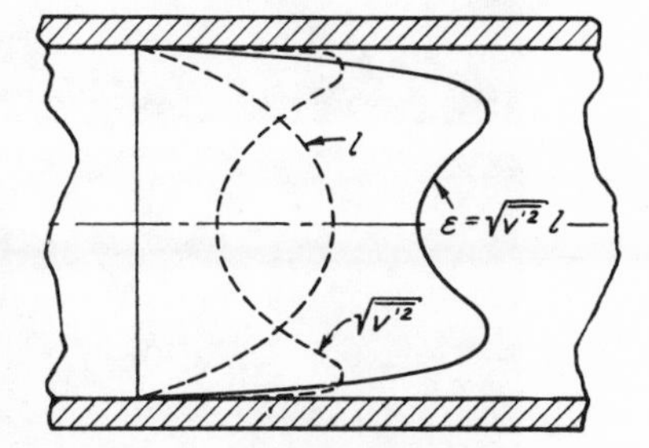

FIG. 94. Variation in eddy viscosity across a circular pipe.

Just as the mean size of the eddies in turbulent flow is governed by the scale of the boundary geometry, the mean velocity of fluctuation for given boundary conditions and a given fluid depends upon the mean velocity of the flow itself. In other words, the average mixing coefficient for the entire flow section, ϵ_m, must vary with VL. Dividing both quantities by the kinematic viscosity, it follows that the ratio of the eddy viscosity to the molecular viscosity in turbulent flow must vary with the Reynolds number:

$$\frac{\epsilon_m}{\nu} \sim \mathbf{R} \tag{135}$$

In general, therefore, not only will the relative effectiveness of the mixing process increase with increasing Reynolds number, but also two states of turbulent flow may be said to be dynamically similar if the boundaries are geometrically similar and the Reynolds numbers are the same.

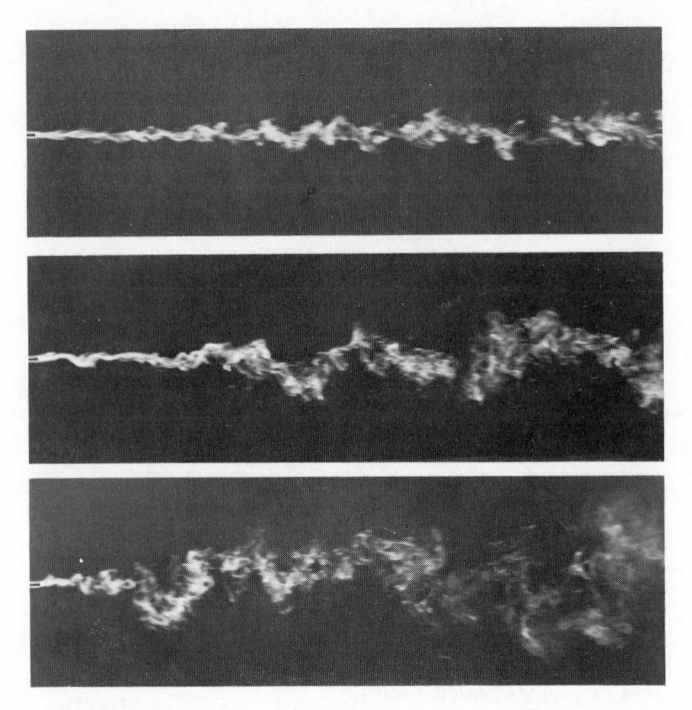

PLATE XIII. The diffusion of smoke in winds having the same velocity but different degrees of turbulence.

Example 34. Flow of a gas through a horizontal 2-inch pipe yields a measured pressure drop of 20 pounds per square foot per 1000 feet of pipe when the mean velocity is 12 feet per second; the density and the kinematic viscosity of the gas are, respectively, 15×10^{-4} slug per cubic foot and 2×10^{-4} square foot per second. What percentage error would result were the rate of discharge computed from the measured pressure gradient according to the equation of Poiseuille?

Since

$$\mathbf{R} = \frac{VD}{\nu} = \frac{12 \times \frac{2}{12}}{2 \times 10^{-4}} = 10{,}000$$

the flow is well beyond the practical limit ($\mathbf{R} = 2000$) of the laminar regime. For the given velocity, the actual rate of energy dissipation is, therefore, far greater than that computed by the equation of Poiseuille; consequently, the velocity indicated by Eq. (121) for the given pressure drop should be considerably in excess of that which actually occurs at this stage of turbulent flow. That is, solving Eq. (121) for V,

$$V = \frac{D^2(p_1 - p_2)}{32\mu L} = \frac{(\frac{2}{12})^2 \times 20}{32 \times (2 \times 10^{-4} \times 15 \times 10^{-4}) \times 1000} = 57.8 \text{ fps}$$

which represents an error of

$$\frac{57.8 - 12}{12} \times 100 = 381\%$$

PROBLEMS

210. Crude oil is pumped through a 6-inch pipeline which is subject to seasonal changes in temperature. At the maximum temperature of 100° Fahrenheit, when $\nu = 3 \times 10^{-4}$ square feet per second, a power input of 3 horsepower per 1000 feet of line is required to maintain a flow of 1 cubic foot per second. What power input would be required to maintain the same rate of flow at the minimum temperature of 30° Fahrenheit, if the viscosity of the oil is then 10 times as great? (Note: First compute the Reynolds numbers; assume a specific gravity of 0.9 at both temperatures.)

211. A mixing device used in chemical manufacture is found to require 15 minutes to yield a satisfactory liquid suspension of finely divided solids. If both the size and the rotational speed of the agitator were doubled, how long an interval should be required to reach essentially the same stage of mixing?

212. Measurements of wind velocity against time in the lee of a tower yield the sinusoidal curve shown in the accompanying sketch. Estimate therefrom the magnitudes of l, $\sqrt{\overline{v'^2_x}}$, and ϵ for the eddies shed by the tower.

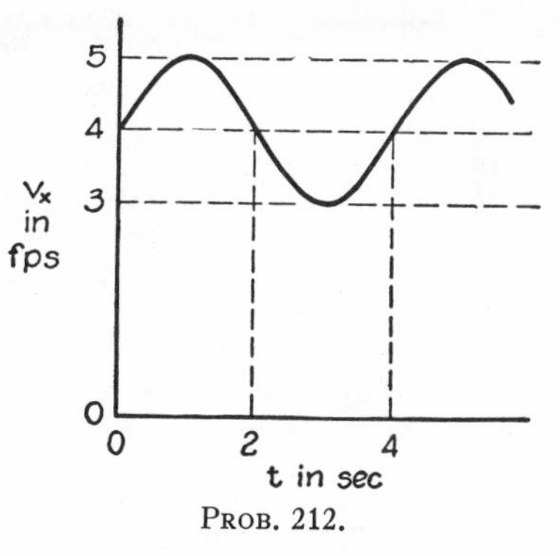

PROB. 212.

213. Estimate the relative magnitudes of the eddy viscosity for the three photographs of smoke diffusion shown in Plate XIII.

214. Assuming that the wind velocity varies with the logarithm of elevation above the surface of the earth (i.e., $v = C_1 + C_2 \log_{10} y$), estimate the velocity at an elevation of 1000 feet when measurements at 10 feet and 100 feet yield velocities of 7 and 11 feet per second, respectively.

215. As a first approximation, the size of the eddies at the end of the hydraulic jump may be assumed proportional to the depth y_2 and the velocity fluctuations proportional to the mean velocity V_2. Compare the values of ϵ resulting from jumps at Froude numbers of 5 and 10, the depth y_1 being the same in each case.

216. In a given wind the smoke from a factory chimney is found to spread 15 feet laterally in a distance of 100 feet. If the eddy size is assumed proportional to that of the boundary (i.e., building) irregularities and the eddy velocity proportional to the velocity of the wind, what change in the lateral diffusion of the smoke would result from a 50 per cent increase in wind velocity?

217. It is sometimes assumed, for convenience, that the kinematic eddy viscosity ϵ is constant across the normal section of a pipe or open channel. To what form of the velocity distribution curve would this necessarily correspond?

QUESTIONS FOR CLASS DISCUSSION

1. Does the viscosity of (*a*) liquids, and (*b*) gases, increase or decrease with temperature? Suggest reasons for the difference in behavior.

2. Describe the distribution of shear in a cylindrical vessel of fluid rotating about its axis at constant angular speed. Discuss the variation in shear which would be caused by an abrupt change in the rotational speed of the tank.

3. At what rate is work done by the shear on the vertical faces of the element of Fig. 76?

4. Show that the relationship preceding Eq. (124) may be obtained from a summation of the rates at which work is done by the several forces shown in Fig. 76.

5. If fluid pressure is not a form of energy, why is the pressure gradient a measure of the rate of energy dissipation in steady, uniform flow?

6. Cite instances of laminar flow encountered in everyday life.

7. Suggest means of measuring both dynamic and kinematic viscosity based on the various examples of laminar flow discussed in this chapter.

8. With reference to Fig. 81, in what zone of laminar flow through a tube would the temperature tend to rise at the greatest rate? How would the resulting heat be transferred to the remaining zones of flow?

9. Heat is supplied to fluid passing through a conduit by means of resistance coils in an external jacket. How will the heat reach the innermost zone of flow in (*a*) laminar motion, and (*b*) turbulent motion? Will the transfer of heat be more rapid in (*a*) or in (*b*)?

10. With reference to Fig. 94, where in the pipe cross section would the rate of heat transfer in Question 9*b* be the smallest?

11. Show that the Reynolds number may be considered to represent the ratio of a typical inertial reaction to a typical viscous force.

12. The dynamic viscosity of water is approximately 70 times as great as that of air at normal temperature. Is the kinematic viscosity of water greater or less than that of air? Of what significance is this fact in connection with the dynamically similar flow of air and water through the same conduit?

13. In model studies of gravitational effects, velocities are reduced in proportion to the square root of the scale ratio. What velocity change is required in model studies of viscous effects? What practical difficulty would such a relationship introduce in model studies involving both gravitational and viscous effects?

14. Why does the irregular filament of dye in the Reynolds apparatus (Fig. 88) not represent a stream filament?

15. Describe examples of flow in which turbulence is (*a*) advantageous and (*b*) disadvantageous.

SELECTED REFERENCES

PRANDTL, L. "The Mechanics of Viscous Fluids." Vol. III, *Aerodynamic Theory*, Springer, 1935.

HERSEY, M. D. *Theory of Lubrication*. Wiley, 1936.

MUSKAT, M. *Flow of Fluids through Porous Media*. McGraw-Hill, 1937.

KÁRMÁN, TH. VON. "Turbulence" (Twenty-Fifth Wilbur Wright Memorial Lecture) *Journal of the Royal Aeronautical Society*, December, 1937.

CHAPTER VII

SURFACE RESISTANCE

32. THEORY OF THE BOUNDARY LAYER

Essence of the Prandtl theory. In the preliminary analysis of fluid motion without regard to the effect of viscosity, it was found that the pattern of motion was governed primarily by the geometry of the boundaries between which or around which the fluid moved. Since, however, the velocity of a viscous fluid at the zone of contact with a solid boundary is invariably the same as the velocity of the boundary, it is evident that the pattern of motion of a viscous fluid must differ appreciably—at least in the boundary neighborhood—from that indicated by the flow net. It is the purpose of this and the succeeding chapter to investigate the influence of such viscous effects upon the velocity distribution, and to evaluate the magnitude of the resulting forces exerted by the fluid upon the boundary.

As must be concluded from the discussion in the foregoing chapter, the extent to which viscous action may be expected to modify the basic pattern obtained from the flow net will depend upon the magnitude of a characteristic Reynolds number. The smaller the Reynolds number, that is, the larger will be the viscous influence. The larger this parameter, on the other hand, the smaller will be the importance of viscosity in determining the ultimate pattern of motion. Indeed, although viscous resistance to deformation must, mathematically speaking, extend throughout the moving fluid in all cases, it so happens that at high values of **R** such resistance will be appreciable only in the immediate vicinity of the boundary. Recognition of this fact early in the present century by the German scientist Prandtl may be considered to mark the foundation of modern fluid mechanics.

Consider, for example, the motion of a fluid at the velocity v_0 past the streamlined strut shown in Fig. 95. If only viscous effects are taken into account, the velocity distribution will follow curve I, v being zero at the boundary and gradually increasing toward the limit v_0 with distance from the boundary. If only accelerative effects are taken to account, however, the lateral velocity distribution will be as indicated by curve IV, v being a maximum at the boundary and gradually decreasing toward the limit v_0 with distance from the boundary. Since curve I evidently corresponds to an extremely low and curve IV to an extremely high Reynolds number, it is apparent that intermediate values of **R** should yield intermediate distribution curves such as II and III.

So long as the velocity varies according to curves I and II, it will be seen that the flow pattern around a body of this form cannot even be approximated by means of the flow net. Curve IV, of course, corresponds exactly to the flow net, but curve III differs perceptibly

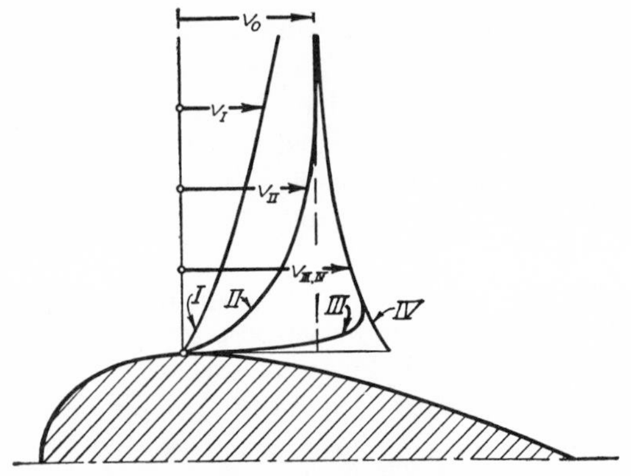

FIG. 95. Modification of boundary influence with increasing Reynolds number.

therefrom only in the neighborhood of the boundary. In the latter case, therefore, it may conveniently be assumed that the major part of the viscous deformation is confined to what is known, after Prandtl, as the *boundary layer*—a relatively thin zone of flow in the immediate boundary vicinity. If, therefore, the Reynolds number is sufficiently high (as it is in the majority of engineering problems), the flow net may still be used to indicate the accelerative aspects of the fluid motion, while the viscous aspects may be evaluated through an analysis of the motion in the boundary layer alone.

Characteristics of the laminar boundary layer. Since the true boundary layer is, from definition, very thin in comparison with other boundary dimensions, for the sake of convenience it may, as a first approximation, be assumed to develop along either side of a flat plate held longitudinally in a moving fluid. At any arbitrary section along the plate the velocity will vary essentially as shown in Fig. 96, having a magnitude of zero at the plate and rapidly approaching the limit v_0 with normal distance y. If the distance $y = \delta$ (delta) is defined as that at which the velocity is, say, within 1 per cent of its asymptotic limit, δ thus becomes a nominal measure of the thickness of the boundary layer—i.e., of the region in which the major portion of the viscous deformation takes place.

According to the foregoing discussion, the relative magnitude of δ should vary inversely with a characteristic Reynolds number. In other words, if δ is expressed as a ratio to the distance x from the leading edge of the plate, and if the Reynolds number is composed of the length x, of the velocity v_0, and of the kinematic viscosity ν, then δ/x should be a function of $\mathbf{R} = v_0x/\nu$. Indeed, it has been shown both analytically (through use of the momentum principle) and experi-

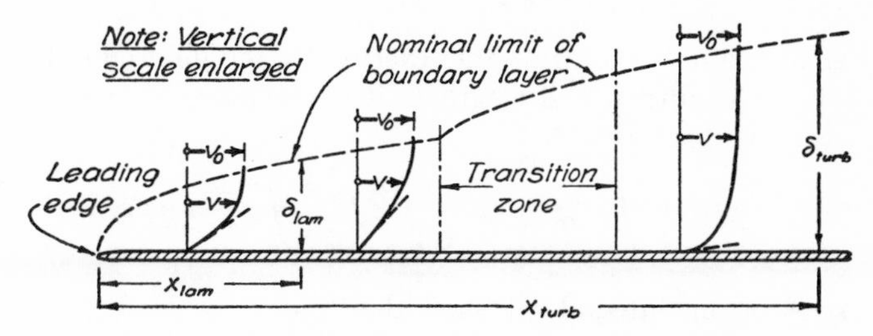

FIG. 96. Development of the boundary layer along a flat plate.

mentally that, so long as the flow in this region is laminar, the velocity distribution will be approximately parabolic and, roughly,

$$\frac{\delta}{x} = \frac{5}{\sqrt{v_0x/\nu}} = \frac{5}{\mathbf{R}^{1/2}} \tag{136}$$

Evidently, for a given distance x from the leading edge of the plate, the nominal boundary-layer thickness will decrease with increasing velocity, and will increase with increasing kinematic viscosity. Similarly, for a given fluid, a given plate, and a given velocity of flow, the nominal boundary-layer thickness will increase with increasing distance x from the leading edge—much as shown at the left of Fig. 96.

If the velocity distribution is to remain approximately parabolic in form while the thickness of the boundary layer increases, it is apparent that the velocity gradient at the plate will accordingly decrease in magnitude, as in the two curves at the left of Fig. 96. From the relationship $\tau = \mu\, dv/dy$ it follows that the intensity of shear τ_0 at the surface of the plate will also be reduced as the distance x increases. In fact,

$$\tau_0 = \mu\left(\frac{dv}{dy}\right)_{y=0} = \mu\left(C\frac{v_0}{\delta}\right) = \frac{\mu C v_0 \sqrt{v_0x/\nu}}{5x} = \frac{2C}{5\sqrt{v_0x/\nu}}\,\frac{\rho {v_0}^2}{2}$$

from which it will be seen that τ_0 may be expressed in terms of the familiar quantity $\rho v_0^2/2$ and a *local drag coefficient* c_f; that is,

$$\tau_0 = c_f \frac{\rho v_0^2}{2} \tag{137}$$

in which c_f evidently consists only of the numerical constant $2C/5$ and the Reynolds number. The constant has been shown both analytically and experimentally to have the magnitude 0.664, whence

$$c_f = \frac{0.664}{\mathbf{R}^{1/2}} \tag{138}$$

In order to evaluate the total drag exerted by the fluid on either side of a plate of breadth B and length L, it is necessary to integrate Eq. (137) over the distance $x = L$:

$$F = B\int_0^L \tau_0\, dx = \frac{0.664\mathrm{B}}{\sqrt{v_0/\nu}} \frac{\rho v_0^2}{2} \int_0^L x^{-1/2}\, dx$$

Evaluation of this integral will show that

$$F = C_f BL \frac{\rho v_0^2}{2} \tag{139}$$

in which C_f is a *mean drag coefficient* which, like c_f, depends only upon the Reynolds number (written now in terms of the length L), as plotted in Fig. 97:

$$C_f = \frac{1.328}{\mathbf{R}^{1/2}} \tag{140}$$

Attention might well be called at this point to the structure of the coefficients c_f and C_f. From Eqs. (137) and (139) it will be seen that

$$c_f = \frac{\tau_0}{\rho v_0^2/2} \quad \text{and} \quad C_f = \frac{F/BL}{\rho v_0^2/2}$$

Comparison with Section 13 will show that these quantities, which are similar in structure to the Euler number, have the same significance in problems of tangential stress as the Euler number has in problems of normal stress. In the following section the two types of stress will be found to be interdependent. In the present case, however, the extreme thinness of the boundary layer results in essential constancy of the normal stress—i.e., the pressure intensity—at all points. Flow past a thin plate therefore represents one of the very few problems in which the Euler number as such is of no significance.

Characteristics of the turbulent boundary layer. As in the case of flow through circular tubes, flow in a laminar boundary layer will eventually become unstable as the Reynolds number (more specifically, the local stability parameter $\chi = y^2(dv/dy)/\nu$) is increased. From the composition of the number $\mathbf{R} = v_0x/\nu$, it will be seen that an increase in the magnitude of **R** might represent a decrease in viscosity, an increase in velocity, or an increase in distance from the leading edge. In other words, if the plate is sufficiently long, at some point (see Fig. 96) the laminar flow may be expected to develop eddies which rapidly

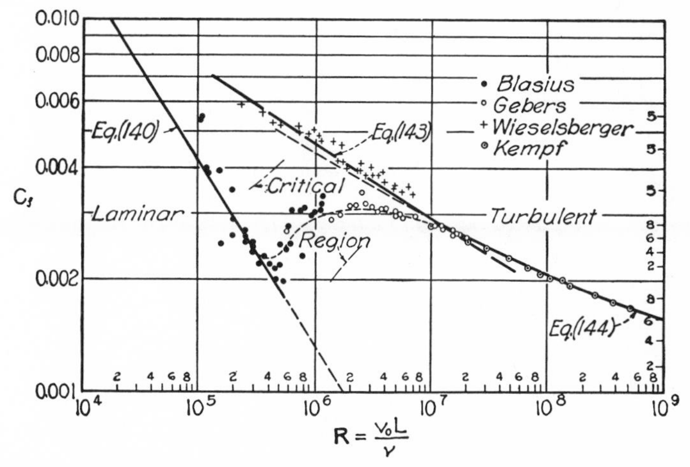

FIG. 97. The mean drag coefficient for a plane boundary as a function of the Reynolds number.

give rise to turbulence within the boundary layer, the critical Reynolds number having the approximate magnitude of 400,000 but varying considerably with the state of the oncoming flow and the condition of the leading edge of the boundary. If the total boundary length is not very great, the combined drag of the laminar and turbulent zones must be evaluated from an empirical transition curve such as that shown in Fig. 97. If, on the other hand, the boundary is very long, the length of the laminar layer may be neglected, and the drag computed as though the layer became turbulent at the outset.

The presence of turbulence in the boundary layer not only increases the rapidity with which the zone of the disturbance expands but also considerably modifies the velocity distribution and the surface drag. In other words, the resulting process of mixing tends to make the velocity more uniform throughout the greater part of the layer, yet at the same time produces a very rapid change in velocity near the wall. Indeed, if the wall is very smooth, laminar motion will still persist over a small zone known as the *laminar sublayer*, as shown in Fig. 98*b*, the velocity gradient at the boundary determining the surface drag just as in the foregoing case. The resulting conditions are approximately described by the following relationships for Reynolds

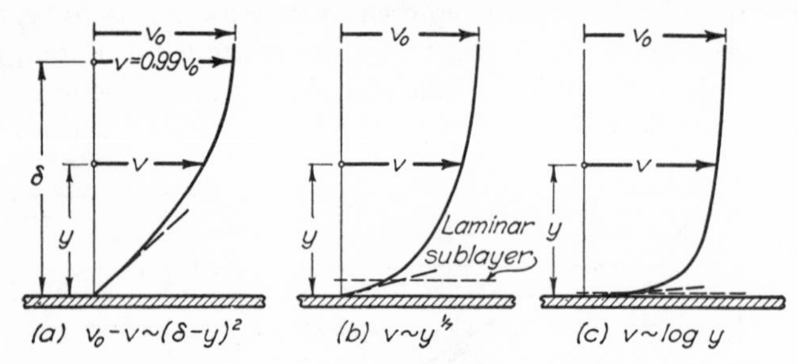

FIG. 98. Distribution of velocity in laminar and turbulent boundary layers.

numbers as high as 20,000,000 (refer to Fig. 97), corresponding to the seventh-root variation of the velocity indicated in Fig. 98*b*:

$$\frac{\delta}{x} = \frac{0.377}{\mathbf{R}^{1/5}} \tag{141}$$

$$c_f = \frac{0.059}{\mathbf{R}^{1/5}} \tag{142}$$

$$C_f = \frac{0.074}{\mathbf{R}^{1/5}} \tag{143}$$

The local and mean coefficients of drag have the same significance as in Eqs. (137) and (139). Beyond the limit of validity of these relationships ($\mathbf{R}$ = 20,000,000), a more refined method of analysis, based on the logarithmic velocity distribution of Fig. 98*c*, yields for C_f the Kármán-Schoenherr equation

$$\frac{1}{\sqrt{C_f}} = 4.13 \log_{10} (\mathbf{R}C_f) \tag{144}$$

which is plotted at the right of Fig. 97. Since the basis of this analysis is essentially the same as that of the following sections, further comment will not be made at this point, except to note that von Kármán, once a student of Prandtl's, is now a leader in the field of aerodynamics in this country, as is Schoenherr in the field of naval architecture.

While the simple illustration of flow along a flat plate has been used to introduce the boundary-layer concept, it must be realized that boundary-layer phenomena play an important role in practically every case of flow transition from one boundary form to another. As indicated in Fig. 99*a*, for example, a laminar boundary layer begins to develop at the entrance—or "leading edge"—of the Reynolds demonstration tube, where the velocity is essentially constant from wall to wall, gradually expanding until the layers from opposite sides approach the center—approximately at the section

$$\frac{x}{D} = 0.06 \frac{VD}{\nu} \tag{145}$$

Only from this section on is the flow nearly uniform, with the parabolic velocity curve of established laminar motion. If the Reynolds

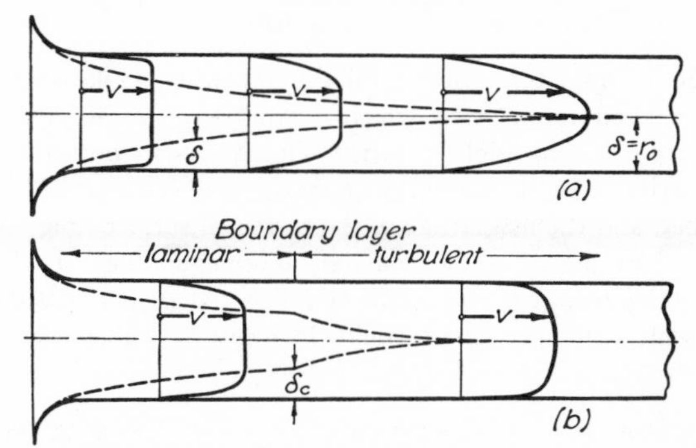

FIG. 99. Development of the boundary layer in a circular pipe.

number is sufficiently high, however, the laminar flow in the boundary layer will become unstable before this section is reached (Fig. 99*b*), further growth taking place in accordance with the principles of the turbulent boundary layer; under such conditions the ultimately uniform flow (beyond $x/D \approx 50$) will be turbulent, except for the thin laminar sublayer along the walls.

Example 35. A thin plate 3 feet wide and 7 feet long is towed through water ($\nu = 1.6 \times 10^{-5}$ square feet per second) at a velocity of 5 feet per second. Compute the drag on the two sides of the plate, assuming that (*a*) the boundary layer remains laminar, (*b*) the boundary layer becomes unstable at $\mathbf{R} = 400{,}000$ and (*c*) the boundary layer becomes turbulent at the leading edge.

The Reynolds number for the distance $x = L$ will be

$$\mathbf{R} = \frac{v_0 L}{\nu} = \frac{5 \times 7}{1.6 \times 10^{-5}} = 2.19 \times 10^6$$

From Eq. (139)

$$F = 2\left(C_f B L \frac{\rho v_0^2}{2}\right) = 2 \times C_f \times 3 \times 7 \times \frac{1.94 \times 5^2}{2} = 1020 C_f$$

Taking values of C_f from Fig. 97 for the three given conditions,

(a) $$F = 1020 \times 0.0009 = 0.92 \text{ lb}$$

(b) $$F = 1020 \times 0.0032 = 3.26 \text{ lb}$$

(c) $$F = 1020 \times 0.004 = 4.08 \text{ lb}$$

Of further interest is a comparison of the boundary-layer thickness at the end of the plate under conditions (a) and (c) as computed from Eqs. (136) and (141):

(a) $$\delta_{\max} = \frac{5L}{\mathbf{R}^{1/2}} = \frac{5 \times 7}{(2.19 \times 10^6)^{1/2}} = 0.024 \text{ ft}$$

(c) $$\delta_{\max} = \frac{0.377L}{\mathbf{R}^{1/5}} = \frac{0.377 \times 7}{(2.19 \times 10^6)^{1/5}} = 0.143 \text{ ft}$$

Example 36. Drag tests are made in a towing tank on a 1:25 scale model of a proposed ocean vessel which is expected to attain a peak speed of 16 knots. Since it is impossible to satisfy simultaneously both the Froude and the Reynolds criteria for similarity, the tests are made at identical values of the Froude number, and the magnitude of the corrected surface drag for the prototype is deduced by means of the boundary-layer equations. If the prototype has a length of 300 feet at the waterline and a wetted area of 18,000 square feet, to what full-scale peak-speed drag would the measured value of 3.45 pounds at model scale correspond?

Assuming that $\rho_m = 1.94$ slugs per cubic foot, $\rho_p = 1.99$ slugs per cubic foot, $\nu_m = 1.2 \times 10^{-5}$ square foot per second, and $\nu_p = 1.3 \times 10^{-5}$ square foot per second, and noting that 16 knots represents a velocity of $1.69 \times 16 = 27$ feet per second,

$$\mathbf{F}_p = \mathbf{F}_m = \frac{27}{\sqrt{32.2 \times 300}} = \frac{v_m}{\sqrt{32.2 \times \frac{300}{25}}}; \quad v_m = \frac{27}{\sqrt{25}} = 5.40 \text{ fps}$$

$$\mathbf{R}_p = \frac{27 \times 300}{1.3 \times 10^{-5}} = 6.23 \times 10^8$$

$$\mathbf{R}_m = \frac{5.40 \times \frac{300}{25}}{1.2 \times 10^{-5}} = 5.40 \times 10^6$$

From Fig. 97 or Eq. (143), the value of C_f corresponding to $\mathbf{R}_m$ is found to be 0.0033, for which the surface drag becomes, according to Eq. (139),

$$(F_s)_m = C_f A \frac{\rho v^2}{2} = 0.0033 \times \frac{18{,}000}{\overline{25^2}} \times \frac{1.94 \times \overline{5.40^2}}{2} = 2.69 \text{ lb}$$

Assuming, then, that the resistance due to the waves is the difference between the total drag and the surface drag, the force required by the model to overcome the wave resistance is $3.45 - 2.69 = 0.76$ pound. The corresponding force for the prototype (see Example 21) should then be

$$(F_w)_p = (F_w)_m \frac{\rho_p}{\rho_m} \left(\frac{L_p}{L_m}\right)^3 = 0.76 \times \frac{1.99}{1.94} \times \overline{25^3} = 12{,}200 \text{ lb}$$

But for the corresponding value of $\mathbf{R}_p$, $C_f = 0.00166$, whence the prototype surface drag becomes

$$(F_s)_p = 0.00166 \times 18{,}000 \times \frac{1.99 \times \overline{27}^2}{2} = 21{,}700 \text{ lb}$$

The total drag of the prototype should therefore be

$$F_p = (F_s)_p + (F_w)_p = 21{,}700 + 12{,}200 = 33{,}900 \text{ lb}$$

PROBLEMS

218. Plot to scale the variation in τ_0 and δ along the plate of Example 35 for conditions (*a*) and (*c*), and sketch the approximate forms of the corresponding curves or condition (*b*).

219. If a plate 5 feet square is towed through water at a constant velocity of 3 feet per second, it is found to attain a maximum drag of 253 pounds when held at right angles to its direction of motion. What fraction of this drag would be encountered when the plate is towed edgewise at the same velocity? (Assume a water temperature of 60° Fahrenheit.)

220. Estimate the surface drag encountered by the hull of a submarine when traveling fully submerged at a speed of 8 knots, if the length of the hull is 200 feet and its surface area 18,000 square feet; use, as a first approximation, the density and viscosity of fresh water at 50° Fahrenheit.

221. The hull of a dirigible has a length of 300 feet and a surface area of 60,000 square feet. Estimate the surface drag on the hull when its speed is 50 miles per hour in air at a temperature of 40° Fahrenheit.

222. A streamlined train is 750 feet long, a typical cross section having a perimeter of 25 feet above the wheels. Evaluate the approximate surface drag of the train when its speed is 60 miles per hour and the temperature of the atmosphere is 70° Fahrenheit.

223. If the wetted area of the model ship of Problem 127 is 8 square feet, what surface drag would be encountered at the given speed in fresh water at 60° Fahrenheit? What total drag would be encountered by the prototype, assuming the salt-water characteristics given in Example 36?

224. Oil having a kinematic viscosity of 5×10^{-4} square foot per second is to be discharged from a large pressure tank through a 1-inch pipe, and the rate of discharge is to be determined from piezometer readings made in the zone of uniform motion well beyond the inlet. If the inlet is rounded to prevent disturbances due to separation, how far downstream may the flow be assumed to have become truly uniform (*a*) when $V = 5$ feet per second, (*b*) when $V = 10$ feet per second, and (*c*) when $V = 20$ feet per second? How far from the inlet should the first piezometer connection be placed?

33. VELOCITY DISTRIBUTION NEAR SMOOTH AND ROUGH BOUNDARIES

Velocity variation with distance from the boundary. In the case of laminar flow near a plane boundary, it was found in Chapter VI that an expression for the velocity distribution could be obtained by integration of the following form of Eq. (110):

$$\frac{dv}{dy} = \frac{\tau}{\mu} \tag{146}$$

From the parallel relationship for turbulent flow [see Eq. (133)]

$$\frac{d\bar{v}}{dy} = \frac{\bar{\tau}}{\eta} \tag{147}$$

one might therefore seek to obtain a corresponding relationship for the velocity distribution in the case of turbulence. During the integration of Eq. (146), however, the molecular viscosity μ could be considered constant for a given state of flow, while in Eq. (147) the eddy viscosity is known to vary with distance from the boundary (see Fig. 94). Fortunately, experiments indicate that in turbulent flow the intensity of shear and the eddy viscosity vary in such a manner that their ratio is, as a first approximation, directly proportional to $\sqrt{\tau_0/\rho}$ (a parameter having the dimension L/T and known as the *shear velocity*), and inversely proportional to the distance y from the boundary. That is, henceforth omitting the bars denoting temporal means, Eq. (147) may be rewritten in the readily integrable form

$$\frac{dv}{dy} = \frac{\tau}{\eta} = \frac{2.5}{y}\sqrt{\frac{\tau_0}{\rho}}$$

in which the factor 2.5 is the constant of proportionality determined from experiment.

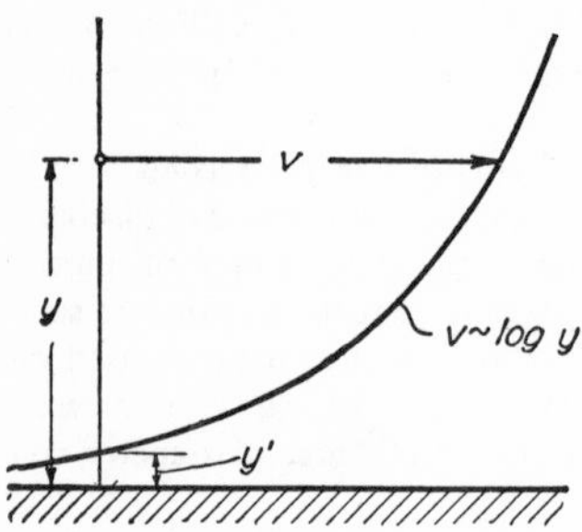

FIG. 100. Characteristics of the logarithmic velocity curve.

Integration of this expression leads at once to the relationship

$$v = 2.5\sqrt{\frac{\tau_0}{\rho}}\log_e y + C$$

which states that in turbulent flow the velocity will vary directly with the logarithm of the distance from the boundary, as shown in Fig. 100. From reference to this diagram it will be seen that the constant of integration may be evaluated from the condition that $y = y'$ when $v = 0$, whence $C = -2.5\sqrt{\tau_0/\rho}\log_e y'$. Introducing this value and dividing by $\sqrt{\tau_0/\rho}$, the ratio of the local velocity of flow to the shear velocity will be found to depend only upon the ratio y/y':

$$\frac{v}{\sqrt{\tau_0/\rho}} = 2.5 \log_e \frac{y}{y'} = 5.75 \log_{10} \frac{y}{y'} \tag{148}$$

Derivation of the Kármán-Prandtl velocity equations. At first glance, the inherent tendency of the logarithmic curve to indicate a zero velocity a finite distance from the boundary would appear to be in disagreement with physical fact. It must be recalled, however, that the flow in the immediate vicinity of a smooth boundary is invariably laminar. Inspection of Fig. 101 will show, moreover, that this zone of laminar motion must extend well beyond the distance y' if a smooth transition is to exist between the velocity curves for the laminar and turbulent zones. Indeed, if one arbitrarily selects the intersection of the parabolic and the logarithmic curves as the nominal borderline $y = \delta'$ between the two types of motion (δ' then representing the thickness of the laminar sublayer), the parameters y' and δ' may reasonably be expected to be interdependent.

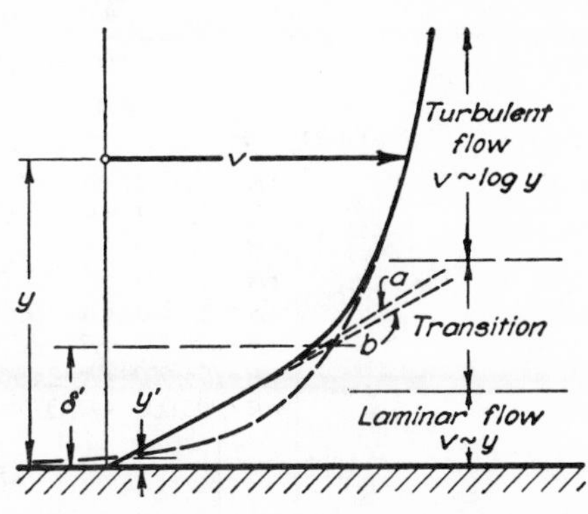

FIG. 101. Definition sketch for evaluating the relative thickness of the laminar sublayer.

Since the distance $y = \delta'$ marks the stability limit of the laminar sublayer, it should, according to Section 30, be represented by a constant magnitude of the stability parameter $\chi = y^2\,(dv/dy)/\nu$. Replacing, for simplicity, the parabolic segment a in Fig. 101 by the straight line b, which it very closely approximates in this zone, Eq. (146) takes the form

$$\frac{dv}{dy} = \frac{\tau_0}{\mu}$$

whence

$$\chi = \frac{y^2\,dv/dy}{\nu} = \frac{\delta'^2\tau_0}{\nu\mu} = C$$

or

$$\delta' = \frac{C^{1/2}\nu}{\sqrt{\tau_0/\rho}} = C'y'$$

That this hypothesis is correct has been demonstrated by various experimental studies, in particular by those of Prandtl's student Nikuradse on closed conduits. The latter yield, for the constants C and C', the values 135 and 107, whereupon

$$\delta' = \frac{11.6\nu}{\sqrt{\tau_0/\rho}} \tag{149}$$

and

$$y' = \frac{\delta'}{107} = \frac{0.108\nu}{\sqrt{\tau_0/\rho}}$$

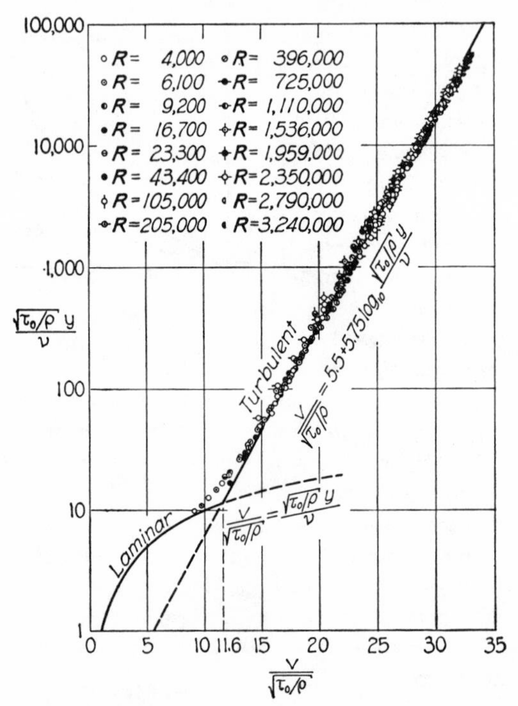

FIG. 102. Experimental verification of the Kármán-Prandtl equation for the velocity distribution near a smooth boundary.

Introduction of the latter expression in Eq. (148) then results in what is known as the Kármán-Prandtl equation for the velocity distribution in turbulent flow near smooth boundaries,

$$\frac{v}{\sqrt{\tau_0/\rho}} = 5.75 \log_{10} \frac{\sqrt{\tau_0/\rho}\, y}{\nu} + 5.5 \tag{150}$$

while in the laminar sublayer

$$\frac{v}{\sqrt{\tau_0/\rho}} = \frac{\sqrt{\tau_0/\rho}\, y}{\nu} \tag{151}$$

These equations are plotted in Fig. 102, the semi-logarithmic form being used both to show the nature of Eq. (150) and to give comparable emphasis to all zones of flow. The experimental data of Nikuradse, which include measurements from the boundary layer to the centerline of a series of circular pipes at various Reynolds numbers, are seen to follow closely the trend of the two equations.

In the case of boundary roughness (see Fig. 103), it is hardly to be presumed that a laminar sublayer will exist at the boundary if the

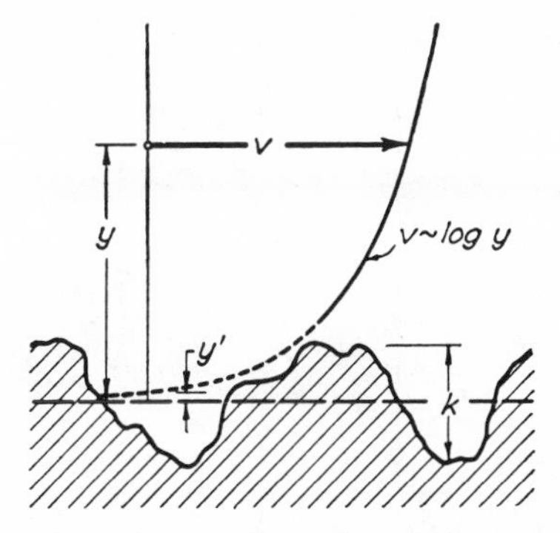

FIG. 103. Definition sketch for evaluating the relative magnitude of boundary roughness.

roughness magnitude k is actually greater than the value of δ' computed from Eq. (149). On the other hand, one might now reasonably expect the parameter y' to be directly proportional to k. As a matter of fact, experiments by Nikuradse and others, using pipes artificially roughened by cemented coatings of sand grains of diameter k, indicate that if $k > 10\delta'$

$$y' = \frac{k}{30}$$

Introduction of this value in Eq. (148) then leads to the following counterpart of Eq. (150)—the Kármán-Prandtl equation for the velocity distribution in turbulent flow near rough boundaries,

$$\frac{v}{\sqrt{\tau_0/\rho}} = 5.75 \log_{10} \frac{y}{k} + 8.5 \tag{152}$$

which is plotted in Fig. 104, together with Nikuradse's test data. It is again to be noted that the experimental points represent many series of measurements extending from very near the boundary to the centerline of the pipe.

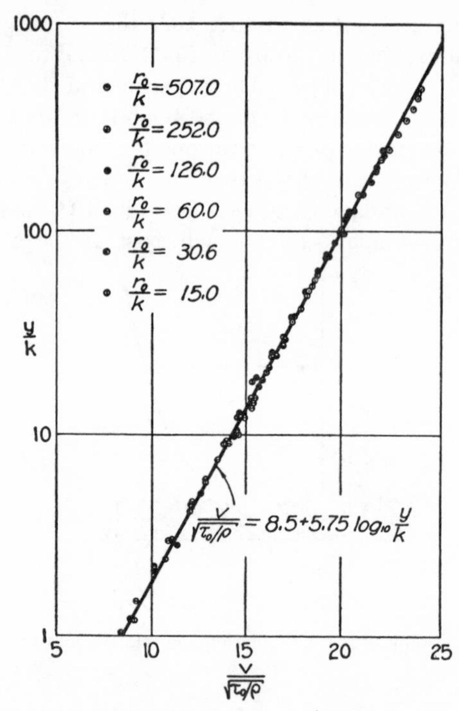

FIG. 104. Experimental verification of the Kármán-Prandtl equation for the velocity distribution near a rough boundary.

The mean velocity as a common denominator. Although the logarithmic curve has an infinite slope at the limit $y = 0$ and continues to have a finite slope for all values of y however great, such lack of agreement with actual conditions both at the boundary and very far away will introduce relatively little error if Eqs. (150) and (152) are integrated across a normal section to determine the corresponding rate of flow. Thus, from Eq. (6), for the case of a circular tube

$$Q = \int_0^{r_0} 2\pi r\, v\, dr = 2\pi\sqrt{\tau_0/\rho}\int_0^{r_0} r\left(2.5\log_e\frac{\sqrt{\tau_0/\rho}\,(r_0 - r)}{\nu} + 5.5\right)dr$$

$$= \pi r_0^2\sqrt{\tau_0/\rho}\left(5.75\log_{10}\frac{\sqrt{\tau_0/\rho}\,r_0}{\nu} + 1.75\right)$$

The ratio of the mean velocity $V = Q/A$ to the shear velocity then becomes

$$\frac{V}{\sqrt{\tau_0/\rho}} = 5.75 \log_{10} \frac{\sqrt{\tau_0/\rho}\, r_0}{\nu} + 1.75 \tag{153}$$

A similar procedure for the case of rough pipes will be found to yield the relationship

$$\frac{V}{\sqrt{\tau_0/\rho}} = 5.75 \log_{10} \frac{r_0}{k} + 4.75 \tag{154}$$

If, now, Eq. (153) is subtracted from Eq. (150), or Eq. (154) from Eq. (152), there will result identical expressions having the form

$$\frac{v - V}{\sqrt{\tau_0/\rho}} = 5.75 \log_{10} \frac{y}{r_0} + 3.75$$

In other words, when referred to the mean velocity the Kármán-Prandtl expressions for the velocity distribution near smooth and rough boundaries become identical.

That the velocity curve in reality varies only with conditions of boundary drag may be seen by introducing a coefficient of surface resistance f similar to the factor c_f of the foregoing section; that is, rewriting the shear velocity in the form $\sqrt{\tau_0/\rho} = V\sqrt{f/8}$ (which will be discussed in the following pages),

$$\frac{v - V}{V\sqrt{f}} = 2 \log_{10} \frac{y}{r_0} + 1.32$$

which indicates that the relative velocity distribution in a circular pipe is a function only of y/r_0 for a given value of the resistance coefficient f, regardless of whether the boundary is rough or smooth.

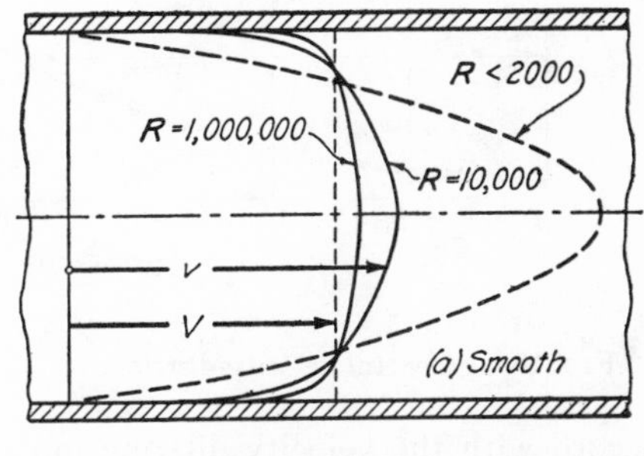

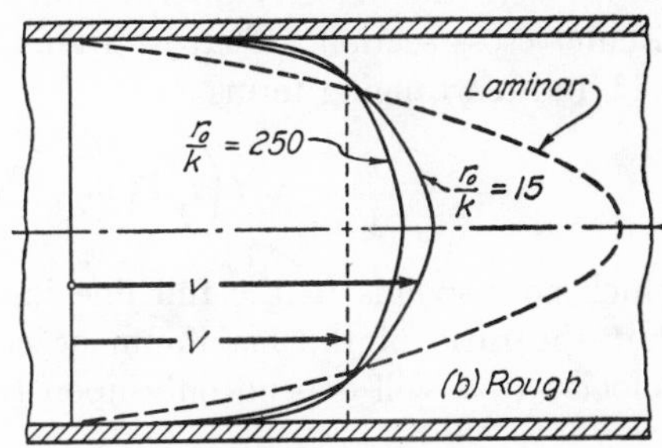

FIG. 105. Typical curves of velocity distribution in smooth and rough pipes.

The two families of distribution curves of Fig. 105 should therefore reduce to a single curve (shown as a broken line in Fig. 106) when plotted according to this equation. The slight discrepancy between the curve and typical experimental values taken from Figs. 102 and 104 is seen to be surprisingly small when one recalls that a relationship derived specifically for the boundary vicinity [Eq. (148)] has now been applied to the entire zone of flow. As a matter of fact, only a small adjustment in numerical coefficients will yield an expression which is in close

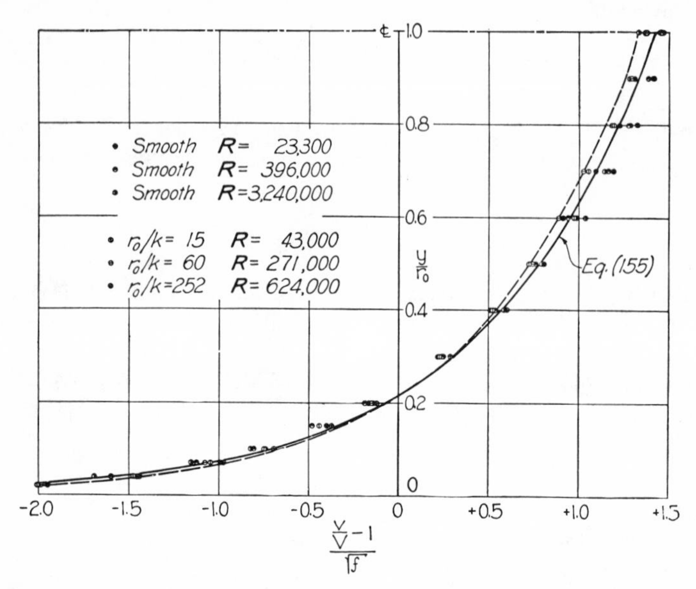

FIG. 106. General velocity-distribution diagram for smooth and rough pipes.

accord with the velocity distribution for uniform flow in any pipe of circular cross section. That is, after changing 2 and 1.32 to 2.15 and 1.43 and rearranging terms,

$$\frac{v}{V} = \sqrt{f}\left(2.15 \log_{10} \frac{y}{r_0} + 1.43\right) + 1 \tag{155}$$

which corresponds to the full line in Fig. 106. It follows herefrom that the ratio of the maximum or centerline velocity to the mean velocity Q/A will depend only upon f:

$$\frac{v_{\max}}{V} = 1.43\sqrt{f} + 1 \tag{156}$$

Although the experimental evidence upon which the foregoing discussion is based is taken largely from studies on circular conduits, the fact must be emphasized that the logarithmic relationships for velocity distribution are by no means limited to such boundary conditions. Indeed, it is only when integrated over different forms of cross section that the relationships may be expected to vary. In the case of flow at the uniform depth y_0 in a very wide channel, for instance, it would be found that

$$\frac{v - V}{V\sqrt{f}} = 2 \log_{10} \frac{y}{y_0} + 0.88 \tag{157}$$

This differs from the corresponding equation for circular conduits only in the factor 0.88, which evidently embodies the effect of cross-sectional form.

Example 37. Measurements with a Pitot tube in a 6-foot water main provided the following velocities:

r in ft	v in fps
0.00	6.56
0.60	6.45
1.20	6.13
1.80	5.70
2.40	4.96
2.70	4.42

Determine the corresponding magnitudes of V and f.

If the foregoing measurements were sufficiently accurate, the required values of V and f could be obtained through simultaneous solution of Eq. (155) written for any two points. Since individual field measurements are seldom exact, however, it is advisable to plot the measured velocities against y/r_0 on semilogarithmic paper, as shown in the accompanying diagram, and to determine V and f from the characteristics of the straight line drawn through the plotted points.

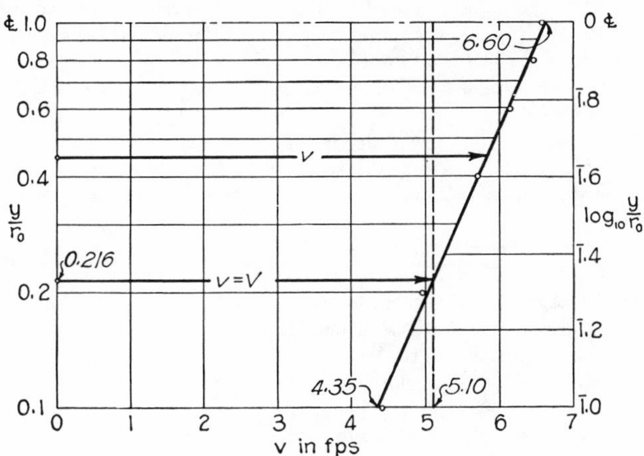

Equation (155), when rewritten as follows,

$$v = 2.15\, V\sqrt{f}\, \log_{10} \frac{y}{r_0} + 1.43\, V\sqrt{f} + V$$

is of the form $x = ay + b$, in which $a = 2.15\, V\sqrt{f}$ and $b = 1.43\, V\sqrt{f} + V$. The coefficient a evidently corresponds to the slope of the straight line in the diagram (i.e., the change in velocity per unit change in $\log_{10} y/r_0$), and the coefficient b to the velocity intercept at $\log_{10} y/r_0 = 0$. Since it is seen from the diagram that $a = 6.60 - 4.35 = 2.25$ fps and $b = 6.60$ fps,

$$2.15\, V\sqrt{f} = 2.25 \qquad \text{and} \qquad 1.43\, V\sqrt{f} + V = 6.60$$

Simultaneous solution of these equations then yields the results desired:

$$V = 5.10 \text{ fps} \qquad \text{and} \qquad f = 0.042$$

It might be noted in passing that the velocity-distribution curve corresponding to Eq. (155) passes through the point ($v/V = 1.0$, $y/r_0 = 0.216$); as a result, the mean velocity $V = 5.10$ fps may be read directly from the diagram at the ordinate $y/r_0 = 0.216$.

PROBLEMS

225. Wind-velocity measurements at heights of 5 feet and 10 feet above the ground in a level meadow yielded values of 8 and 9 feet per second, respectively. What is the corresponding magnitude of the parameter y'? What velocity should prevail at a height of 50 feet under the same conditions?

226. Oil having a kinematic viscosity of 5×10^{-4} square foot per second and a specific gravity of 0.9 flows at the rate of 0.5 cubic foot per second through a smooth 4-inch pipe. What is the nominal thickness of the laminar sublayer? What is the velocity at the distance δ' from the pipe wall?

227. The velocity of flow in a badly corroded 3-inch pipe is found to increase 15 per cent as a Pitot tube is moved from a point ½ inch from the wall to a point 1 inch from the wall. Estimate the mean height of the corrosion tubercles.

228. Determine, for the conditions of Example 37, the following: (*a*) the magnitude of y'; (b) the corresponding magnitudes of δ' and k; (*c*) the magnitude of δ' corresponding to Eq. (149). Is the pipe rough or smooth?

229. Measurements in a very wide river at a section having a depth of 10 feet yield a surface velocity of 8 feet per second when the rate of flow is 60 cubic feet per second per foot width. What intensity of bed shear is indicated by these measurements?

230. In order to determine the rate of flow of air through a ventilating duct of circular cross section, Pitot tubes are installed at the midpoint and at one quarter point of the 7-foot diameter. What rate of flow is indicated by simultaneous readings of 18 and 15 feet per second at the two points?

231. In the determination of the mean velocity of flow in wide rivers, it is common practice to estimate the mean velocity in a vertical section by making a single measurement at a distance below the surface equal to six-tenths of the depth. Assuming Eq. (157) to represent the true distribution in the vertical, determine the actual distance below the surface at which $v = V$.

34. RESISTANCE OF SMOOTH AND ARTIFICIALLY ROUGHENED PIPE

The Darcy-Weisbach resistance coefficient. Since uniform flow in a conduit may be regarded as a limiting case of boundary-layer development, the intensity of shear at the boundary of a conduit should be expressible in a relationship analogous to Eq. (137) of Section 32. That is, replacing v_0 by V,

$$\tau_0 = c_f \frac{\rho V^2}{2} = \frac{f}{4} \frac{\rho V^2}{2} \tag{158}$$

in which f, the *Darcy-Weisbach resistance coefficient* used in conduit analysis, is simply a multiple of c_f. In the case of steady flow along a flat plate, the intensity of boundary shear is determined from the longitudinal force required to hold the plate in position. In the case of steady flow through a conduit, on the other hand, τ_0 must be evaluated from the force which is required to maintain the flow. For instance, since the boundary shear in a horizontal pipe is necessarily in equilibrium with the force due to the pressure gradient, the product of the intensity of shear and the boundary area in a given length of pipe should equal the product of the cross-sectional area and the reduction in pressure intensity in that distance:

$$\tau_0 \pi D L = -\frac{dp}{dx} L \frac{\pi D^2}{4}$$

Substituting $p_1 - p_2$ for $(-dp/dx)L$ and solving for τ_0,

$$\tau_0 = (p_1 - p_2) \frac{D}{4L}$$

Upon introducing the latter value in Eq. (158), one obtains the following basic equation for pipe resistance:

$$p_1 - p_2 = f \frac{L}{D} \frac{\rho V^2}{2} \tag{159}$$

In the case of sloping pipes—in particular if the fluid in question is a liquid and the resistance is evaluated in terms of change in piezometric head—comparison with Section 28 will show that Eq. (159) may be written in the form

$$h_f = f \frac{L}{D} \frac{V^2}{2g} \tag{160}$$

in which h_f represents the decrease in piezometric head over the distance L. It still remains, of course, to evaluate the resistance coefficient f in one or the other of these expressions.

Variation of f with the Reynolds number. If Eq. (159) is solved for f, the result will be seen to represent the product of a geometrical ratio and the reciprocal of the Euler number squared:

$$f = \frac{D}{L}\frac{\Delta p}{\rho V^2/2} = \frac{D}{L}\frac{1}{\mathbf{E}^2}$$

According to Eq. (131), if the boundary geometry is given, the Euler number for the flow of a viscous fluid will be a function of the Reynolds number; it should therefore follow that

$$f = \phi(\mathbf{R})$$

A similar operation on Eq. (160) will be found to yield the expression

$$f = \frac{D}{L}\frac{h_f}{V^2/2g}$$

While the product $V^2/2gh_f$ would appear to be a form of the Froude number, it is in reality merely the result of dividing numerator and denominator of the Euler number by γ. Indeed, since the only free surface present in such flow is that in open piezometer columns, gravity can have no influence whatever upon the flow pattern, and the Froude number therefore has no bearing upon this phase of boundary resistance. In other words, if experimental values of f are determined from either of the foregoing equations and plotted against the corresponding values of $\mathbf{R} = VD/\nu$, the resistance coefficient for flow in geometrically similar pipes should be found to depend only upon the Reynolds number.

The correctness of this conclusion will be seen from Fig. 107, which includes resistance measurements on smooth pipes over a very wide Reynolds-number range. At values of $\mathbf{R}$ below 2000, the points will be seen to follow, on a logarithmic plot, the straight line

$$f = \frac{64}{\mathbf{R}} \tag{161}$$

That this corresponds to the equation of Poiseuille will be apparent from the following amplification of Eq. (121):

$$p_1 - p_2 = \frac{32\mu VL}{D^2} \times \frac{2\rho V}{2\rho V} = \frac{64\mu}{VD\rho}\frac{L}{D}\frac{\rho V^2}{2}$$

To the right of the approximate critical limit $\mathbf{R} = 2000$, the empirical relationship of Blasius,

$$f = \frac{0.316}{\mathbf{R}^{1/4}} \tag{162}$$

(which corresponds to Eq. (142) for the boundary layer), is seen to indicate the experimental trend as high as **R** = 100,000, beyond which a deviation from the straight line is quite apparent. However large this deviation may ultimately become, nevertheless, it appears from experimental evidence now available that the limit $f = 0$ would be approached asymptotically as **R** became infinitely great. Since

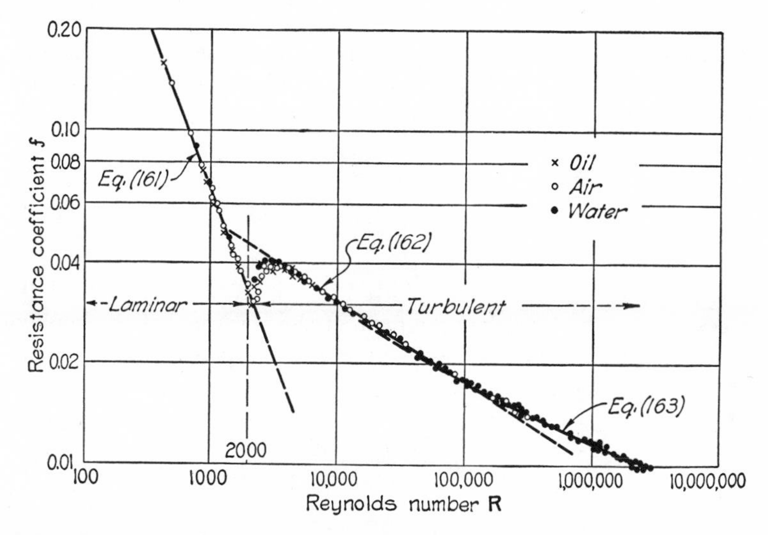

FIG. 107. Variation of the resistance coefficient with the Reynolds number for smooth pipes.

the limit $\mathbf{R} = \infty$ corresponds to a zero fluid viscosity, and since $f = 1/\mathbf{E}^2$, the hypothesis that **E** should become independent of **R** at this limit appears fully justified. It now remains to seek a rational expression for f which will embody this function.

Derivation of the Kármán-Prandtl resistance equations. Since both the velocity distribution and the pressure gradient are related to the boundary shear, it should be possible to derive an expression for f in terms of the same parameter which was found to determine the velocity distribution near a smooth boundary—namely, the relative thickness of the laminar boundary film. Thus, from Eq. (158) for τ_0, it is evident that the shear velocity may be written in the form

$$\sqrt{\frac{\tau_0}{\rho}} = V\sqrt{\frac{f}{8}}$$

introduction of which in Eq. (153) at once yields

$$\frac{V}{V\sqrt{f}} = \frac{1}{\sqrt{8}}\left(5.75 \log_{10} \frac{V\sqrt{f}\, r_0}{\sqrt{8}\,\nu} + 1.75\right)$$

whence

$$\frac{1}{\sqrt{f}} = 2.03 \log_{10}\left(\frac{VD}{\nu}\sqrt{f}\right) - 0.91$$

With a slight modification of numerical factors to conform with experimental measurement, this then becomes [compare with Eq. (144) for the boundary layer] the Kármán-Prandtl resistance equation for turbulent flow in smooth pipes,

$$\frac{1}{\sqrt{f}} = 2 \log_{10} \mathbf{R}\sqrt{f} - 0.8 \tag{163}$$

which is the third relationship plotted in Fig. 107. From Eq. (149), however, it will be seen that the ratio of the thickness of the laminar sublayer to the pipe radius will be inversely proportional to the quantity $\mathbf{R}\sqrt{f}$ appearing in the foregoing equation:

$$\frac{\delta'}{r_0} = \frac{11.6\nu}{V\sqrt{f/8}\, D/2} = \frac{65.6}{\mathbf{R}\sqrt{f}} \tag{164}$$

In other words, the resistance coefficient for smooth conduits is dependent only upon the relative thickness of the laminar sublayer.

Since the artificially roughened pipes used by Nikuradse in the velocity investigations already mentioned cannot be considered geometrically similar either to smooth pipes or to one another, it is to be expected that a plot of f against $\mathbf{R}$ for each particular value of the relative roughness k/r_0 would yield a different functional trend. In other words, if a smooth pipe is assumed to represent the limit $k/r_0 = 0$, a composite plot of data for various relative roughnesses should consist of a family of curves deviating systematically from that of Fig. 107 with increasing values of k/r_0. That such is the case is shown by Fig. 108, which includes Nikuradse's measured values on pipes covering a thirty-fold variation in the ratio of sand-grain diameter to pipe radius, together with the three curves of Fig. 107. The trend of the Poiseuille relationship for laminar flow is seen to be followed closely by all points, regardless of the relative roughness, below the critical limit $\mathbf{R} = 2000$, which indicates that boundary roughness has no influence upon the resistance to laminar flow. Beyond the critical limit, moreover, the curves continue to follow the smooth-boundary relationship to a Reynolds number which increases with increasing "relative smoothness" r_0/k. Nevertheless, each of the curves is seen eventually to become horizontal, whereafter no influence of the Reynolds number (i.e., of the fluid viscosity) appears to exist.

Since this horizontal limit evidently corresponds to conditions in which the velocity distribution depends only upon the boundary roughness, it would be reasonable to expect that an expression for the

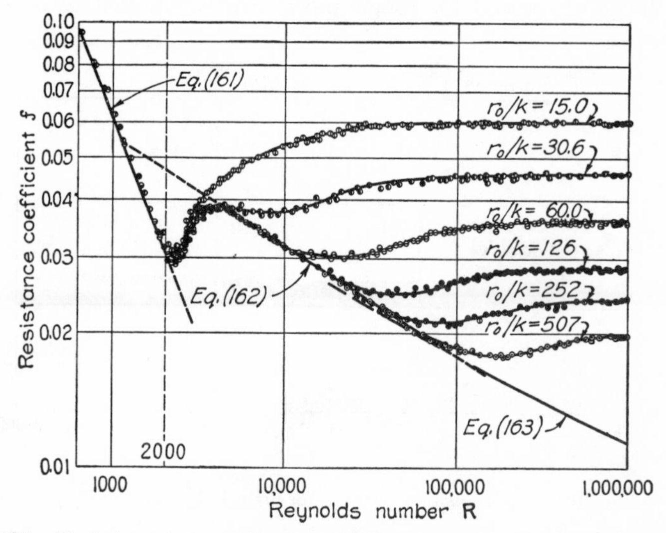

FIG. 108. Variation of the resistance coefficient with the Reynolds number for artificially roughened pipes.

ultimate resistance coefficient of rough pipes might be obtained from Eq. (154) through elimination of the shear velocity. Thus, since $\sqrt{\tau_0/\rho} = V\sqrt{f/8}$, this equation takes the form

$$\frac{V}{V\sqrt{f}} = \frac{1}{\sqrt{8}}\left(5.75 \log_{10} \frac{r_0}{k} + 4.75\right)$$

and

$$\frac{1}{\sqrt{f}} = 2.03 \log_{10} \frac{r_0}{k} + 1.68$$

With a slight modification of numerical values, this relationship will be found to indicate the limiting magnitude of f for each of the roughness curves of Fig. 108; that is, the Kármán-Prandtl resistance equation for turbulent flow in rough pipes becomes

$$\frac{1}{\sqrt{f}} = 2 \log_{10} \frac{r_0}{k} + 1.74 \qquad (165)$$

Boundary roughness versus the laminar sublayer. Since a pipe of any given roughness will evidently function as a smooth pipe at low Reynolds numbers and as a rough pipe at high Reynolds numbers, an intermediate zone must exist in which viscous and roughness effects together determine the boundary resistance. Inasmuch as the relative thickness of the laminar sublayer is the sole resistance criterion for smooth pipes and the relative magnitude of the boundary roughness the sole criterion for rough pipes, one is tempted to conclude

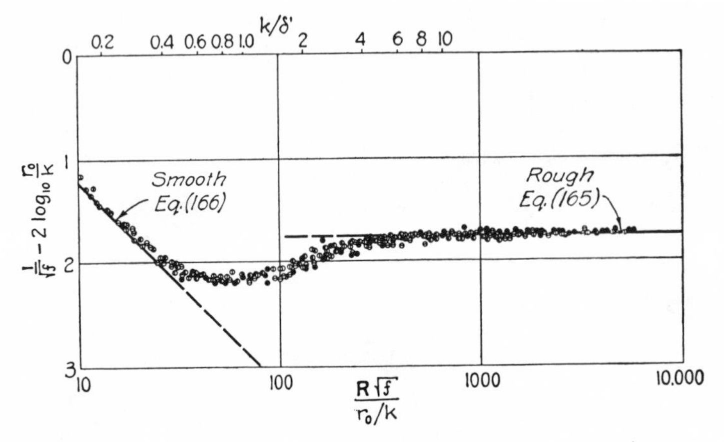

FIG. 109. The transition function for artificially roughened pipes.

that it must be the ratio of the magnitudes δ' and k which governs this transition region. In other words, so long as the boundary irregularities are wholly enclosed by the laminar boundary flow, they should remain ineffective; but once the laminar film becomes thin enough to be made unstable by the roughness disturbances, the resistance should rise accordingly. If it is recalled that δ'/r_0 is inversely proportional to $\mathbf{R}\sqrt{f}$, it will be seen that the quantity $(\mathbf{R}\sqrt{f})/(r_0/k)$ will be proportional to the ratio of the boundary roughness to the computed thickness of the laminar film. Each of the curves in Fig. 108 should therefore begin to deviate from the smooth-pipe relationship at the same value of k/δ', and should likewise approach its horizontal asymptote in the same systematic manner.

This hypothesis may be tested by superposing all curves on a diagram having as abscissa scale the parameter $(\mathbf{R}\sqrt{f})/(r_0/k)$, and as ordinate scale the quantity $1/\sqrt{f} - 2\log r_0/k$, which, according to Eq. (165), has the constant value 1.74 as each curve becomes horizontal. Such a diagram is shown in Fig. 109, which obviously displays a single functional trend from the smooth-pipe curve obtained from Eq. (163),

$$\frac{1}{\sqrt{f}} - 2 \log_{10} \frac{r_0}{k} = 2 \log_{10} \frac{\mathbf{R}\sqrt{f}}{r_0/k} - 0.8 \qquad (166)$$

to the rough-pipe asymptote 1.74. It is noteworthy that the transition begins when the computed thickness δ' of the laminar sublayer is approximately 4 times the roughness magnitude k, and ends when k is approximately 6 times the computed magnitude of δ'—i.e., when $(\mathbf{R}\sqrt{f})/(r_0/k) \approx 400$.

Example 38. What horsepower per mile of line would be required to maintain the flow in Example 37? What is the apparent relative roughness? What per cent error would have resulted if the power required for the given rate of flow had been estimated from the smooth-pipe equations of (*a*) Kármán-Prandtl, (*b*) Blasius, and (*c*) Poiseuille?

As determined in the foregoing example,

$$f = 0.042 \qquad \text{and} \qquad V = 5.10 \text{ fps}$$

Therefore, from Eq. (160), the drop in head per mile will be

$$h_f = f \frac{L}{D} \frac{V^2}{2g} = 0.042 \times \frac{5280}{6} \times \frac{\overline{5.10}^2}{2 \times 32.2} = 14.8 \text{ ft}$$

The required power is then

$$\frac{Q\gamma h_f}{550} = \frac{5.10 \times \frac{\pi}{4} \times 6^2 \times 62.4 \times 14.8}{550} = 242 \text{ hp}$$

Assuming that $\nu = 1.5 \times 10^{-5}$ square foot per second, the Reynolds number will be found to be

$$\mathbf{R} = \frac{VD}{\nu} = \frac{5.10 \times 6}{1.5 \times 10^{-5}} = 2{,}040{,}000$$

Inspection of Fig. 108 will show that flow at this Reynolds number with a resistance coefficient as great as $f = 0.042$ is not only well beyond the range of viscous influence but also high on the scale of relative roughness. In terms of Nikuradse's sand roughness, the corresponding magnitude of r_0/k may be computed from Eq. (165) as follows:

$$\frac{1}{\sqrt{0.042}} = 2 \log_{10} \frac{r_0}{k} + 1.74$$

$$\log_{10} \frac{r_0}{k} = \frac{4.88 - 1.74}{2} = 1.57 \qquad \text{and} \qquad \frac{r_0}{k} = 37$$

Evidently, if roughness equivalent to that of the water main were to be produced artificially, one would have to cement pebbles approximately 1 inch in diameter to the wall of the 6-foot pipe. Many years of service are generally required to produce such a degree of natural tuberculation.

From the head and power expressions it is apparent that the only variable in the third part of the question is the magnitude of f as computed from the three equations. As obtained from Fig. 108, or from the corresponding equations, $f_a = 0.01$, $f_b = 0.0084$, and $f_c = 3.1 \times 10^{-5}$. Therefore, the following errors will result:

(a)
$$\frac{0.01 - 0.042}{0.042} \times 100 = -76\%$$

(b)
$$\frac{0.0084 - 0.042}{0.042} \times 100 = -80\%$$

(c)
$$\frac{3.1 \times 10^{-5} - 0.042}{0.042} \times 100 \approx -100\%$$

PROBLEMS

232. If, on a very hot day, the increase in temperature caused the viscosity of the oil flowing in the smooth pipe of Problem 189 to decrease to one-tenth the value given, what would be the resulting difference in pressure intensity between points A and B for the same rate of flow?

233. An 8-inch smooth pipe 10,000 feet long transmits per minute 600 gallons of oil having a specific gravity of 0.90 and a dynamic viscosity of 0.0008 pound-second per square foot. If a decrease in temperature causes the viscosity to increase to 0.008 pound-second per square foot and if the power input remains the same, what will be the change in the rate of flow?

234. Natural gas is pumped through a smooth 10-inch pipe with a pressure drop of ½ inch of water per 1000 feet. The dynamic viscosity of the gas is 2.1×10^{-7} pound-second per square foot, and the gas weighs 0.06 pound per cubic foot. Determine the rate of flow.

235. Submit on logarithmic paper a plot of the ratio of the thickness δ' of the laminar sublayer to the radius r_0 as a function of the Reynolds number for turbulent flow through a smooth pipe.

236. If a conduit 6 feet in diameter is to carry a flow of 150 cubic feet of water per second with a minimum expenditure of energy, what is the order of magnitude of the permissible surface irregularities?

237. Compare the cost of pumping the same fluid at the same volumetric rate through 6-inch and 8-inch pipes having the same absolute roughness $k = 0.001$ foot; assume that the Reynolds number is sufficiently high for viscous effects to be negligible.

238. If sand grains 0.03 inch in diameter are cemented on the inner surface of a 4-inch pipe for test purposes, at what velocity of water at 60° Fahrenheit will the surface roughness (a) just begin to disturb the laminar sublayer, and (b) completely eliminate the influence of viscosity upon the boundary drag?

35. RESISTANCE OF COMMERCIAL PIPE

Development and use of a general resistance diagram. Were the actual roughness of all materials geometrically similar to the uniform sand roughness used by Nikuradse, Figs. 108 and 109 would be immediately useful in evaluating the resistance of all kinds of pipe. However, even a non-uniform sand roughness, as shown by Colebrook and White, will yield quite a different transition curve between the smooth-pipe and rough-pipe asymptotes, indicating that the coarsest irregularities of the boundary will begin to disturb the laminar sublayer long

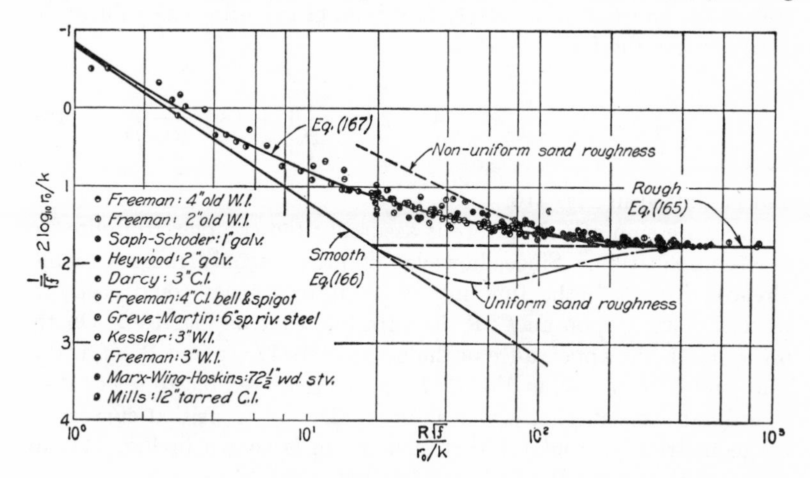

FIG. 110. The transition function for commercial pipe materials.

before the finest irregularities become effective. It is obvious, furthermore, that such commercial materials as metal, wood, and masonry will vary considerably in average height, form, and pattern of irregularities, so that it would be impossible to describe them completely in terms of a single length corresponding to the sand-grain diameter. Nevertheless, the *equivalent sand roughness* of any surface may arbitrarily be evaluated in terms of the sand roughness k which would yield the same limiting value of f in the same diameter of pipe. If, when reduced to a common basis in this manner, the transition curves of most commercial materials proved to be of essentially the same form —whatever that form might be—a single resistance function would then suffice for all.

In Fig. 110 is shown a series of such transitions for various materials in common use. It would appear from the location of the points with reference to the plotted curves of Nikuradse and Colebrook-White that natural roughness is definitely not uniform like the sanded surfaces used by Nikuradse, and yet not of such pronounced irregularity as the non-uniform sand of Colebrook and White, since all values lie between the two curves. Although there is considerable scatter of the individual points, a consistent systematic deviation is not greatly in evidence. As a first approximation, therefore, it appears safe to assume that the transition function for many—if not all—of the most important commercial materials may satisfactorily be represented by a single curve. Indeed, a semi-empirical equation proposed by Colebrook will be found to be asymptotic to both the smooth-pipe and the rough-pipe equations, and to follow closely the trend of experimental values; this equation has the form

$$\frac{1}{\sqrt{f}} - 2\log_{10}\frac{r_0}{k} = 1.74 - 2\log_{10}\left(1 + 18.7\,\frac{r_0/k}{\mathbf{R}\sqrt{f}}\right) \qquad (167)$$

and is plotted in Fig. 110.

An expression of this nature is assuredly too complex to invite frequent application. Since, however, it includes only those variables already discussed, the function which it represents could easily be used if plotted upon an $f : \mathbf{R}$ diagram like that of Fig. 108. On the other hand, the appearance of the parameters $1/\sqrt{f}$ and $\mathbf{R}\sqrt{f}$, rather than simply f and $\mathbf{R}$, in Eq. (167) makes it preferable to select these as ordinate and abscissa scales, in order that all transition curves will be geometrically similar. Such a diagram is shown in Fig. 111, together with alternative scales of f and $\mathbf{R}$. Since

$$f = \frac{2gh_f}{V^2}\frac{D}{L}, \quad \mathbf{R} = \frac{VD}{\nu}, \quad \text{and} \quad \mathbf{R}\sqrt{f} = \frac{D^{3/2}}{\nu}\sqrt{\frac{2gh_f}{L}}$$

proper selection of scales will permit direct calculation of any one of the variables involved, depending upon the form of the problem.

Mention must be made at this point of the fact that the values of equivalent sand roughness k tabulated on this diagram correspond to materials in new condition. Practically every such material will increase in roughness with use, due to corrosion or incrustation, at a rate depending upon the nature of the material and of the fluid. Since this will correspond, for a given pipe, simply to an upward progression across the curves of Fig. 111, knowledge of the effective roughness k at any stage of deterioration will still permit evaluation of the resistance coefficient for any Reynolds number. Colebrook and White have, moreover, shown that boundary roughness may be expected to increase

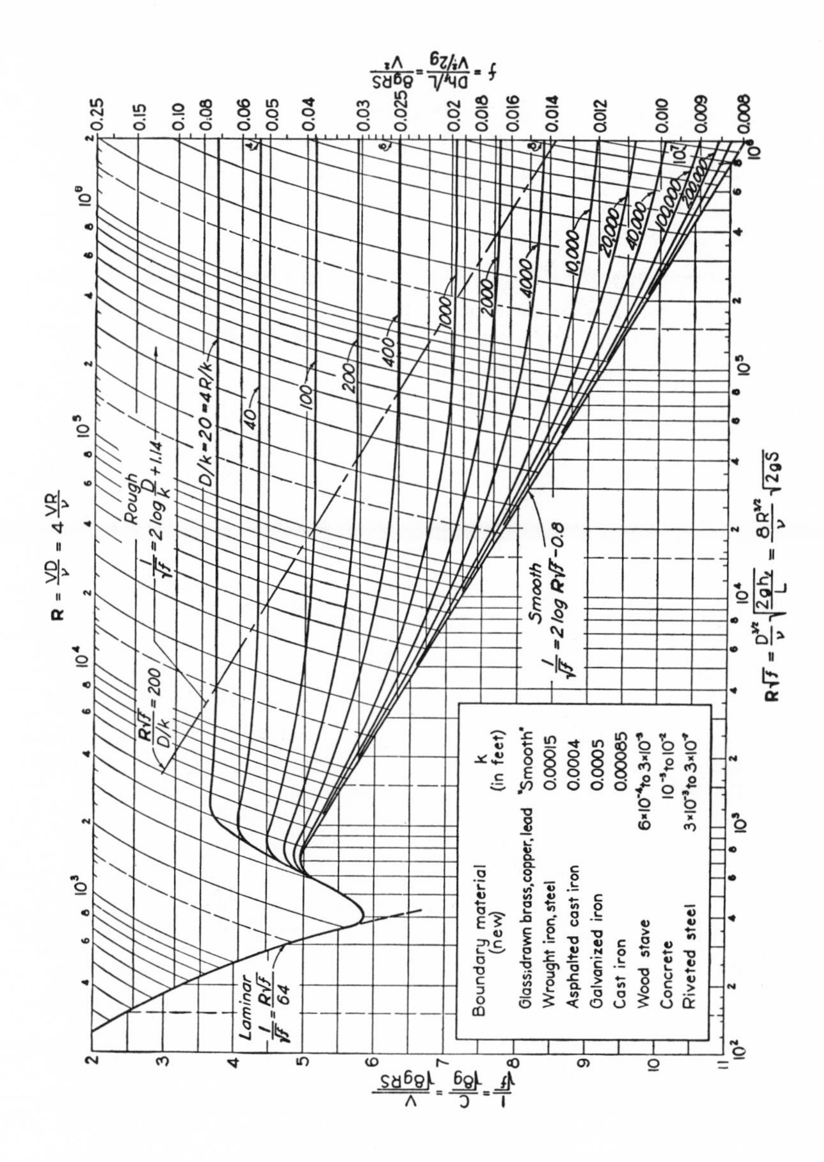

FIG. 111. General resistance diagram for uniform flow in conduits.

with time in approximate accord with the expression

$$k = k_0 + \alpha t \tag{168}$$

in which k_0 is the roughness of the new material, k the roughness at any time t, and α the time rate of roughness increase. Two resistance measurements at different times will evidently permit evaluation of the constants k_0 and α, whereupon a fair estimate may be made of the future variation in pipe capacity.

Example 39. The 24-inch main of a city water-supply system requires, after 1 year of service, 12 horsepower per 1000 feet of line to maintain a flow of 20 cubic feet per second. At the end of 3 years of service, the power requirement for the same rate of flow is found to have increased 7 per cent. What power should be necessary after 10 years of service, assuming that the main must then deliver 25 cubic feet per second?

After 1 year of service, in 1000 feet of line the loss of head will be

$$h_f = \frac{P}{Q\gamma} = \frac{12 \times 550}{20 \times 62.4} = 5.28 \text{ ft}$$

whence

$$f = \frac{2gh_fD}{V^2L} = \frac{64.4 \times 5.28 \times 2}{(20/\pi)^2 \times 1000} = 0.017$$

The corresponding Reynolds number (assuming $\nu = 1.2 \times 10^{-5}$ square foot per second) is

$$\mathbf{R} = \frac{VD}{\nu} = \frac{(20/\pi) \times 2}{1.2 \times 10^{-5}} = 1{,}060{,}000$$

From Fig. 111, for these values of f and $\mathbf{R}$,

$$\frac{D}{k_1} = 2000 \quad \text{and} \quad k_1 = 0.001 \text{ ft}$$

The rate of flow remaining constant, a 7 per cent increase in power corresponds to a 7 per cent increase in f. From Fig. 111, for $f = 1.07 \times 0.017 = 0.018$ and the same Reynolds number,

$$\frac{D}{k_3} = 1600 \quad \text{and} \quad k_3 = 0.00125 \text{ ft}$$

Therefore, from Eq. (168),

$$0.001 = k_0 + \alpha \times 1 \quad \text{and} \quad 0.00125 = k_0 + \alpha \times 3$$

whence, by solution of these simultaneous equations,

$$k_0 = 0.000875 \text{ ft} \quad \text{and} \quad \alpha = 0.000125 \text{ ft/yr}$$

Hence, after 10 years of service

$$k_{10} = k_0 + \alpha t_{10} = 0.000875 + 0.000125 \times 10 = 0.00213 \text{ ft}$$

From Fig. 111, for the parameters

$$\frac{D}{k_{10}} = \frac{2}{0.00213} = 938 \quad \text{and} \quad \mathbf{R} = \frac{VD}{\nu} = \frac{(25/\pi) \times 2}{1.2 \times 10^{-5}} = 1{,}330{,}000$$

it will be found that $f = 0.020$. After 10 years of service, therefore, in 1000 feet of line the loss of head should be

$$h_f = f \frac{L}{D} \frac{V^2}{2g} = 0.020 \times \frac{1000}{2} \times \frac{(25/\pi)^2}{64.4} = 9.8 \text{ ft}$$

which would require

$$\frac{P}{550} = \frac{Q\gamma h_f}{550} = \frac{25 \times 62.4 \times 9.8}{550} = 28 \text{ hp}$$

PROBLEMS

239. The "Big Inch" is a welded steel pipeline, 24 inches in diameter, designed to carry a flow of 12,600,000 gallons of oil per day across the country, pumping stations being located every 50 miles. If the pump efficiency is 85 per cent, determine the power input required at each station when the dynamic viscosity and specific weight of the oil are 5×10^{-4} pound-second per square foot and 55 pounds per cubic foot, respectively.

240. What size of galvanized pipe would be required to carry a flow of 0.5 cubic foot of water per second at 40° Fahrenheit without exceeding a head loss of more than 3 feet per 100 feet of pipe?

241. A wrought-iron sewer line 8 inches in diameter is laid on a grade of 1:500. Assuming liquid properties similar to those of water at 40° Fahrenheit, determine the normal capacity of the line—i.e., the rate of flow which will yield a gradient of piezometric head equal to the slope of the sewer.

242. If a flow of 150 cubic feet of water per second through a 5-foot wood-stave penstock at 60° Fahrenheit results in a head loss of 3 feet in 1000 feet of line, what is the magnitude of the equivalent sand roughness k?

243. Tests on the flow of air through a 3-foot circular ventilating duct of a vehicular tunnel yield a pressure drop of 1.5 inches of water in 1500 feet of duct when the rate of flow is 9000 cubic feet per minute. What is the magnitude of the equivalent sand roughness k?

244. A 12-inch steel pipe carries crude oil having a kinematic viscosity of 5×10^{-4} square foot per second and a specific gravity of 0.9. What rate of flow will correspond to a power input of 20 horsepower per mile of line?

245. Water is to be pumped through a 3-foot riveted-steel conduit for which the equivalent sand roughness has been found to have the magnitude $k = 0.01$ foot. If the highest point in the conduit is 75 feet above the pump at a section 3000 feet away, what discharge pressure must be maintained at the pump during a flow of 60 cubic feet per second at 50° Fahrenheit to prevent the pressure intensity from becoming less than atmospheric at the point of maximum elevation?

246. Field tests on a 15-inch cast-iron water main indicate that the wall roughness has increased to 0.005 foot after many years of service. If the main now carries a

peak flow of 8 cubic feet per second, what increase in flow at the same power input would result from replacing the line with new cast-iron pipe of the same diameter?

247. Gasoline is pumped through a vertical 2-inch steel pipe at a temperature such that the kinematic viscosity is 10^{-5} square foot per second and the specific gravity 0.75. For purposes of estimating the rate of flow, an inverted U-tube is connected to piezometer taps located 20 feet apart on the vertical pipe. If the space above the liquid columns in the U-tube is filled with air under pressure, what rate of flow would be indicated by a differential head of 5 inches?

248. After 10 years of service, an asphalted cast-iron water main 18 inches in diameter is found to require 30 per cent more power to deliver the 8 cubic feet of water per second for which it was originally designed. What is the corresponding magnitude of the rate of roughness increase α?

36. UNIFORM CONDUITS OF NON-CIRCULAR CROSS SECTION

Effects of cross-sectional form; the hydraulic radius. Since the velocity distribution is again the same at all cross sections, the resistance to uniform flow between parallel boundaries (the two-dimensional counterpart of flow in pipes of circular cross section) may again be evaluated by integration of Eqs. (150) and (152) and subsequent elimination of the shear velocity. However, since the cross section is no longer circular, it is only to be expected that different constants of integration will be obtained. Thus, for smooth boundaries it will be found that

$$\frac{1}{\sqrt{f}} = 2.03 \log_{10} \left(\frac{2BV}{\nu} \sqrt{f} \right) - 1.08$$

while for rough boundaries

$$\frac{1}{\sqrt{f}} = 2.03 \log_{10} \frac{B/2}{k} + 2.11$$

The change in the factors -1.08 and 2.11 from the values of -0.91 and 1.67 for smooth and rough pipe evidently represents the entire effect of cross-sectional form upon the resistance relationships, since the coefficient 2.03 is the same in every case. There is, unfortunately, insufficient experimental evidence to permit a correction of these constants of integration similar to that found necessary in the case of pipes. On the other hand, laboratory studies of uniform flow with a free surface indicate that the foregoing relationships for parallel boundaries are, as might be surmised, quite applicable to free-surface conditions, provided that the boundary spacing B be replaced by the uniform depth y_0 (i.e., $y_0 = B/2$). The resistance diagram for such flow should then be quite similar to that for pipes, shown in Fig. 111, except for a slight displacement of all curves in accordance with the change in form coefficients.

In the case of conduits of other cross-sectional forms, however, it cannot be assumed that the velocity distribution is the same at all parts of a normal section, in particular if the boundary is composed of planes intersecting at an angle—for instance (see Fig. 112), if the cross section is rectangular. It is therefore evident that the resulting intensity of shear τ_0 will no longer be the same at all points of the

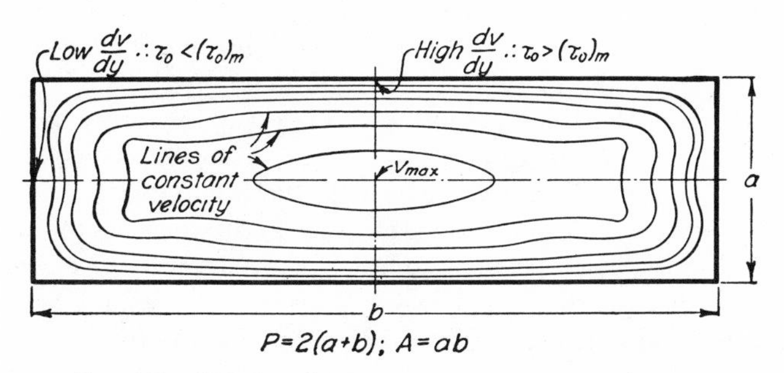

FIG. 112. Velocity distribution in a non-circular conduit.

boundary, so that an integration of the general velocity equations should not yield more than a qualitative indication of the resistance function.

Nevertheless, it is still reasonable to assume that the *mean* intensity of boundary shear will vary according to the same type of expression as Eq. (158); that is,

$$(\tau_0)_m = \frac{f}{4}\frac{\rho V^2}{2} \tag{169}$$

in which f will differ from that for conduits of circular cross section only in the numerical factor embodying the effect of form. In a manner similar to the development of Eqs. (159) and (160), the tangential stress along the conduit boundary may now be equated to the force maintaining flow; in other words, denoting by P the so-called *wetted perimeter* of the cross section (i.e., the perimetric length of contact between fluid and boundary),

$$(\tau_0)_m PL = -\gamma \frac{dh}{dx} LA = \gamma h_f A$$

Through elimination of $(\tau_0)_m$ and replacement of the ratio A/P by the symbol R, these two relationships then combine in the form

$$h_f = f \frac{L}{4R} \frac{V^2}{2g} \tag{170}$$

The quantity R is known as the *hydraulic radius* of the cross section. If the cross section is circular, $R = A/P = (\pi D^2/4)/(\pi D) = D/4$, and Eq. (170) becomes identical with Eq. (160) for pipes; if the width of the conduit is very great, on the other hand, R will be found to approach as a limit either $B/2$ or y_0, depending upon whether the fluid is confined between two parallel boundaries or flows along a single boundary with a parallel free surface.

Flow in a very wide or a very deep conduit may be considered to represent the maximum deviation of a normal section from the circular. The effect of cross-sectional form upon the resistance coefficient should then also be a maximum, the magnitude of which may be judged from the deviation of the form coefficients noted earlier in this section. Although the actual magnitude of the form factor for any shape of section may be approximated by a method developed by Keulegan, the variation of this factor from that of the circular section will be negligible so long as the width-depth ratio of the conduit is not excessively great or small. In other words, Fig. 111 for pipe resistance may safely be used in estimating the resistance of other types of conduit, provided these do not depart too greatly from circular proportions. With this in view, the various parameters of this diagram have also been written in terms of the hydraulic radius.

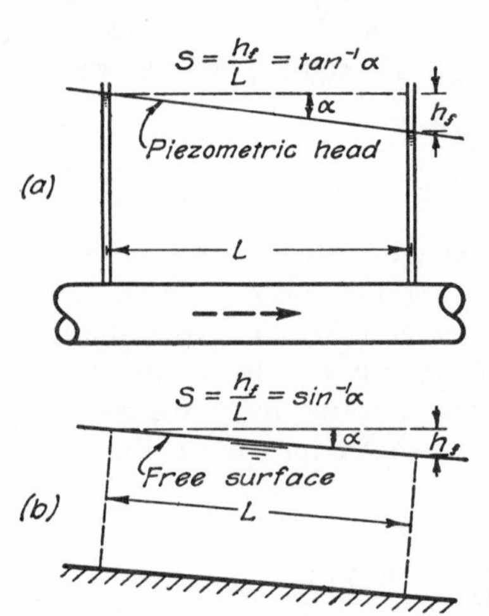

FIG. 113. Definition sketches for the piezometric slope in closed and open conduits.

Significance of the Chézy discharge coefficient. Further significance of the parameter $1/\sqrt{f}$ of this diagram will become apparent if Eq. (170) is solved for the mean velocity:

$$V = \sqrt{\frac{8g}{f}}\sqrt{R\frac{h_f}{L}}$$

In the case of a horizontal closed conduit the ratio $h_f/L = -\Delta h/L$ represents the *slope* of the line of piezometric head, and hence is customarily given the symbol S, the negative sign being eliminated by defining a downward slope (i.e., a decreasing head) as positive. In the case of uniform free-surface flow (Fig. 113), since the boundary itself must then slope with the free surface or line of piezometric head, S will be seen to represent the sine of the angle between either the boundary or the free surface and the horizontal. Finally, upon replacing the radical $\sqrt{8g/f}$ by the coefficient C, the foregoing relationship becomes simply

$$V = C\sqrt{RS} \tag{171}$$

or, since $Q = VA$,

$$Q = AC\sqrt{RS} \tag{172}$$

This form of the resistance expression is known as the *Chézy equation*, and C as the *Chézy coefficient*, after a French engineer of the eighteenth century. Unlike f, which is directly proportional to the *resistance*, C is evidently a *discharge* coefficient, since it is proportional to the rate of flow. Again unlike f, which is a pure number, C obviously has the dimension of $\sqrt{g}$; g, however, is essentially constant, so that the Chézy coefficient may be considered simply a multiple of the parameter $1/\sqrt{f}$ which was found to play a predominant role in the Kármán-Prandtl resistance equations. In other words, since

$$\frac{C}{\sqrt{8g}} = \frac{1}{\sqrt{f}} \tag{173}$$

the ordinate scale of Fig. 111 will, in American units, be found to represent almost exactly 1/16 the magnitude of C. This diagram may thus be used as conveniently with the Chézy equation as with the Darcy-Weisbach equation for which it was developed.

The Manning formula and its limitations. Long before the derivation of the Kármán-Prandtl equations of velocity distribution and resistance, it was necessary to evaluate the probable loss of head h_f in a given conduit from empirical relationships which took into consideration only the effect of boundary roughness. Since some of these are still commonly used in hydraulic design—particularly in the case of open channels—and since the simplicity of one, the *Manning formula*, continues to warrant its use in first approximations, the accuracy of this formula might well be tested in the light of the foregoing analysis.

The Manning formula is generally written as

$$Q = \frac{1.49AR^{2/3}S^{1/2}}{n} \tag{174}$$

in which n is a characteristic of the boundary material, as indicated by the table on page 219. Solution of Eqs. (172) and (174) for the Chézy C, however, will show that

$$C = 1.49\frac{R^{1/6}}{n}$$

whence, in terms of the dimensionless coefficient f of Eq. (165),

$$\frac{1}{\sqrt{f}} = \frac{1.49}{\sqrt{8g}} \frac{R^{1/6}}{n}$$

As n varies only with the boundary roughness, the factor 1.49 must have the dimension of $\sqrt{g}$. Therefore, since the hydraulic radius R

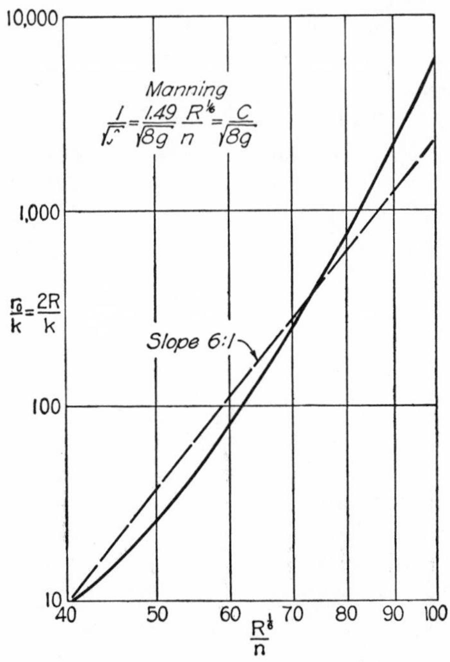

FIG. 114. Relationship between absolute boundary roughness and the Manning n.

is a linear factor, it follows that n must have the dimension $L^{1/6}$, R/n^6 then representing a ratio of two lengths similar to the relative-roughness parameter r_0/k used in the preceding section. Indeed, since both roughness ratios are now related to f, it is possible to evaluate the Manning parameter $R^{1/6}/n$ in terms of the parameter $r_0/k = 2R/k$, as shown in Fig. 114. At once apparent is the fact that a *threefold* increase in n (i.e., between the practical limits of 0.01 and 0.03) corresponds roughly to a *thousandfold* increase in absolute roughness k. Were the empirical formula of Manning compatible with the more nearly correct relationship of Eq. (165), the curve of Fig. 114 would be a straight line having the slope 6:1—the actual deviation of the curve from this line being an indication of the error involved in

Eq. (174); it would appear from this plot that the Manning formula is most dependable for intermediate values of relative roughness. Also to be emphasized is the failure of an expression of this type to include the influence of viscosity; it may therefore be expected to be least dependable at low values of the Reynolds number.

TABLE II

VALUES OF THE MANNING ROUGHNESS FACTOR FOR VARIOUS BOUNDARY MATERIALS

Boundary Surface	Manning n (feet$^{1/6}$)
Planed wood	0.010–0.014
Unplaned wood	0.011–0.015
Finished concrete	0.011–0.013
Unfinished concrete	0.013–0.016
Cast iron	0.013–0.017
Riveted steel	0.017–0.020
Brick	0.012–0.020
Rubble	0.020–0.030
Earth	0.020–0.030
Gravel	0.022–0.035
Earth with weeds	0.025–0.040

Example 40. An open channel of the cross section shown is to be constructed of unfinished concrete to carry 125 cubic feet of water per second. Determine (*a*) the characteristics of the section, (*b*) the approximate roughness k, and (*c*) the required slope.

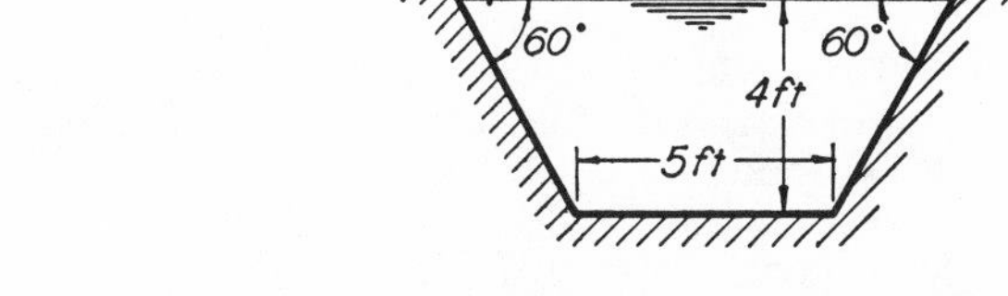

(*a*) Evaluating the cross-sectional area, the wetted perimeter, and the hydraulic radius,

$$A = 4 \times \frac{5 + (5 + 2 \times 4 \times 0.577)}{2} = 29.2 \text{ ft}^2$$

$$P = 5 + 2 \times \frac{4}{0.866} = 14.2 \text{ ft}$$

$$R = \frac{A}{P} = \frac{29.2}{14.2} = 2.05 \text{ ft}$$

Then, from the relationship following Eq. (174),

$$C = 1.49 \frac{R^{1/6}}{n} = 1.49 \times \frac{(\overline{2.05^{1/3}})^{1/2}}{0.014} = 1.49 \times 80 = 120 \text{ ft}^{1/2}/\text{sec}$$

(*b*) From Fig. 114, when $R^{1/6}/n = 80$, $R/k \approx 350$, whence

$$k \approx \frac{R}{350} \approx 0.006 \text{ ft}$$

(*c*) From Eq. (172),

$$S = \frac{Q^2}{C^2A^2R} = \frac{\overline{125}^2}{\overline{120}^2 \times \overline{29.2}^2 \times 2.05} = 0.00062$$

PROBLEMS

249. Determine the rate of flow of oil ($\mu = 3 \times 10^{-4}$ pound-second per square foot, $\gamma = 55$ pounds per cubic foot) through the annular space between the walls of two coaxial brass pipes under a gradient of piezometric head of -0.025, the inside diameter of the outer pipe being 6 inches and the outside diameter of the inner pipe 3 inches.

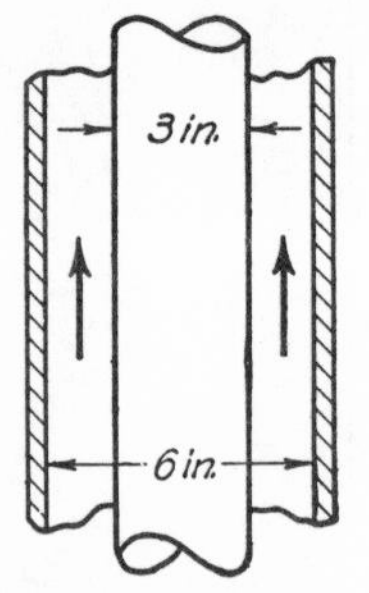

PROB. 249.

250. A ventilating duct of galvanized sheet metal has a cross section 6 feet wide and 4 feet high. What power per 100 feet would be required to maintain a flow of 25,000 cubic feet of air per minute at 60° Fahrenheit?

251. For a flow of 200 gallons of water per minute through a horizontal 3-inch rough pipe the pressure drop is 12.5 pounds per square inch per 100 feet of pipe. Determine (*a*) the corresponding value of the Manning *n*, and (*b*) the equivalent sand roughness *k*.

252. What width-depth ratio of a rectangular open channel will yield the largest possible hydraulic radius for a given cross-sectional area?

253. Water flows down a chute having a slope of 30° and a rectangular cross section 5 feet wide, the bottom being roughened to an extreme degree by transverse cleats to reduce the tendency of such flow to become unstable at high velocities. If the average vertical depth is 2 feet, determine the average magnitudes of the pressure intensity and the intensity of shear along the bottom. If the chute is supported at 10-foot intervals along the slope, what resultant load must each support carry? (Note that the pressure distribution must be hydrostatic across a normal section, and that the pressure head is therefore not equal to the vertical depth below the free surface.)

254. A laboratory flume for research on open-channel flow is constructed of polished sheet brass in the form of a 60° trough, in order that the cross sections of the flow will be geometrically similar (i.e., equilateral triangles) at all depths. What slope should this flume have in order to carry 1 cubic foot of water per second at a uniform depth of 1 foot? (Assume a water temperature of 60°.)

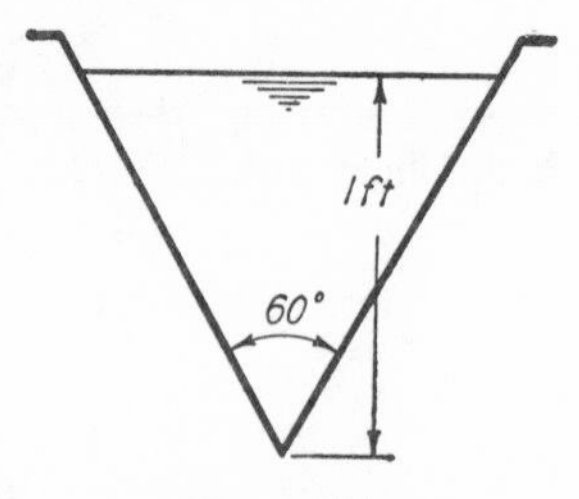

PROB. 254.

255. At what rate will water flow at a uniform depth of 2 feet in a rectangular open channel 4 feet wide if the channel is of unfinished concrete and slopes 5 feet per mile?

256. Of what material should the channel shown in cross section be constructed in order to deliver 230 cubic feet of water per second at a depth of 6 feet and a slope of 1 foot per mile?

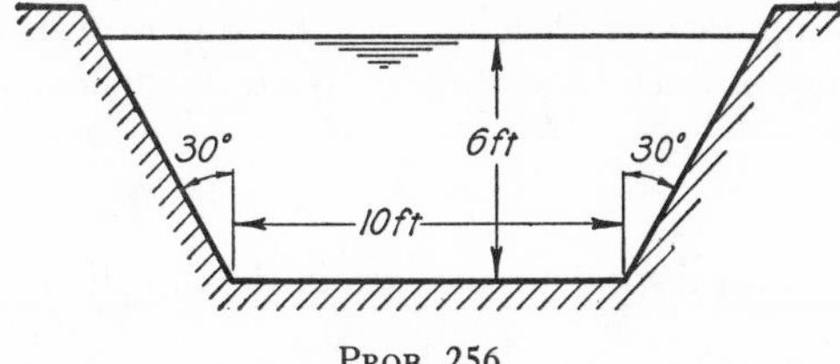

PROB. 256.

257. A sewer line of circular cross section 5 feet in diameter is to carry 35 cubic feet of water per second when flowing half full. If the inner surface is of finished concrete, on what slope should the line be laid to insure uniform flow?

258. It is frequently assumed as a first approximation that flow in a channel having an irregular cross section may be analyzed by dividing the section into a series

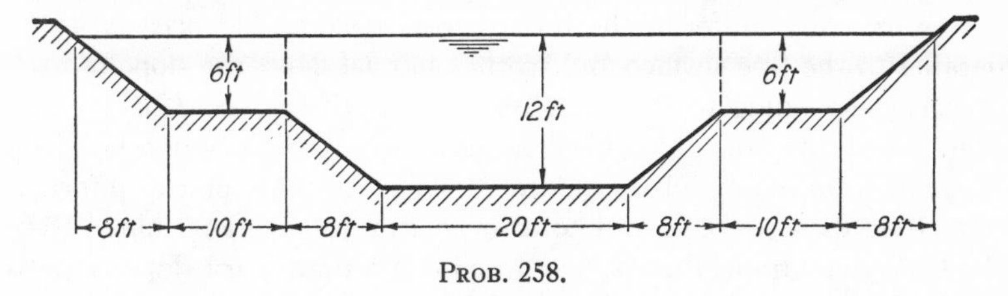

PROB. 258.

of elements, the wetted perimeter of each element including only the channel boundary. If the channel shown in cross section is of earth ($n = 0.025$) and has a slope of 1 foot per mile, what rate of flow should it carry at the given depth?

37. GRADUALLY VARIED FLOW IN OPEN CHANNELS

Step computation of surface profiles. Although the foregoing section discussed the resistance accompanying uniform flow in open channels, it will become apparent from the present section that uniform flow is actually a limiting condition which may be approached, but never truly attained. This is due to the fact that boundary resistance tends to transmit over a very great distance the effects of local non-uniformity caused by a boundary transition. Long before the uniform limit is approached, however, the variation in depth will be so gradual that accelerative effects may be completely ignored. Evidently, just as local transitions involving rapid acceleration were studied in Chapters IV and V without regard to boundary resistance, the analysis of depth variation under conditions of negligible acceleration may now be treated as a phase of the resistance problem. In order that accelerative effects be truly negligible, of course, it is essential that the analysis be restricted to cases in which the non-uniformity of motion is very slight; such motion is commonly described as *gradually varied flow.*

So far as computations of depth changes due to resistance are concerned, one may proceed most directly from the definition equation of specific head:

$$H_0 = H - z_0$$

Differentiation with respect to distance in the direction of flow at once yields

$$\frac{dH_0}{dx} = \frac{dH}{dx} - \frac{dz_0}{dx}$$

The first term evidently signifies the rate of change of the specific head, $H_0 = V^2/2g + y = Q^2/2gA^2 + y$, along the channel. The second term likewise represents the rate of change in total head. If the flow were truly uniform, the line of total head would necessarily be parallel to the free surface and to the channel floor, the slope of each then corresponding to the quantity $S = V^2/C^2R$ of the Chézy equation. Since the flow under consideration is only slightly non-uniform, it would seem reasonable to assume, for lack of more precise information, that the rate of loss of total head can still be determined from the Chézy equation; that is, again defining a downward slope as positive, $dH/dx = -S_H \approx -V^2/C^2R$. The last term, $-dz_0/dx$, is evidently the slope of the channel bottom, S_0. Upon introduction of the latter quantities, the relationship takes the form

$$\frac{dH_0}{dx} = -\frac{V^2}{C^2R} + S_0$$

If, now, the exact derivative dH_0/dx is replaced by the ratio of the change ΔH_0 in specific head to the corresponding finite length ΔL of channel, the foregoing expression may be written approximately as

$$\Delta L \approx \frac{\Delta H_0}{S_0 - (V^2/C^2R)_m} \tag{175}$$

which is the form most convenient for step-by-step integration.

The computation procedure is as follows: Starting from a section of known depth, one determines, for the given channel characteristics and rate of flow, the change $\Delta H_0 = \Delta(V^2/2g) + \Delta y$ for an arbitrary increment Δy (or decrement $-\Delta y$) of depth; the average value of V^2/C^2R for the depths y and $y + \Delta y$ is then assumed to represent the mean value over the distance ΔL, and the latter distance is evaluated according to Eq. (175). This process is repeated for successive increments of depth, and the summation $\Sigma\Delta L$ at any time during the computations will evidently represent the distance from the initial section to that at which the depth equals $y + \Sigma\Delta y$. As in all cases of step-by-step integration, of course, the accuracy of results will depend to a great extent upon the relative magnitude of the depth increment which is chosen.

Example 41. In order to decrease the mean velocity of flow in a rectangular channel, a low overflow dam is installed near the downstream end. The channel is of unfinished concrete ($n = 0.014$), and has a width of 10 feet and a bottom slope of 0.001. If the depth of flow just before the dam is 7.5 feet during a peak discharge of 300 cubic feet per second, how far upstream will backwater cause a velocity reduction of at least 20 per cent?

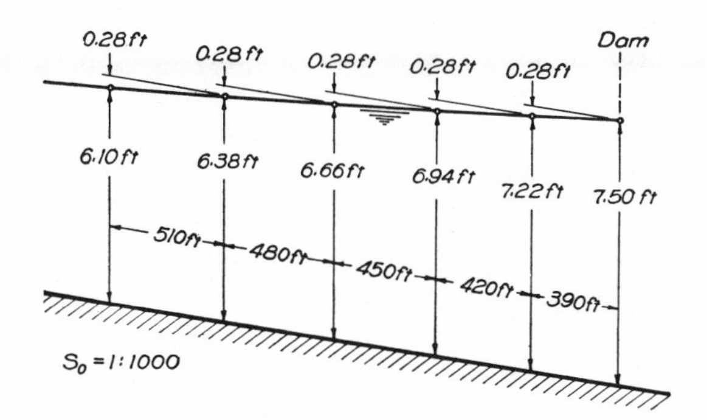

The uniform depth y_0 of the undisturbed flow a great distance upstream may be determined from the Manning relationship,

$$Q = \frac{1.49AR^{2/3}S^{1/2}}{n} = \frac{1.49}{0.014} \times 10y_0 \times \left(\frac{10y_0}{10 + 2y_0}\right)^{2/3} \times (0.001)^{1/2} = 300 \text{ cfs}$$

trial solution of which yields

$$y_0 = 4.88 \text{ ft}$$

Since a velocity reduction of 20 per cent corresponds to a depth increase of 25 per cent (i.e., Vy = constant), it is necessary to determine the distance $\Sigma\Delta L$ to the section at which $7.50 + \Sigma\Delta y = 1.25y_0 = 6.10$ feet. This is most easily accomplished in the following tabular form, in which the values required for calculating ΔL according to Eq. (175) are obtained in systematic steps. As indicated in the accompanying sketch, $\Delta y = \Sigma\Delta y/5 = (6.10 - 7.50)/5 = -0.28$ foot—that is, the depth decreases in the upstream (i.e., negative) direction.

y	A	V	$V^2/2g$	H_0	P	R	C	V^2/C^2R	$(V^2/C^2R)_m$	ΔH_0	ΔL	$\Sigma\Delta L$
7.50	75.0	4.00	0.25	7.75	25.0	3.00	128	0.00033				0
									0.00034	−0.26	−390	
7.22	72.2	4.15	0.27	7.49	24.4	2.95	128	0.00036				−390
									0.00038	−0.26	−420	
6.94	69.4	4.32	0.29	7.23	23.9	2.90	127	0.00040				−810
									0.00042	−0.26	−450	
6.66	66.6	4.51	0.31	6.97	23.3	2.85	127	0.00044				−1260
									0.00047	−0.25	−480	
6.38	63.8	4.71	0.34	6.72	22.8	2.80	126	0.00050				−1740
									0.00053	−0.24	−510	
6.10	61.0	4.92	0.38	6.48	22.2	2.75	126	0.00055				−2250

Evidently, the weir will cause a velocity reduction of at least 20 per cent for a distance of more than 2000 feet upstream.

Differential equation of gradually varied flow. As in earlier examples of the one-dimensional analysis of open-channel flow, it is again apparent that the form of equation best suited to quantitative computation is also the least effective in giving a qualitative picture of depth variation in general. For the latter purpose, again limiting the discussion to the case of flow in a rectangular channel of very great width, the derivative must be taken with respect to x of the expression

$$H = \frac{q^2}{2gy^2} + y + z_0$$

with the following result:

$$\frac{dH}{dx} = -\frac{q^2}{gy^3}\frac{dy}{dx} + \frac{dy}{dx} + \frac{dz_0}{dx}$$

Replacing, as before, dH/dx by $-V^2/C^2R = -V^2/C^2y$ and dz_0/dx by $-S_0$, and solving for dy/dx,

$$\frac{dy}{dx} = \frac{S_0 - V^2/C^2y}{1 - q^2/gy^3}$$

It will be recalled from Section 24 that $q^2/g = y_c^3$, whence $1 - q^2/gy^3 = 1 - (y_c/y)^3$. Moreover, from the Manning formula,

$$S_0 = \frac{V_0^2}{C_0^2 y_0} = \frac{q^2}{\left(1.49\,\frac{y_0^{1/6}}{n}\right)^2 y_0^3} \quad \text{and} \quad \frac{V^2}{C^2y} = \frac{q^2}{\left(1.49\,\frac{y^{1/6}}{n}\right)^2 y^3}$$

whence

$$S_0 - \frac{V^2}{C^2y} = S_0\left[1 - \frac{1}{S_0}\frac{V^2}{C^2y}\right] = S_0\left[1 - \left(\frac{y_0}{y}\right)^{10/3}\right]$$

Upon introduction of these quantities, the foregoing relationship for dy/dx becomes, significantly,

$$\frac{dy}{dx} = S_0 \frac{1 - \left(\frac{y_0}{y}\right)^{10/3}}{1 - \left(\frac{y_c}{y}\right)^3} \tag{176}$$

The longitudinal rate of change in depth is thus seen to depend upon three significant dimensionless parameters: the bottom slope, the ratio of the uniform depth to the actual depth, and the ratio of the critical depth to the actual depth.

Classification of surface profiles. Of initial interest in qualitative analysis of the surface profile is the sign of dy/dx, for if it is positive the depth will increase, and if it is negative the depth will decrease, in the direction of flow. Evidently, this sign will depend upon whether S_0 is positive or negative, and whether y_0/y and y_c/y are greater or less than 1. Consider, for instance, a case of flow similar to that of the foregoing example, in which $S_0 > 0$ and $y > y_0 > y_c$; since the numerator and denominator at the right of Eq. (176), as well as S_0, will then be positive, dy/dx must also be positive, indicating that the depth must increase in the direction of flow. Were the weir producing the backwater extremely high, the numerator and denominator would both approach unity as y became great; hence as a limit $dy/dx = S_0$, from which it is evident that the downstream asymptote of such a surface profile would be a horizontal line, as indicated in Fig. 115. The upstream asymptote, of course, is the line of uniform depth, since $dy/dx = 0$ in Eq. (176) when $y = y_0$.

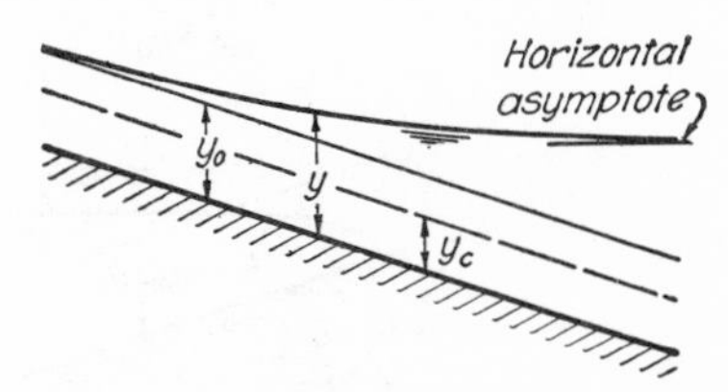

FIG. 115. Characteristics of a backwater profile.

If each of the various possible positive and negative combinations of the three parameters is investigated in this manner, it will be found that twelve different forms of surface profile will result, as shown schematically in Fig. 116. These profiles are conveniently classified according to slope and depth in the following manner: If S_0 is negative, the channel slope is called *adverse* (A); if $S_0 = 0$, the channel is *horizontal* (H); if S_0 is positive, the channel slope is termed *mild* (M) when $y_0 > y_c$, *critical* (C) when $y_0 = y_c$, and *steep* (S) when $y_0 < y_c$. If the surface profile lies above both the normal and critical-depth lines, it is of type 1; if between these lines, it is of type 2; and if below both lines, it is of type 3. The twelve profile forms are labeled accordingly in Fig. 116.

Although the qualitative verification of each profile form will be left to the reader, certain general remarks are still in order. First, the scale of all profiles sketched in Fig. 116 is greatly reduced in the horizontal direction, for if plotted to an undistorted scale the rate of change in depth would be scarcely noticeable. Second, since even a slope

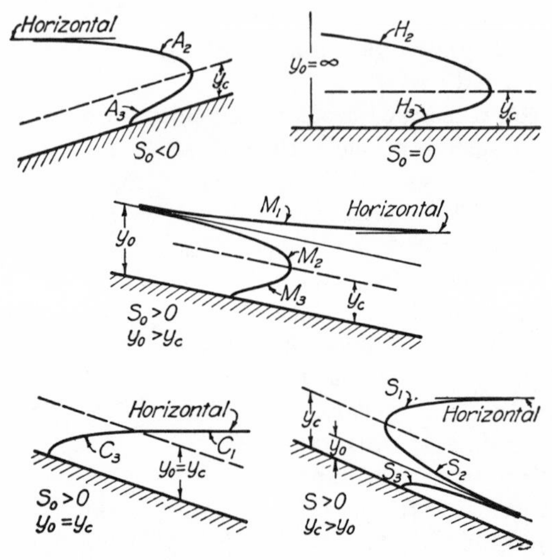

FIG. 116. Surface profiles on adverse, horizontal, mild, critical, and steep slopes.

which is hydraulically steep can deviate only a few degrees from the horizontal if the flow is to be truly of the gradually varied type, it actually matters little whether depths are measured in the vertical (as plotted in the figure) or at right angles to the channel bottom. Finally, despite the fact that Eq. (176) indicates an infinite slope of the free surface when $y = y_c$ and when $y = 0$, these sections must be regarded merely as limiting conditions at which the assumption of negligible acceleration (i.e., hydrostatic pressure distribution) is no longer fulfilled; since the actual zones of appreciable acceleration (Chapters IV and V) are relatively short, they would obviously be of insignificant length in comparison with the profiles of Fig. 116.

Control sections; profile analysis. The qualitative analysis of depth variation in a given channel proceeds in the following manner: The parameters y_0 and y_c are computed for the given rate of flow and the given channel characteristics, and the lines of uniform and critical depth are plotted to scale on a foreshortened diagram of the channel profile. Depths are then evaluated for existing *control* sections (i.e., inlets, weirs, sluice gates, falls, and breaks in channel slope), with due regard to the fact that profiles lying *above* the critical-depth line are determined by a *downstream* control; and profiles *below* the critical-depth line by an *upstream* control; in other words, the disturbance produced at a control section can travel upstream only if the velocity of flow is less than the wave celerity (i.e., only if the depth is greater than the critical). The appropriate portions of the surface profiles shown in Fig. 116 are then sketched upon the channel diagram. Should a surface discontinuity appear to exist in any open reach (that is, should the depth have to change from $y < y_c$ to $y > y_c$ between control sections), this will invariably indicate the formation of a hydraulic jump at a location commensurate with the required depth relationship of Eq. (109); it is otherwise physically impossible for a surface profile to cross either the uniform-depth or the critical-depth line without a change in boundary configuration. Once the general characteristics of the entire profile have been determined in this manner, evaluation of the actual depth variation in any reach may safely proceed according to the step-by-step process of integration.

The foregoing qualitative analysis of free-surface form was, for the sake of simplicity, restricted to flow in a rectangular channel of very great width. The curves of Fig. 116 will, however, be representative of those in any channel of uniform cross section, if y_0 is evaluated for the condition $V_0^2/C_0^2R_0 = S_0$, and y_c for the condition $V_c = \sqrt{gy_m}$, in which the quantity y_m is the mean depth—i.e., the ratio of the cross-sectional area to the free-surface width. Indeed, these curves may be used qualitatively in the analysis of surface variation in natural streams as well, provided that local variations in slope, cross-sectional form, and boundary resistance are properly taken into account; since the step-by-step method of integration was not developed specifically for uniform channels, it is evidently suited to all such backwater computations.

Despite the fact that the open-channel problems with which the engineer is concerned are still restricted to the flow of water in canals and rivers, emphasis must again be laid upon the close analogy between the gravity flow of a liquid in an open channel and the gravity flow of air masses or water masses which are slightly heavier or lighter than the surrounding air or water because of temperature differences or the presence of material in solution or suspension. Meteorologists and oceanographers are thus as vitally concerned with the resistance problem in gravity flows as are hydraulic engineers. Their problem is, needless to say, considerably more difficult than that discussed herein, owing to the additional drag at the interface between the slightly lighter and heavier mediums. Nevertheless, in its basic essentials the phenomenon of flow under gravitational attraction is universally the same.

Example 42. A channel of considerable length leads from a large reservoir and terminates in an abrupt fall; a sluice gate is located approximately midway between the two ends. Compare the general variation in depth for conditions of (*a*) mild and (*b*) steep slope.

In case (*a*) the uniform-depth line will lie above the critical-depth line, as shown in the sketch. Since the sluice gate produces an upstream depth greater

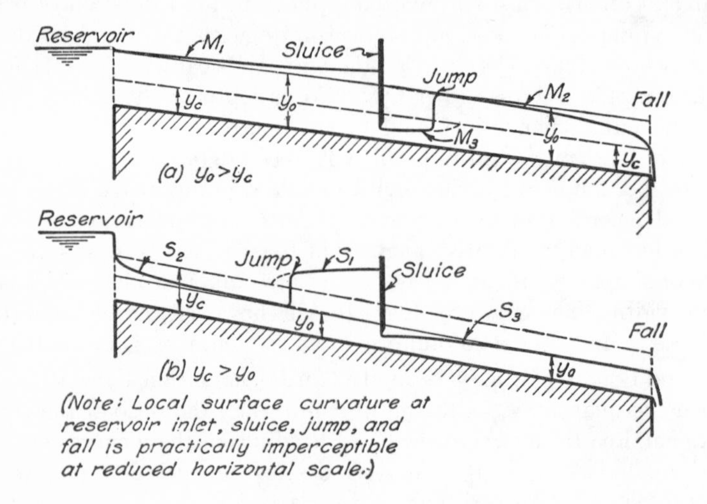

than y_c and a downstream depth less than y_c, it will control the surface profile in both directions. The *M*–1 profile will extend to the channel inlet, which will therefore play no role in the problem. But if the *M*–3 profile continued to its limit y_c, the flow beyond would be physically impossible. The fall hence serves as the control for an *M*–2 profile which extends far enough upstream to permit a hydraulic jump to form between depths satisfying the momentum relationship. In case (*b*) these conditions are reversed. Although the sluice gate again serves as a double control, if the distance from the inlet is sufficiently great the *S*–1 profile would—like the *M*–3 profile of case (*a*)—reach its limit y_c at some intermediate point. The channel inlet therefore controls an *S*–2 profile which extends up to the section at which the hydraulic jump forms. Evidently, it is now the inlet rather than the sluice gate which determines the rate of flow. The fall is no longer a control, since the *S*–3 profile lies below the line of critical depth.

PROBLEMS

259. Determine the error which would result from computing in a single step the length of transition obtained in five steps in Example 41.

260. How far upstream from the dam of Example 41 will the backwater effect extend, assuming that its practical limit is reached when the depth of flow is within 1 per cent of the uniform depth?

261. If the water flowing over the dam of Example 41 leaves the apron with a depth of 1.5 feet, and if the abrupt fall at the end of the channel is 75 feet downstream, at what depth will the water approach the fall?

262. If the dam of Example 41 had not been installed, the depth at all points along the channel would have been below rather than above the uniform depth. Assuming that the critical depth prevails just upstream from the fall at the end of the channel, determine and plot the surface profile for a distance upstream of at least 1000 feet.

263. The uniform depth of flow in a very long channel of rectangular cross section is 6 feet when the rate of flow is 30 cubic feet per second per foot width. A sluice gate which controls the inflow at the upstream end produces under these conditions a sheet of water 1 foot deep, which gradually expands to yield a profile of the M-3 type. If the computed coordinates of this profile are as follows, how far downstream from the sluice gate will the depth be such that a stable hydraulic jump will form?

x	y
0 ft	1.00 ft
100 ft	1.18 ft
200 ft	1.36 ft
300 ft	1.55 ft
400 ft	1.75 ft
500 ft	1.98 ft
600 ft	2.29 ft

264. A long rectangular channel of unfinished concrete 12 feet wide carries a flow of 250 cubic feet of water per second, the bottom slope changing abruptly from 0.01 to 0.0001 at an intermediate section. Show that the uniform depth of the flow approaching the break in grade is less than the critical depth, while that beyond is greater than the critical depth. A hydraulic jump must evidently form in the vicinity of the transition; will the jump occur on the steep or on the mild slope?

265. Sketch to approximate scale the form of the free surface and the line of total head at an abrupt change in channel slope, assuming that the critical depth is 5 feet, the uniform depth for the upstream slope is 7 feet, and the uniform depth for the downstream slope is 3 feet.

266. A channel of rectangular cross section and moderate length leads on a mild slope from a reservoir to the forebay of a power plant, the rate of flow through the channel being controlled by the water level in the forebay, which in turn is controlled by the gate opening of the turbines. At zero flow the water surface in the channel is horizontal, and at peak flow the forebay level is below the critical-depth level at the end of the channel. Submit sketches showing (*a*) the sequence of surface profiles, and (*b*) the variation in rate of flow, as the water in the forebay changes from maximum to minimum elevation.

QUESTIONS FOR CLASS DISCUSSION

1. Boundary-layer theory is based upon the premise that the thickness δ is very small compared with other linear dimensions. Does this premise eventually fail to be fulfilled (*a*) near to or far from the leading edge, (*b*) with increasing or decreasing viscosity, and (*c*) at high or low velocity?

2. Cite cases previously discussed in which the extent of viscous influence is too great for the boundary-layer theory to apply.

3. The mass density does not appear in the resistance equation for established laminar flow in a tube. Why must the density appear in the resistance equation for the laminar boundary layer?

4. The turbulent boundary layer and the laminar sublayer correspond to what zones of flow through a conduit?

5. In what localities and under what boundary circumstances would a laminar sublayer exist in atmospheric motion?

6. Does the ratio of the maximum to the mean velocity in a pipe increase or decrease (*a*) with increasing Reynolds number, and (*b*) with increasing relative roughness?

7. Show that the equation

$$\frac{v_{max} - v}{\sqrt{\tau_0/\rho}} = 5.75 \log_{10} \frac{r_0}{y}$$

may be obtained for smooth and rough pipes alike. What is the significance of this relationship?

8. Why does the equation $f = 64\mu/VD\rho$ for laminar flow in pipes *not* indicate that the resistance depends upon the density?

9. Distinguish between the actual roughness and the effective roughness of a conduit boundary.

10. Is a perfectly smooth surface physically possible? What bearing does this fact have upon the limit of validity of the resistance equation for smooth conduits?

11. Why is it more advantageous in the long run to know the equivalent sand roughness of a given boundary material than to obtain from a handbook the approximate loss in head for a particular rate of flow and diameter of pipe?

12. The resistance characteristics herein discussed refer to continuous lengths of pipe. Are joints apt to cause large or small variations in overall resistance?

13. Enumerate the similarities and dissimilarities of the coefficients c_f, C_f, f, and C.

14. What factors contribute to the "aging" of pipes?

15. What assumption is made in applying the Chézy equation to the case of gradually varied flow?

16. The flow pattern of uniform flow in an open channel was shown to be independent of the Froude number. Show that the form of a surface profile in gradually varied flow does depend upon the Froude number.

17. Under what circumstances do control sections govern (*a*) upstream and (*b*) downstream conditions of flow?

18. Verify the trend of each surface profile in Fig. 116.

19. Of what significance is the line of uniform depth in a horizontal channel?

20. Show that in uniform flow down a steep chute the pressure head is not equal to the vertical depth below the free surface.

21. Discuss the significance of the parameter y_c in the case of gravity flow of air masses in the atmosphere.

SELECTED REFERENCES

PRANDTL, L. "The Mechanics of Viscous Fluids." Vol. III, *Aerodynamic Theory*, Springer, 1935.

GOLDSTEIN, S. *Modern Developments in Fluid Dynamics.* Vols. I and II, Oxford, 1938.

KEULEGAN, G. H. "Laws of Turbulent Flow in Open Channels." *Journal of Research of the National Bureau of Standards*, Vol. 21, 1938.

WOODWARD, S. M., and POSEY, C. J. *Hydraulics of Steady Flow in Open Channels* Wiley, 1941.

CHAPTER VIII

FORM RESISTANCE

38. BOUNDARY-LAYER SEPARATION

Restrictions of the boundary-layer theory. Irrotational flow and laminar flow have repeatedly been stated to represent the two extremes of fluid motion involving both mass acceleration and viscous resistance to deformation. In the mathematical analysis of steady irrotational flow, the assumed absence of viscous effects is found to result in the complete lack of resistance to motion, however much the fluid may be deformed. In steady laminar flow, on the other hand, the preponderance of viscous effects leads to a distribution of pressure and velocity wholly unlike that which would be indicated by the flow net; in fact, the designation *deformation drag* to characterize the latter condition of motion aptly implies a widespread distortion of the basic flow pattern. Nevertheless, in the foregoing pages it was shown that over an intermediate range between these two extreme cases the zone of appreciable viscous deformation of the flow is confined to a relatively thin layer of fluid next to the boundary. It should be reasonable, under such conditions, to expect the flow net to indicate the general distribution of velocity and pressure throughout the moving fluid, the actual resistance to motion being attributable to *surface drag*—i.e., to viscous shear in the boundary layer alone.

Such a conclusion is, indeed, fully warranted—provided that one bear in mind three closely related restrictions: First, construction of the flow net according to the boundary geometry requires that the boundary layer be of negligible thickness at all points. Second, so long as the boundary layer remains as thin as assumed, the pressure must be essentially the same at the boundary as it is just outside the boundary layer. Third, since the velocity of the fluid must be zero at all points along the boundary, any deceleration in this zone which might otherwise be produced by boundary curvature is physically impossible. In other words, whereas acceleration tends to minimize viscous effects, the reduction in velocity and the corresponding increase in pressure indicated by a divergence of the stream lines near a boundary are not compatible with normal boundary-layer conditions.

The mechanism of separation. In the left half of Fig. 117 is shown to a greatly enlarged radial scale the distribution of velocity in the neighborhood of a boundary which curves in such a way as to produce a general increase in velocity (as shown by the converging stream lines) and a corresponding decrease in pressure. The boundary-layer thickness, it will be recalled, tends to increase with distance in the direction of flow, but in this case such a tendency is more or less counterbalanced by the convergence of the stream lines and the corresponding acceleration of the flow in general. Except for the thin layer of retarded fluid near the boundary, therefore, the velocity distribution of the flow as a

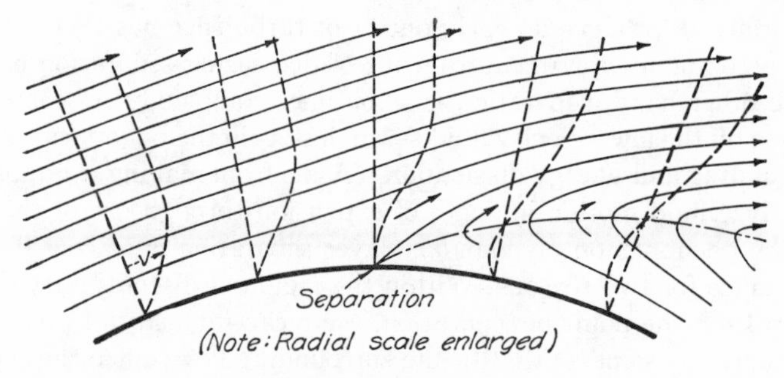

FIG. 117. Boundary-layer separation in a zone of deceleration.

whole may be obtained from the configuration of the flow net, which in turn permits evaluation of the pressure distribution at all points of the flow and of the boundary as well.

The right half of this figure, on the contrary, illustrates the behavior of the flow as the boundary curves in such a way as to produce deceleration of the fluid—deceleration, at least, as would be indicated by the diverging lines of the customary flow net. In this case the normal tendency of the boundary layer to expand with distance in the direction of flow is strengthened by the tendency of the stream lines to diverge from the boundary as the flow decelerates. But, since the velocity at the boundary is already zero, further deceleration at the zone of contact is not physically possible; the flow as a whole can continue beyond this zone only if a discontinuity at the boundary is produced. As indicated in the figure, such discontinuity involves *separation* of the flow from the boundary, the stream line which abruptly leaves the boundary dividing the oncoming flow from a region of reverse flow on the downstream side.

Were it possible to define the form of this stream line of separation, the flow net might still be used to determine the resulting distribution of velocity and pressure. The phenomenon of separation is not an easy one to analyze quantitatively, however, for the location of the point of separation depends in general upon the form and roughness of the boundary and upon the Reynolds number of the flow. Indeed, only if the boundary is angular in form will the separation point remain fixed in position. In the case of a boundary of easy curvature, the zone of separation will advance upstream as the Reynolds number is increased—and then suddenly shift downstream when the boundary layer becomes turbulent, owing to the fact that the lateral mixing of the turbulent fluid makes the velocity distribution more uniform and thus reduces the tendency toward separation. Roughening the boundary surface to produce an early outset of turbulence has therefore become a common experimental means of decreasing separation effects; on the other hand, separation is sometimes completely prevented by drawing off the low-velocity fluid through slots in the boundary surface.

Form drag and energy dissipation. Aside from making it impossible to analyze fluid motion in zones of expansion through use of the flow net, the phenomenon of boundary-layer separation is of considerable importance for two reasons: Within the region of discontinuity downstream from the point of separation, the mean intensity of pressure is essentially the same as that of the surrounding flow; since the separation generally occurs at a point of increased velocity—and hence of decreased pressure—a low pressure will prevail throughout the region of reverse flow. Separation is, in addition, a source of instability, the backflow along the boundary usually giving rise to eddies, which in turn lead to fully developed turbulence in the wake of the boundary transition. Thus, not only does this lowered pressure on the downstream side produce an additional boundary force known as *form drag*, but the generation of eddies on a relatively large scale greatly augments the drain upon the energy of the flow. In other words, the occurrence of separation necessarily increases both the resistance to motion and the rate of dissipation of mechanical energy through turbulence and viscous shear.

Example 43. In what zones can separation be expected to occur during flow over the spillway shown in profile? Suggest changes in design which would reduce or eliminate the separation tendency.

Since the stream lines near the boundary diverge locally at points a, b, c, and d, separation will occur in each of these zones as indicated. That at a, which is in a region of low velocity and hence not of great importance, could be reduced by eliminating the 90° angle of the boundary. Local deceleration leading to separation at b and c, which may be dangerous to the stability of the structure at high flow, should be avoided by patterning the spillway profile after the lower surface of a weir nappe for flow at the same head. No separation will take place in turbulent flow at d if the radius of the bucket is large in comparison with the thickness of the sheet.

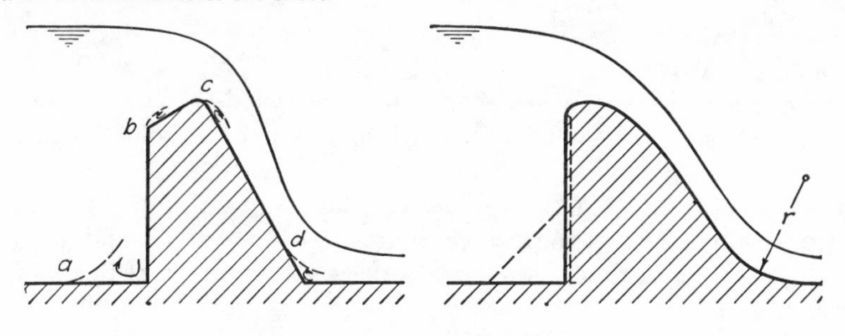

PROBLEMS

267. Separation generally occurs along the boundary of a pipe just upstream from an orifice (see Fig. 28). What effect would this have (a) upon the pressure intensity at the base of the orifice and (b) upon the coefficient of contraction? Suggest a boundary form which would eliminate such separation.

268. Would the pattern of irrotational flow around the streamlined strut shown in Fig. 22 be more closely approached in laminar or in turbulent flow? Why?

269. Cite instances in which separation is a desirable flow phenomenon.

39. DISTRIBUTION OF FLUID PRESSURE ON IMMERSED BODIES

Pressure variation around bodies of revolution. In flow at low Reynolds numbers, the variation in pressure intensity around an immersed body is so nearly hydrostatic as to warrant no further atten-

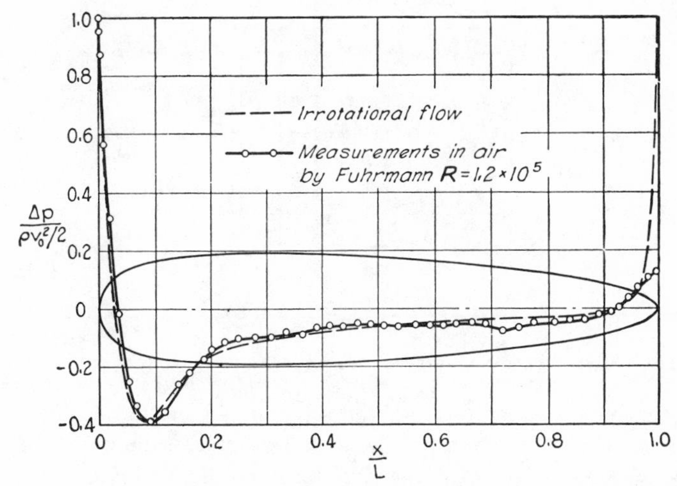

FIG. 118. Pressure distribution for flow past a streamlined body of revolution

tion. At high Reynolds numbers, on the other hand, unbalanced normal stresses may be so great as to require a detailed knowledge of pressure distribution if structural stability is to be attained. (A salient—and very costly—example of the need for accurate information as to pressure variation is found in the 1940 failure of the Tacoma Bridge due to unexpected oscillation in a high wind.)

So long as the form of a body is such that at high Reynolds numbers separation either will not occur or will be of negligible effect, the distribution of pressure may be found to a fair degree of approximation through application of the flow net or its three-dimensional equivalent. In Fig. 118, for instance, is shown the measured variation of the quantity $\Delta p/(\rho v_0^2/2)$ (i.e., the local Euler number) around the hull of a model dirigible, in comparison with that obtained from mathematical principles of irrotational flow; only at the very rear of the body is a perceptible difference apparent, indicating that the gradual boundary curvature and the presence of turbulence in the boundary layer have almost entirely eliminated the zone of discontinuity.

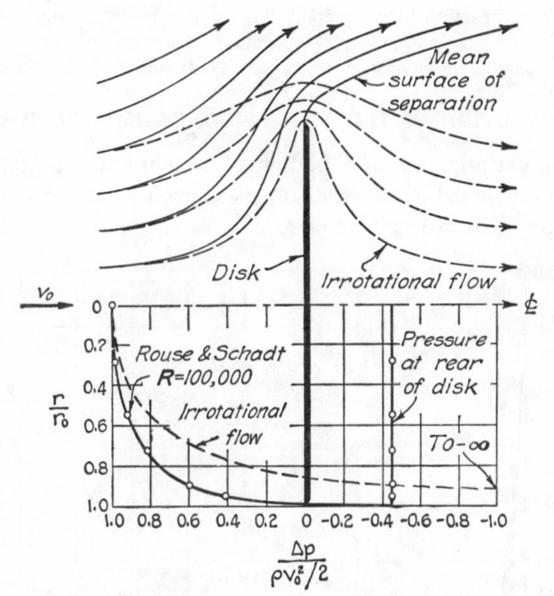

FIG. 119. Stream lines and pressure distribution for flow past a circular disk

An extreme departure from the streamlined profile of Fig. 118 is found in the case of a thin circular disk held at right angles to the flow. Were the flow pattern symmetrical fore and aft, as indicated by the broken lines for irrotational flow in Fig. 119, the same positive intensity of pressure would exist at each stagnation point, but at the edge of the disk a zero radius of curvature would require a pressure intensity of negative infinity. However, since such severe reduction in pressure is physically impossible, it is evident that the fluid must separate from the boundary in this zone. Separation, to be sure, will increase the radius of curvature of the boundary stream lines and thus make the pressure drop at the edge of the plate less extreme; but

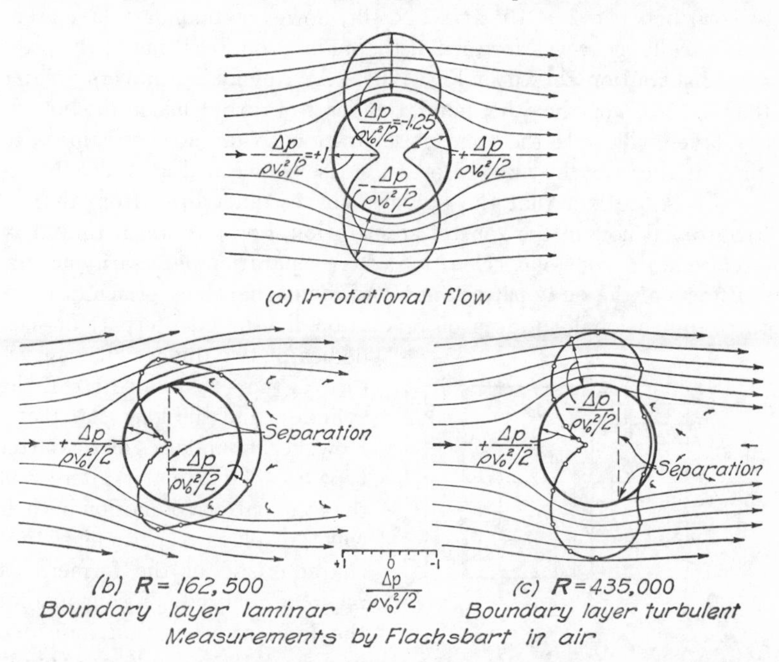

FIG. 120. Stream lines and pressure distribution for flow past a sphere.

this same low intensity of boundary pressure will then prevail throughout the zone of discontinuity, yielding the measured distribution curve shown in the figure. Knowledge of the geometry of the discontinuity surface would, of course, permit solution for the pressure variation by graphical methods; although such geometry may not in itself be determinate, the fact remains that, from moderately low to extremely high Reynolds numbers, it is governed solely by the boundary form.

Between these two limiting conditions of pressure distribution at moderate to high Reynolds numbers is a vast assortment of body shapes for which the relative pressure distribution varies with both the boundary geometry and the Reynolds number. That is,

$$\frac{\Delta p}{\rho {v_0}^2/2} = \frac{1}{\mathbf{E}^2} = \phi(\mathbf{R}, \text{form}) \tag{177}$$

Typical of such boundary forms is the sphere. According to the stream-line pattern for irrotational flow (see Example 5), the distribution of pressure should follow a curve which is symmetrical about the midsection, the highest intensity of pressure occurring at the two stagnation points, and the lowest around the circumference at right angles to the flow, as shown in Fig. 120*a*. At Reynolds numbers (VD/ν) between 2×10^4 and 2×10^5, however, boundary-layer separation will occur well forward of the midsection, resulting in the pressure distribution shown in Fig. 120*b*. At Reynolds numbers greater than 2×10^5, on the other hand, the onset of turbulence in the boundary layer will shift the zone of separation so far downstream as to yield at the rear the marked pressure rise shown in Fig. 120*c*. Noteworthy is the fact that the pressure curves differ little from that of irrotational flow in the zone of acceleration, however much they may deviate in the zone of deceleration where separation necessarily occurs.

Effects of the eddy pattern in two-dimensional flow. Each of these three-dimensional surfaces of revolution has its counterpart in such bodies of two-dimensional curvature as the streamlined strut, the cylinder, and the long, flat plate. However, boundaries of the latter type are of particular interest in that the pattern of motion in their wake displays a periodicity not characteristic of the former. It must be realized, of course, that the lines of separation indicated in Figs. 119 and 120 are simply averages with time, since an appreciable degree of fluctuation is to be expected in almost any zone of discontinuity. Although similar lines of average separation may be drawn for the plate and cylinder, and although the corresponding curves of mean pressure distribution (Figs. 121 and 122) may be obtained from measurement, the deviation from these patterns of two-dimensional motion will proceed with a regularity which is of considerable importance to the stability problem. For instance, visual inspection of the flow pattern just behind a cylinder will disclose the fact that the wake consists of a more or less orderly series of vortices which alternate in position about the centerline. Evidently, the region of discontinuity must pendulate from one side of the centerline to the other as the vortices alternately form and detach themselves from the boundary; as a

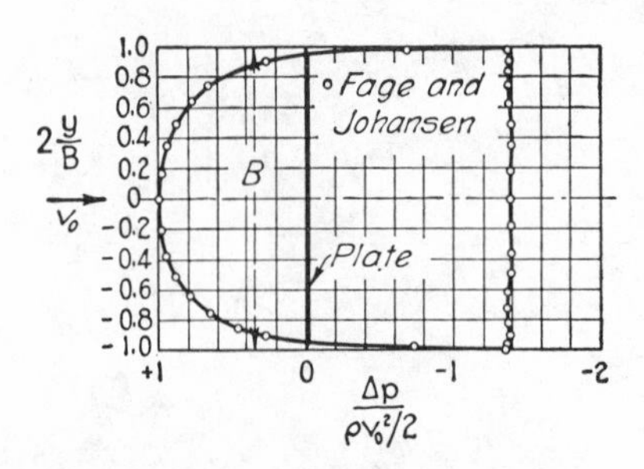

FIG. 121. Distribution of pressure for two-dimensional flow past a plate.

result, the zone of low pressure at the rear shifts from side to side, thereby producing a variable side thrust as well as a longitudinal drag.

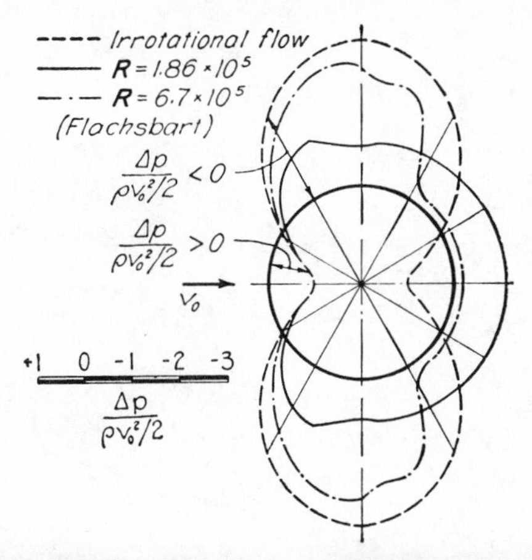

FIG. 122. Distribution of pressure for two-dimensional flow past a cylinder.

The pattern of motion within such a stable trail of vortices has been analyzed by von Kármán as a problem in irrotational flow to which the stream lines of Fig. 123 correspond. This analysis led to the conclusion that the relative spacing a/b of the vortices (see Fig. 124) will

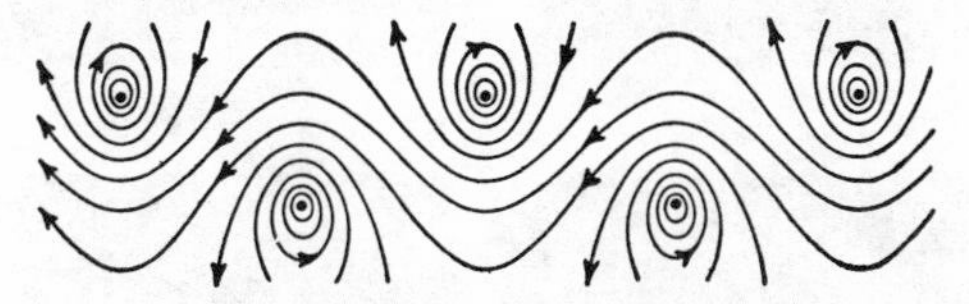

FIG. 123. Instantaneous stream lines of the von Kármán vortex trail.

have the value 0.281, while the vortex system as a whole will move with respect to the fluid at the velocity $v_v = 0.354\ \Gamma/b$, in which Γ (gamma) is a measure of the vortex intensity. That is, if a body moves through a stationary fluid at a velocity v_0, the vortices will

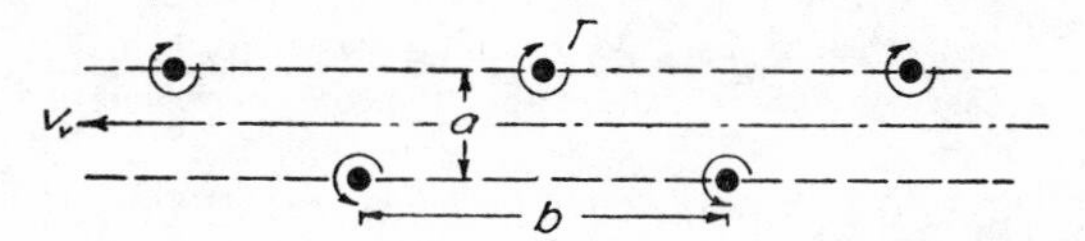

FIG. 124. Definition sketch for the analysis of the vortex trail.

follow the body at the velocity v_v, which is invariably lower than v_0; therefore, if the body is at rest and the fluid passes it at the velocity v_0, the absolute velocity of the vortex system (and hence its velocity relative to the body) will be $v_0 - v_v$. Since the ratio $b/(v_0 - v_v)$ then represents the time required for a pair of vortices to form and pass into the wake, it evidently indicates the period of oscillation of the discontinuity zone at the rear of the body.

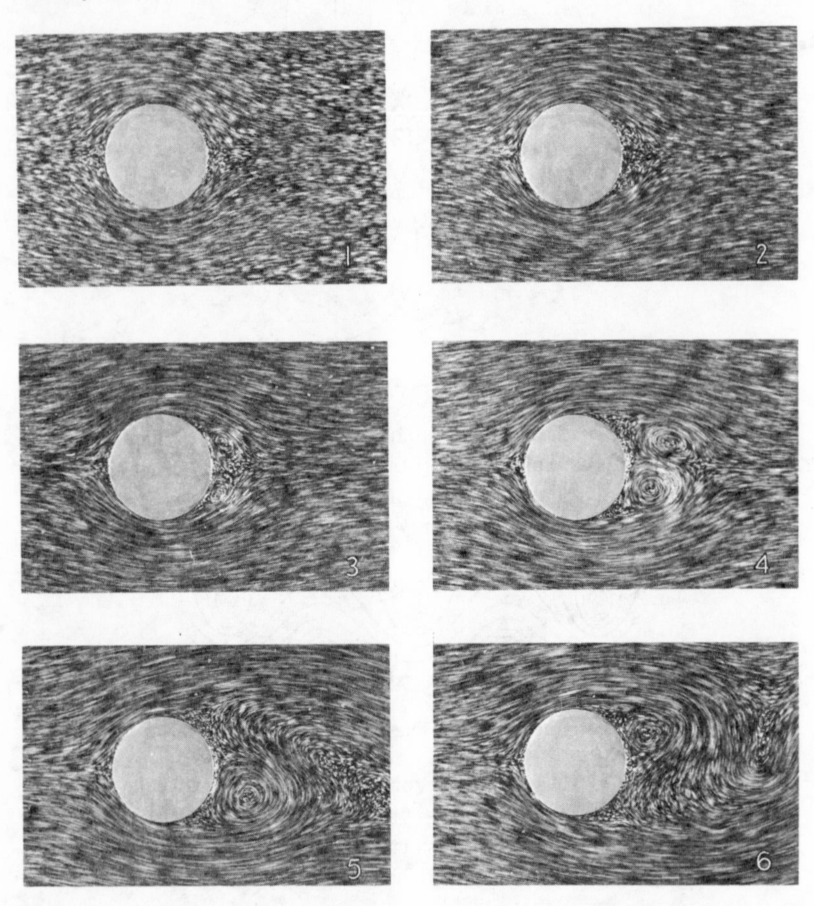

PLATE XIV. Photographs of the wake behind a stationary cylinder, showing successive stages of eddy formation after the commencement of flow.

Although the von Kármán *vortex trail* can thus be analyzed in itself without regard to the viscous effects upon which its formation and ultimate decay depend, the ratio of the characteristic lengths a and b to the dimension of the body, and the ratio of the vortex velocity v_v to the relative velocity v_0 between body and fluid, will necessarily vary with the form of the body and the Reynolds number of the flow. However, for a given body form it will be found that the dimensionless combination of the period of oscillation T, the velocity v_0, and a linear measure L of the body may invariably be expressed as

$$\frac{Tv_0}{L} = \phi(\mathbf{R}) \tag{178}$$

This functional relationship necessarily depends for its determination upon experimental measurement. In the case of the cylinder, it is found that Tv_0/D is approximately 5 for values of $\mathbf{R}$ between 2×10^2

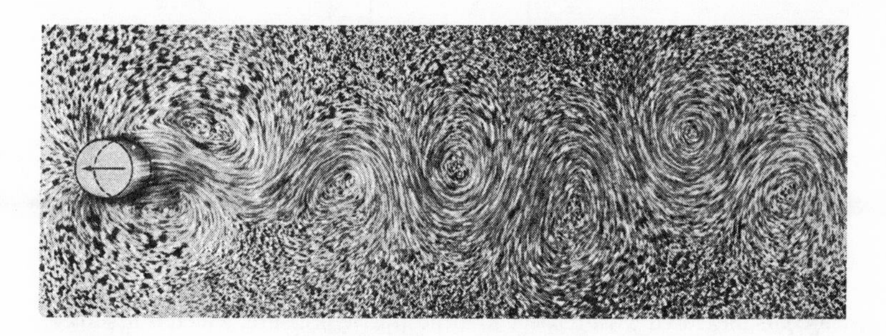

PLATE XV. Vortex trail in the wake of a moving cylinder.

and 2×10^5, decreasing to perhaps half this magnitude as the boundary layer becomes turbulent. Noteworthy is the fact that, in the case of the normal plate, the constant value $Tv_0/B \approx 7$ is applicable so long as the drag is independent of $\mathbf{R}$.

It should by now be evident to the reader that in only a few isolated cases can purely analytical methods yield a quantitative evaluation of the pressure distribution for flow around a body of arbitrary form. Such methods do, however, provide a significant qualitative picture of the general flow characteristics, in that they clarify the causes, indicate the approximate location, and warn of the structural dangers, of boundary-layer separation. Furthermore, through the dimensional correlation of the variables involved, these methods offer a systematic guide for the experimental study of force distribution on a scale model of any projected structure. Indeed, the importance of the wind tunnel and the towing tank to modern structural design cannot be overemphasized.

Example 44. Wind-tunnel tests yielded the accompanying curves of pressure distribution around a model building of elementary form. Estimate the mean intensity of normal force upon the roof of the prototype structure during a 40-

mile-per-hour wind, assuming atmospheric pressure to prevail throughout the interior. What effective change would occur if an open door at the front permitted the internal pressure to attain a maximum magnitude?

According to the diagram, $\Delta p_m/(\rho v_0^2/2) = -0.6$, whence

$$\Delta p_m = -0.6\,\frac{\rho v_0^2}{2} = -0.6 \times \frac{0.0025 \times \left(\dfrac{40 \times 5280}{60 \times 60}\right)^2}{2} = -2.6 \text{ psf}$$

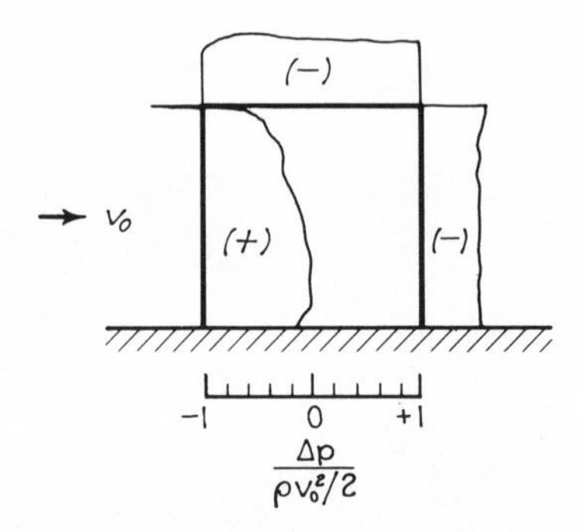

Since Δp is measured with respect to atmospheric pressure, which is assumed to prevail within the building, this represents a normal force in the *upward* direction of 2.6 pounds per square foot of roof. If, on the other hand, the pressure intensity within the building is increased above atmospheric by the amount

$$\Delta p = \frac{\rho v_0^2}{2} = \frac{0.0025 \times \left(\dfrac{40 \times 5280}{60 \times 60}\right)^2}{2} = 4.3 \text{ psf}$$

the force tending to lift the roof will have a mean intensity of $4.3 + 2.6 = 6.9$ pounds per square foot.

PROBLEMS

270. What are the maximum and minimum intensities of pressure on a ¼-inch spherical shot when traveling at a speed of 500 feet per second in air having a temperature of 60° Fahrenheit?

271. Determine from the pressure distribution of Fig. 119 the total force exerted upon a 3-foot circular disk when dragged through water at a velocity of 5 feet per second. (Suggestion: plot the product $2\pi r\,\Delta p$ against r, and integrate graphically.)

272. A horizontal tension member of a hydraulic structure has the form of a T, the 4-inch flange of which is perpendicular to the direction of flow. If this member lies 5 feet below the water surface in a zone of essentially uniform flow, at approximately what velocity may cavitation be expected to develop?

273. Estimate the maximum and minimum intensities of pressure on the 180-foot perisphere of the New York World's Fair of 1939 in a 50-mile-per-hour wind.

274. A billboard having a surface height of 9 feet and a continuous length of 60 feet is so mounted as to be freely exposed to the wind. From the pressure-distribution curve of Fig. 121 estimate the total force exerted upon the structure in a normal wind having a speed of 30 miles per hour.

275. For purposes of velocity measurement, a cylindrical tube ½ inch in diameter is inserted through stuffing boxes at diametrically opposite points on the wall of a 3-foot conduit, upstream and downstream piezometer inlets at a given section of the tube being connected to a differential water gage. Prepare a rating curve of velocity (from 5 to 20 feet per second) against differential head for use of the tube in water at 60° Fahrenheit.

276. Wind-tunnel tests on a model airship hangar indicated that the pressure intensity on the lee side of the hangar in a wind at right angles to the hangar axis corresponded to the magnitude -0.65 of the dimensionless parameter $2\,\Delta p/\rho v_0^2$. If the hangar is covered with sections of roofing material 4 feet by 8 feet in size, what outward force would be exerted on a typical section in a 40-mile-per-hour wind?

277. At what wind velocity would telephone wires ⅛ inch in diameter attain a vibration period of 0.01 second?

278. A rack to collect debris at the entrance of a power plant consists of parallel pipes 2 inches in diameter and 6 inches on center. If the natural period of vibration of the pipes is $\frac{1}{20}$ second, at what water velocities would vibrations of serious magnitude be expected?

40. DRAG OF IMMERSED BODIES

Formulation of the drag equation. The total longitudinal force exerted by a moving fluid upon an immersed body necessarily consists of the summation of the longitudinal components of all normal and tangential stresses upon the boundary surface. In the case of streamlined bodies at high Reynolds numbers, the resultant effect of normal stresses will be a minimum, and the drag can hence be considered due almost entirely to boundary-layer shear. In the case of bodies of angular profile, on the other hand, the reduction of pressure intensity in the region of discontinuity will so outweigh the boundary shear that the drag can be considered due almost entirely to the unbalanced normal forces on the front and rear sides. The former case evidently approaches the limit of pure surface drag, and the latter the limit of pure form drag. Between these limits the proportionate effects of surface and form upon the total drag of a body can be evaluated only by measurement and integration of the pressure distribution.

As a general rule, knowledge of the total drag of a body, rather than its component parts, is required for design purposes. Since both tangential and normal stress are dependent upon the Reynolds number, the term Δp of the Euler number in Eq. (177) may be replaced by the total longitudinal force F divided by the projected area of the body under consideration, the resulting parameter then also being a function of the Reynolds number:

$$\frac{F/A}{\rho v_0^2/2} = C_D = \phi(\mathbf{R}, \text{form})$$

As indicated by this expression, the modified Euler number represents a *coefficient of drag* C_D which, for a given body form, should vary as a function of the Reynolds number. Once the form of this function is known, the drag of the body may be evaluated from the equation

$$F = C_D A \frac{\rho v_0^2}{2} \tag{179}$$

Since C_D is evidently a measure of the relative resistance of bodies of the same cross-sectional area under the same flow conditions, a study of its variation with body form and the Reynolds number will be found of great significance.

Resistance diagram for bodies of revolution. As indicated in Fig. 125, a wealth of experimental data is at hand for the drag coefficient of spheres over a very great Reynolds-number range. At low values of $\mathbf{R}$ (i.e., in the zone of deformation drag) the measurements are seen to follow the straight line $C_D = 24/\mathbf{R}$, which may be shown to correspond to the equation of Stokes [Eq. (122)] by the following operation:

$$F = 3\pi D\mu v_0 \times \frac{8D\rho v_0}{8D\rho v_0} = \frac{24\mu}{v_0 D\rho} \frac{\pi D^2}{4} \frac{\rho v_0^2}{2} = \frac{24}{\mathbf{R}} A \frac{\rho v_0^2}{2} \tag{180}$$

The experimental points begin to deviate from this line as soon as the accelerative effects, ignored by Stokes, begin to become appreciable; a Reynolds number slightly less than unity evidently marks the approximate limit of deformation drag, beyond which the Stokes equation is no longer applicable. With increasing values of $\mathbf{R}$ the zone of appreciable viscous deformation becomes restricted more and more to the immediate boundary vicinity; at the same time, however, accelerative effects become more pronounced, and separation takes place in the zone of deceleration at the rear. By the time a Reynolds number of about 2×10^4 is reached, the viscous shear at the boundary has become so insignificant in comparison with the pressure reduction in the zone of discontinuity that the drag coefficient no longer varies perceptibly with the Reynolds number; this condition corresponds to the pressure distribution shown in Fig. 120*b*.

With the onset of turbulence in the boundary layer, the abrupt contraction in the zone of discontinuity and the accompanying rise in pressure intensity at the rear (Fig. 120c) produce a marked drop in the resistance coefficient, as seen at the right of Fig. 125. If the sphere is very smooth and if the approaching flow is undisturbed, the boundary layer will become unstable at a Reynolds number of approximately 2×10^5, and a reduction in C_D of about 60 per cent will take

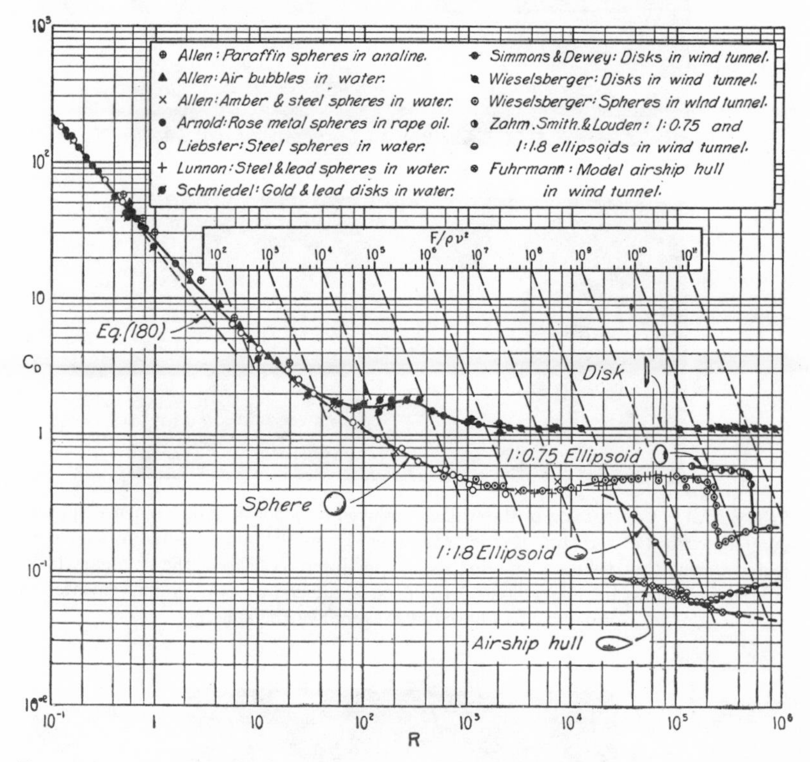

FIG. 125. Coefficients of drag as functions of the Reynolds number for bodies of revolution.

place. Noteworthy is the fact that this critical value of **R** will decrease with increasing boundary roughness or turbulence of the oncoming fluid; boundary roughness, however, is found to limit the reduction of C_D beyond the critical zone.

Unlike the sphere, the circular disk is characterized by a drag coefficient which becomes independent of the Reynolds number beyond the region of deformation drag, owing to the fact that the form of the wake and the accompanying distribution of pressure (see Fig. 119) are governed entirely by the boundary geometry. The other extreme is approached by the streamlined body of Fig. 118, for which the drag

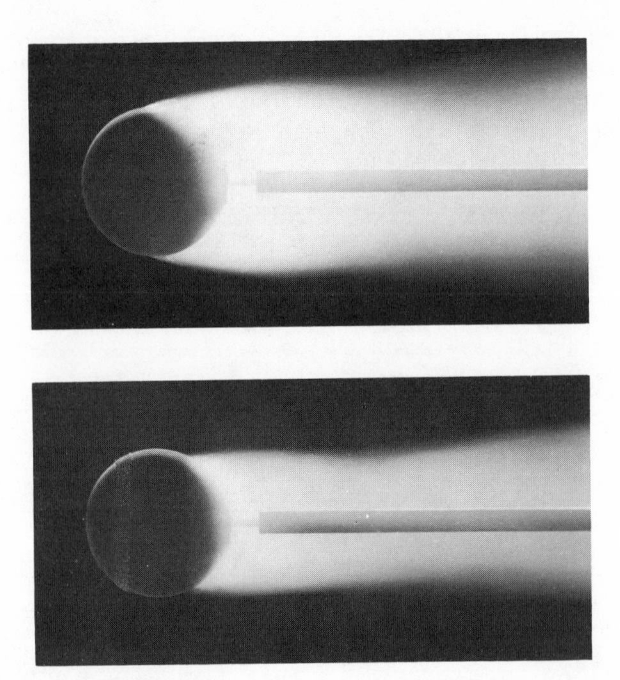

PLATE XVI. Comparison of the wakes produced by smooth and roughened spheres in an air stream at the same Reynolds number ($\mathbf{R} \approx 10^5$), made visible by smoke injected through the tube surrounding the supporting rod. Above, the boundary layer is laminar; below, boundary-layer turbulence is induced by sand grains cemented in a narrow band around the front of the sphere.

PLATE XVII. Region of separation downwind from a circular disk.

at high Reynolds numbers is also shown. It is significant to note that the latter type of body would encounter only about 5 per cent of the resistance of a disk of the same diameter under the same flow conditions.

Such a diagram of C_D against $\mathbf{R}$ at once permits solution for the resistance encountered by a body moving at a given velocity. Certain problems of free fall, on the other hand, require solution for the terminal velocity which would be attained by a body of given weight and size; so long as C_D varies with $\mathbf{R}$, it will be found that the velocity can be determined only by the tedious process of trial and error, unless the resistance diagram is modified as follows: If the parameters

$$C_D = \frac{F}{A\rho v_0^2/2} \quad \text{and} \quad \mathbf{R} = \frac{v_0 D}{\nu}$$

are combined through elimination of the velocity, it will be found that

$$C_D = \frac{1}{\mathbf{R}^2}\frac{2FD^2}{A\rho\nu^2} = \frac{1}{\mathbf{R}^2}\frac{8}{\pi}\frac{F}{\rho\nu^2} \tag{181}$$

A line of $C_D : \mathbf{R}$ for a given value of $F/\rho\nu^2$ on Fig. 125 will then indicate the locus of points for which the resistance is constant, and the intersection of this line with any resistance curve will indicate the Reynolds number and the drag coefficient for the given values of F, ρ, and ν; the

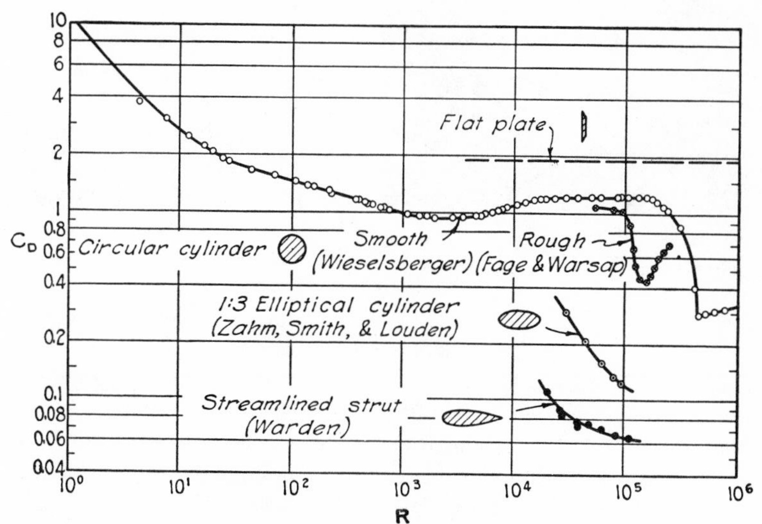

FIG. 126. Coefficients of drag as functions of the Reynolds number for two-dimensional bodies.

corresponding magnitude of v_0 may then be evaluated from either C_D or **R**. A system of these supplementary lines will be found on the diagram for use in fall-velocity evaluation.

Resistance coefficients of miscellaneous bodies. Owing to the close analogy between three-dimensional bodies of revolution and their two-dimensional counterparts, it is only to be expected that the plots of C_D against **R** for the long cylinder, plate, and streamlined strut, as shown in Fig. 126, will be of essentially the same form as those for the sphere, disk, and airship hull of Fig. 125. The fact must be noted, however, that these curves apply only to two-dimensional bodies of very great length, and that bodies of finite length will encounter a lower resistance if three-dimensional flow can occur at the ends.

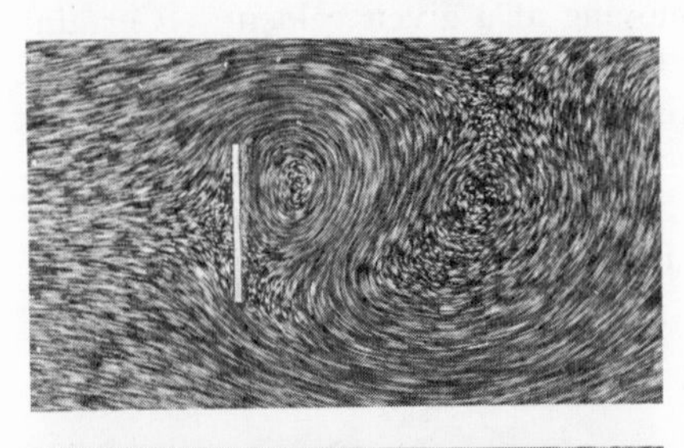

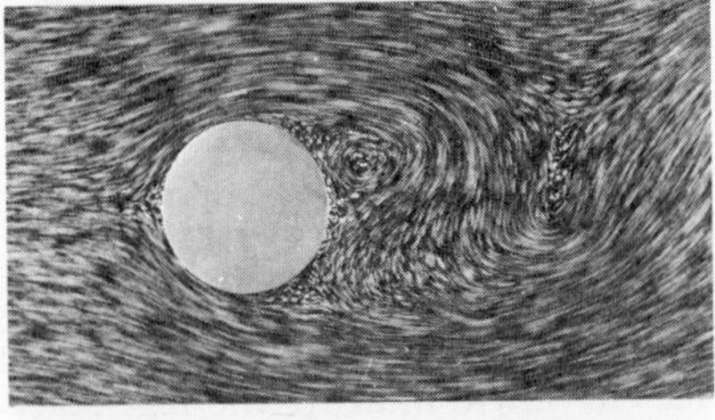

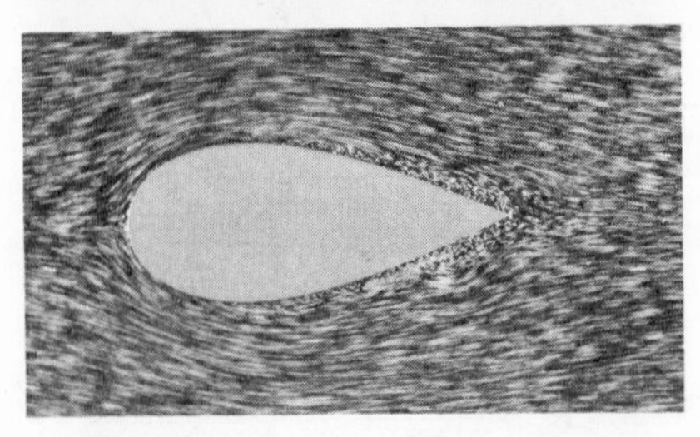

PLATE XVIII. Comparison of the wakes produced by a plate, a cylinder, and a streamlined strut.

The drag of other types of body—whether two- or three-dimensional—will follow the same general trend as the elementary forms herein described. Evidently, the more abrupt the curvature of the sides of the body, the more pronounced the form effect will be in comparison with that of boundary shear. Thus, a thin plate will produce surface drag, yet no form drag whatever, when held parallel to the flow; but the form effect will be extremely great—and the surface effect negligible—when the same plate is turned through 90°. Indeed the process of *streamlining*—i.e., the easement of boundary curvature at the sides and rear—is entirely for the purpose of reducing the form effect upon the total drag. While the ideal limit of streamlining would involve the complete elimination of separation, it should be noted that the surface drag necessarily increases as a body is lengthened; in other words, C_D attains its smallest value when form drag and surface drag together are a minimum.

In the case of spheres, it is probably the lower range of the Reynolds-number scale which is of greatest importance, for this region yields useful information as to the motion of such finely divided material as silt, mist, dust, or soot in gaseous or liquid media. On the other hand, the majority of engineering problems involve flow around bodies at high values of **R**, at which the factor C_D is essentially dependent upon form alone. In many elementary cases, the drag of structures may be

TABLE III

APPROXIMATE VALUES OF THE DRAG COEFFICIENT FOR VARIOUS BODY FORMS

Form of Body	L/D	**R**	C_D
Circular disk		$>10^3$	1.12
Tandem disks (L = spacing)	0	$>10^3$	1.12
	1		0.93
	2		1.04
	3		1.54
Rectangular plate (L = length, D = width)	1	$>10^3$	1.16
	5		1.20
	20		1.50
	∞		1.90
Circular cylinder (axis $\parallel$ to flow)	0	$>10^3$	1.12
	1		0.91
	2		0.85
	4		0.87
	7		0.99
Circular cylinder (axis $\perp$ to flow)	1	10^5	0.63
	5		0.74
	20		0.90
	∞		1.20
	5	$>5 \times 10^5$	0.35
	∞		0.33
Streamlined foil (1:3 airplane strut)	∞	$>4 \times 10^4$	0.07
Hemisphere: Hollow upstream		$>10^3$	1.33
Hollow downstream			0.34
Sphere		10^5	0.50
		$>3 \times 10^5$	0.20
Ellipsoid (1:2, major axis $\parallel$ to flow)		$>2 \times 10^5$	0.07
Airship hull (model)		$>2 \times 10^5$	0.05

estimated through use of the characteristics given in Table III—for example, the force of wind on towers and cables. In more complicated instances (wind effects on bridges or the drag of trains, airships, and undersea craft) the characteristic data must be obtained in the wind tunnel or towing tank through use of scale models tested at comparable values of the Reynolds number.

Example 45. A pilot balloon 3 feet in diameter rises under a net lifting force (buoyancy less weight) of 0.8 pound. Estimate its velocity of ascent.

Assuming that $\nu = 1.4 \times 10^{-4}$ square foot per second and $\rho = 0.0025$ slug per cubic foot,

$$\frac{F}{\rho\nu^2} = \frac{0.8}{0.0025 \times (1.4 \times 10^{-4})^2} = 1.63 \times 10^{10}$$

From the intersection of the corresponding line with the drag curve for spheres on Fig. 125, $C_D = 0.20$ and $\mathbf{R} \approx 5 \times 10^5$. Therefore,

$$v_0 = \sqrt{\frac{2F}{C_D A \rho}} = \sqrt{\frac{2 \times 0.8}{0.20 \times \frac{\pi}{4} \times 3^2 \times 0.0025}} = 21.3 \text{ fps}$$

PROBLEMS

279. The perisphere at the 1939 New York World's Fair had a diameter of 180 feet. Estimate the horizontal force which would have been exerted upon it by a 50-mile-per-hour wind.

280. Improved steamlining produces a 15 per cent reduction in the drag coefficient of a torpedo when traveling fully submerged. What percentage increase in speed would this represent, assuming the driving power to remain the same?

281. The hull of a semi-rigid airship having the profile shown in Fig. 125 has a maximum cross-sectional diameter of 30 feet. Estimate the power required to overcome the resistance of the hull when cruising at a speed of 60 miles per hour in air having a temperature of 40° Fahrenheit?

282. A wind having a velocity of 50 miles per hour strikes a ½-inch transmission cable at right angles. Determine the force exerted upon the cable per foot of length, assuming the temperature of the air to be 50° Fahrenheit.

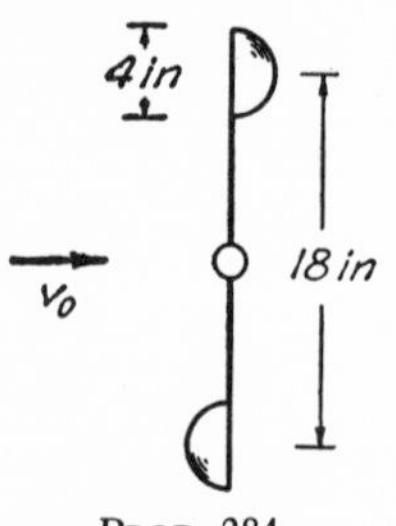

PROB. 284.

283. Determine the terminal velocities of fall of spherical hail stones having diameters of ¼, ½, and 1 inch; assume a specific gravity of 0.9 and an air temperature of 50° Fahrenheit.

284. A cup anemometer, used in measuring wind velocity, consists of two hollow hemispheres mounted in opposite directions at the ends of a horizontal rod which turns freely about a vertical axis, as shown in the accompanying sketch. What torque would be required to hold the rotating member stationary in a 30-mile-per-hour wind?

285. Estimate the maximum load which may be carried by an 18-foot parachute without attaining a velocity of descent of more than 20 feet per second.

286. The ratio of the fall velocity of sand and silt grains to the velocity characteristics of turbulence is of importance in all problems of sedimentation. Assuming such grains to be spherical and to have a specific gravity of 2.56, determine the fall velocities in water at 60° Fahrenheit corresponding to grain diameters of 0.01, 0.1, and 1.0 inch. (Note: the buoyant force of the water must not be ignored.)

287. Solve Problem 271 through use of the corresponding drag coefficient.

288. A stabilizing drag used in shipping consists of two 3-foot disks arranged in tandem 10 feet on center at the end of a long cable. What force would be required to tow the disks through sea water at a speed of 3 miles per hour?

289. Estimate the bending moment at the base of a 50-foot smoke stack 5 feet in diameter during a 60-mile-per-hour wind; assume the air temperature to be 60° Fahrenheit.

290. A lead shot ⅛ inch in diameter is fired upward over a body of open water. (*a*) If the velocity of the shot as it leaves the gun is 500 feet per second, what is the initial air resistance? (*b*) What is the terminal velocity of descent of the shot in air? (*c*) What terminal velocity will be attained by the shot after striking the water?

291. Solve Problem 274 through use of the corresponding drag coefficient.

292. Compare the rate at which a ⅛-inch bubble of air would rise through water with the rate at which a ⅛-inch drop of water would fall through air; assume a temperature of 60° Fahrenheit.

293. Assuming an automobile to have a projected area of 25 square feet, estimate the horsepower required to overcome wind resistance at a speed of 50 miles per hour (*a*) if the car is not streamlined, and (*b*) if the car is well streamlined.

294. From the data given in Table III, estimate the minimum spacing of tandem disks at which the drag of each would be essentially independent of the other.

41. DISTRIBUTION OF PIEZOMETRIC HEAD AT CONDUIT TRANSITIONS

Similarity of flow around bodies and through conduits of comparable form. Nearly every elementary form of immersed body represents the counterpart of a conduit transition, in that the flow characteristics of any such transition are essentially similar to those of the corresponding immersed body. For instance, a plate or diaphragm orifice in a pipe involves form effects comparable to those of a circular disk held at right angles to the flow; a Venturi meter produces surface effects not greatly different from those of an airship hull; or the flow at the front and rear of a circular cylinder held with axis parallel to the direction of motion is not unlike that at a sudden pipe contraction followed by a sudden enlargement. Indeed, such analogous states of motion differ in only two basic ways: first, in the case of immersed bodies the fluid is assumed to extend a considerable distance in all directions, while the centerline of a conduit necessarily marks the transverse limit of the boundary influence; second, in the case of immersed bodies the boundary layer begins to develop at the front of the body itself, while boundary effects in a conduit have already developed far upstream from the transition section.

Since the drag of bodies is most often associated with motion through air, in discussing distribution of boundary pressure in Section 39 little heed was paid to the effect of fluid weight; in any event, of course, the distribution of $\Delta p/(\rho v_0^2/2)$ for gases would be identical to the distribution of $\Delta h/(v_0^2/2g)$ for liquids. In the case of conduit transitions, however, the distribution of piezometric head is of such graphic significance that all distribution plots in the present section will represent liquid flow; needless to say, the parameter $\Delta h/(V^2/2g)$ may likewise be changed to $\Delta p/(\rho V^2/2)$ if the fluid in question is a gas. Owing to the close analogy of flow conditions, it is only to be expected that the distribution of head along the boundaries of any conduit transition will vary with the boundary geometry and the Reynolds number of the flow, just as in the case of immersed bodies. Thus, at low Reynolds numbers viscous effects should again predominate, and at high Reynolds numbers such effects should become subordinate to those of acceleration, the phenomenon of separation then playing a decided role. It is not to be expected, however, that an equivalent effect of boundary-layer turbulence upon the wake will be encountered at very high Reynolds numbers, for the boundary layer of the approaching flow already extends to the centerline of the conduit; on the other hand, the velocity distribution of the approaching flow now becomes of considerable importance.

Variation in head at a local constriction. In Fig. 127 is shown in dimensionless form the distribution of piezometric head at a diaphragm orifice for a high Reynolds number, comparable to that of the disk of Fig. 119. The circumference of the diaphragm is now the zone of stagnation, but it will be seen that separation in the region of deceleration just upstream from the diaphragm does not permit h to attain the full stagnation magnitude. The edge of the orifice opening, similarly, is now equivalent to the circumference of the disk, and the outline of the submerged jet is essentially the same as the separation profile in the wake of the disk; evidently, the low pressure behind the disk also obtains downstream from the diaphragm. Just as the eddies behind the disk gradually spread laterally and eventually decay, the eddies produced by the diaphragm spread toward the centerline of flow; the mixing produced by the spreading eddies simultaneously causes the jet to expand until it finally fills the conduit, with an accompanying rise in pressure. The flow is not yet uniform, however, owing to the presence of turbulence far greater in intensity than that of the approaching flow; in other words, not until the kinetic energy of this excess turbulence has been dissipated through viscous shear over a considerable length of conduit can the effects of the transition upon the velocity and pressure distribution no longer be noted. The great drop in head which is produced by the formation of such turbulence is evident from the diagram.

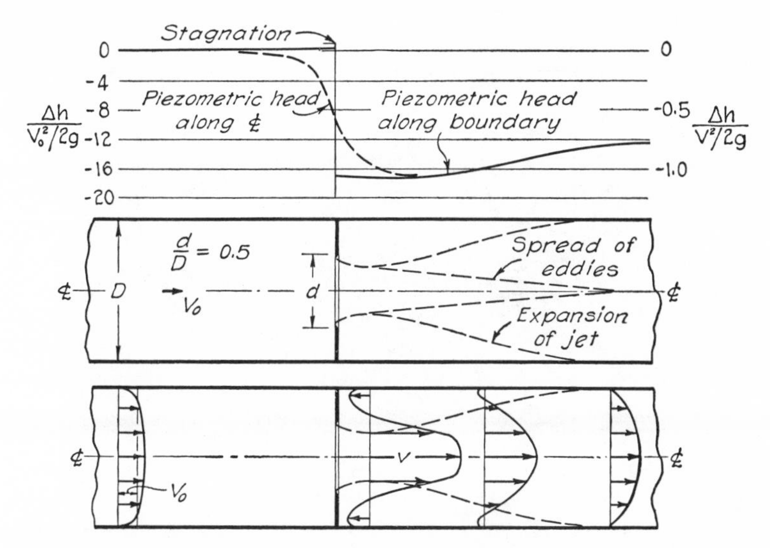

FIG. 127. Distribution of velocity and piezometric head for a pipe orifice.

The other extreme of conduit transition is illustrated by the Venturi meter, which is comparable to the streamlined hull of Fig. 118. However, while the streamlining of an immersed body is generally restricted to the rear portions, the streamlining of a conduit transition

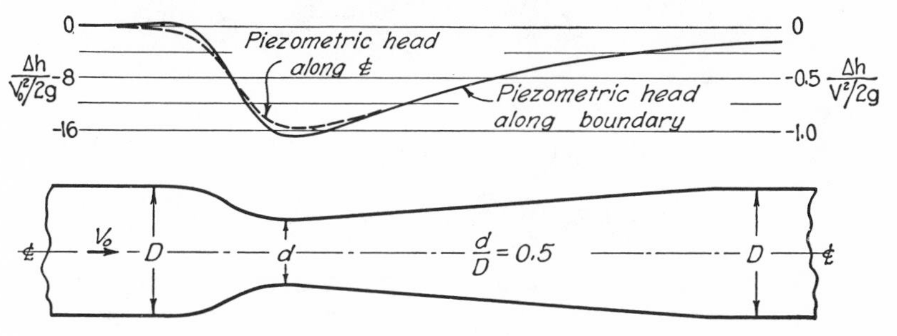

FIG. 128. Distribution of piezometric head for a Venturi meter.

must begin at the very entrance, owing to the low velocity in the laminar sublayer of the approaching flow. Separation at the entrance is therefore avoided by the gradual convergence shown in Fig. 128, which produces only a slight local rise in h. The pressure necessarily drops as the flow section contracts, but at different rates along boundary and centerline; as a result, the pressure distribution can become hydrostatic across the throat section only if the throat is relatively long. Separation in the divergent section, if the flow is turbulent, is practically eliminated when the angle of flare is less than about 10°.

PLATE XIX. Reduction of the disturbance at a conduit constriction through partial streamlining; compare flow patterns with those of Plate XVIII.

The fact that the piezometric head downstream from the transition is nearly as great as that of the approaching flow is an indication of the gain in efficiency which can be obtained by proper streamlining.

Use of head variation in flow measurement. The investigation of pressure distribution at lower Reynolds numbers would be of little practical significance were it not for the fact that transitions of this nature are frequently used for measuring rates of flow. That is, since the change in the quantity $h/(V_0^2/2g)$—or $p/(\rho V_0^2/2)$ if the fluid is a gas—between any two points depends only upon the conduit geometry and the Reynolds number, the same differential head between two such points of a given transition should invariably indicate the same rate of flow for a fluid having the same characteristics. Indeed, from the dimensionless nature of the parameters $\Delta h/(V^2/2g)$ and VD/ν, it is only reasonable to expect that calibration of a given meter with any one fluid would yield a general discharge curve applicable to any other fluid. This curve should then be characteristic of all geometrically similar meters as well, provided only that the approach conditions are also similar and that the piezometers are located at equivalent points. In order to attain uniformity of location and maximum sensitivity of indication, it is customary to place the piezometers for an orifice immediately upstream and downstream from the diaphragm; those for the Venturi meter are usually just upstream from the entrance and at the midpoint of the throat. For the sake of similarity with earlier relationships for orifice discharge [see Eq. (37)], the reference velocity V in the discharge coefficient is again taken as the ratio of the flow rate Q to the minimum cross-sectional area $\pi d^2/4$ of the conduit. Thus

$$\frac{V}{\sqrt{2g\,\Delta h}} = C_d = \phi(\mathbf{R})$$

whereupon the discharge relationship becomes, as before,

$$Q = C_d \frac{\pi d^2}{4} \sqrt{2g\,\Delta h} \qquad (182)$$

The variation of the coefficient of discharge (i.e., the Euler number) with the Reynolds number for pipe orifices and Venturi meters of various proportions is shown in Fig. 129, the Reynolds number, like the Euler number, being based upon the diameter and the mean velocity at the narrowest boundary section. At once evident is the striking similarity of this diagram to Fig. 125 for the drag of immersed bodies of analogous form—similarity, that is, so far as general characteristics are concerned, since C_d is a measure of discharge while C_D is a measure of resistance (i.e., $C_d \sim 1/\sqrt{C_D}$). Thus, so long as the Reynolds number is very low, accelerative effects will be negligible in comparison with those of viscous deformation. Accelerative effects will, however, become appreciable while the approaching flow is still laminar, and the relative influence of the parabolic velocity distribution upon contractions of different relative size is at once evident from the diagram. Ultimately, of course, the viscous effects upon the differential pressure approach a minimum in comparison with accelerative effects hence at high values of $\mathbf{R}$ the several curves approach horizontal asymptotes depending in magnitude only upon the boundary geometry. It is to be noted, however, that boundary roughness is an essential part of such geometry, for, as the relative roughness of the pipe is increased, the increase in the ratio v_{max}/V results in a corresponding rise in the limiting value of C_d.

As in the case of Fig. 125, a plot of this nature may be used conveniently only if the velocity of flow is known. In order to determine

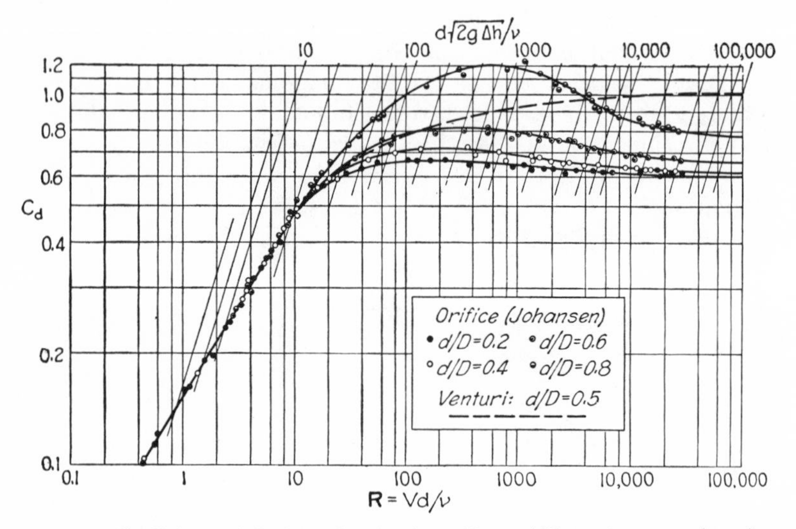

FIG. 129. Coefficients of discharge for the pipe orifice and Venturi meter as functions of the Reynolds number.

the velocity accompanying a given differential pressure, a system of supplementary lines similar to that of Fig. 125 must be added to the diagram. Since

$$C_d = \frac{Q}{\frac{1}{4}\pi d^2\sqrt{2g\ \Delta h}} \quad \text{and} \quad \mathbf{R} = \frac{Qd}{\frac{1}{4}\pi d^2 \nu}$$

elimination of Q will yield the counterpart of Eq. (181):

$$C_d = \mathbf{R}\frac{\nu}{d\sqrt{2g\ \Delta h}}$$

The sloping lines in Fig. 129 represent lines of constant $d\sqrt{2g\ \Delta h}/\nu$; intersection of the proper line with the proper discharge curve evidently yields the corresponding magnitudes of C_d and $\mathbf{R}$, from either of which the rate of flow may be determined.

Abrupt contractions; conduit inlets. Between the two extremes of conduit transition shown in Figs. 127 and 128 there exists a vast assortment of boundary forms. As in the case of immersed bodies, the distribution of boundary pressure will vary with both the Reynolds number and the geometry of the transition in such a complex manner that experimental measurement represents the only accurate means of analysis. In general, the more angular the profile, the greater the drop in pressure intensity or piezometric head in the wake of the transition; and the more perfect the streamlining, the more closely will the downstream conditions approach those upstream. Needless to say,

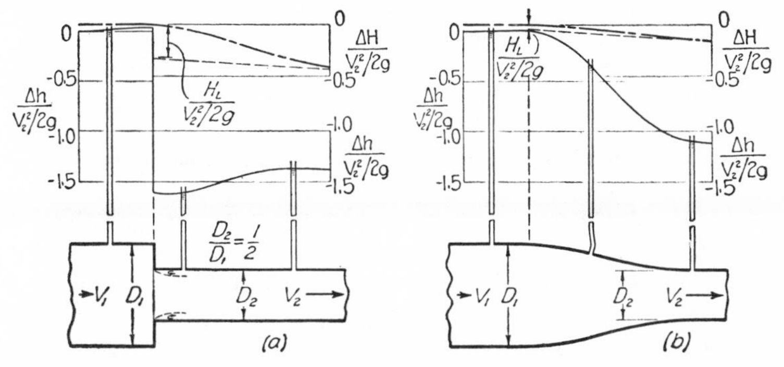

FIG. 130. Variation in head at an abrupt and a gradual reduction in pipe diameter.

however, even complete elimination of boundary-layer separation will not permit complete recovery of piezometric head, owing to the tangential stresses which must invariably exist in the boundary layer. On the other hand, such distribution curves for almost any transition will approach a constant form at moderately high Reynolds numbers since most flows of practical importance are in this region, an investigation of the limiting pressure curves for various elementary transitions will be sufficient for present purposes.

An abrupt contraction in the cross section of a circular conduit, as shown in Fig. 130*a*, will produce two zones of separation. As a result, the piezometric head will not rise to its full stagnation value in the boundary corner, and yet will drop considerably below its ultimate magnitude as the flow continues to contract beyond the transition section. Even though the mixing due to the boundary eddies eventually produces full expansion of the flow some distance downstream, the accompanying rise in piezometric head by no means represents full recovery, as may be seen from comparison with the streamlined transition of Fig. 130*b*.

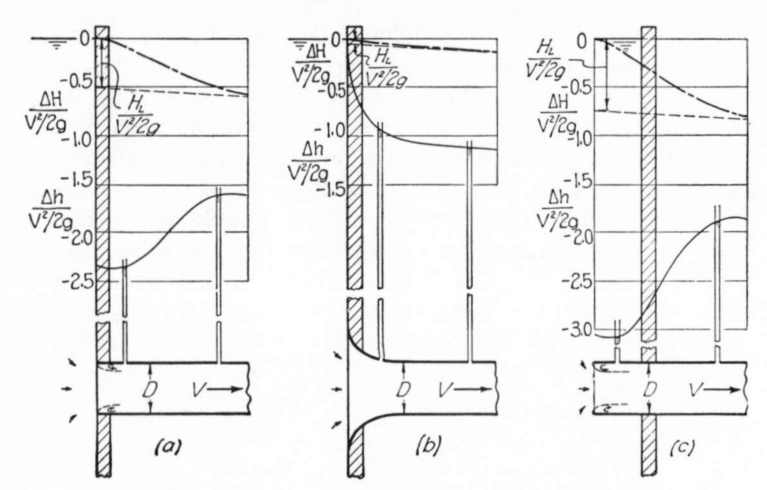

FIG. 131. Variation in head for different forms of pipe inlet.

The ultimate change in h at such a contraction will evidently vary with the contraction ratio. The limiting case, of course, is represented by the inlet to a conduit from a large tank or reservoir, shown in Fig. 131*a*. While this represents the maximum reduction in h of the series of Fig. 130*a*, it will be seen from Fig. 131*c* that projection of the conduit into the reservoir will result in an even greater reduction.

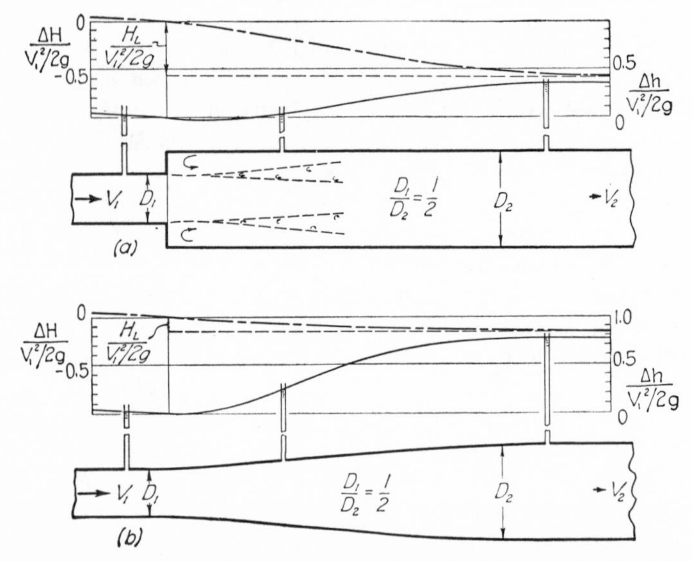

FIG. 132. Variation in head at an abrupt and a gradual increase in pipe diameter

Complete elimination of the separation phenomenon by proper curvature of the boundary, on the other hand, will yield a curve which at no point is lower than that of the ultimately uniform flow (Fig. 131*b*). It is to be noted that high velocities of inflow may reduce the pressure intensity locally to the point of cavitation; proper streamlining of the inlet clearly eliminates all local zones at which cavitation could occur.

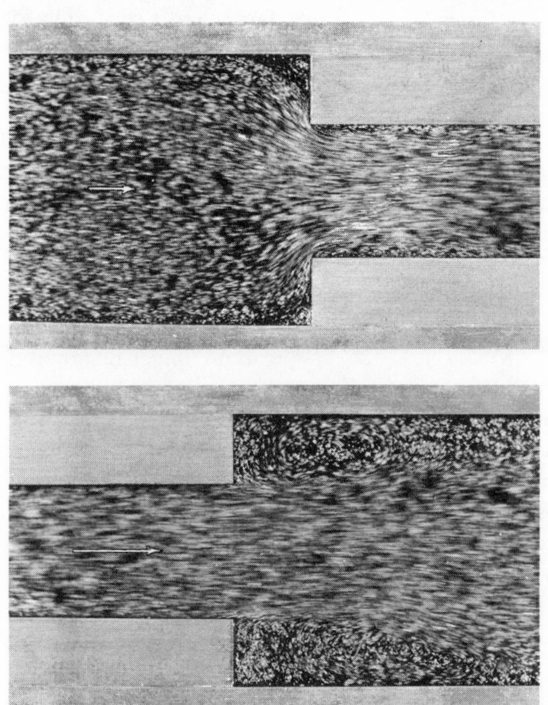

PLATE XX. Comparison of the flow patterns at an abrupt contraction and an abrupt enlargement of cross section.

Abrupt enlargements; conduit outlets. The converse of the abrupt contraction is found in the abrupt enlargement of Fig. 132*a*. Since the pressure distribution just past the transition (i.e., before the stream begins to expand) is essentially the same as that of the approaching flow, such knowledge permits evaluation of the ultimate rise in h through use of the continuity and momentum relationships.

Thus,

$$Q = V_1 A_1 = V_2 A_2$$

and

$$\gamma h_1 A_1 + \gamma h_1 (A_2 - A_1) - \gamma h_2 A_2 = Q\rho(V_2 - V_1)$$

whence

$$\frac{\Delta h}{V_1^2/2g} = 2\frac{A_1}{A_2}\left(1 - \frac{A_1}{A_2}\right) \tag{183}$$

The degree to which the rise in h can be increased by streamlining the enlargement depends upon both the rate of enlargement and the enlargement ratio. That is, while the maximum angle of flare without separation is approximately 10°, the length of divergent boundary evidently increases with the enlargement ratio, thereby decreasing the ultimate rise because of the increased surface drag. Since the variation in h is then governed almost entirely by viscous effects, the influence of the Reynolds number can no longer be ignored. So far as the writer is aware, systematic studies of the problem as a whole have not been made.

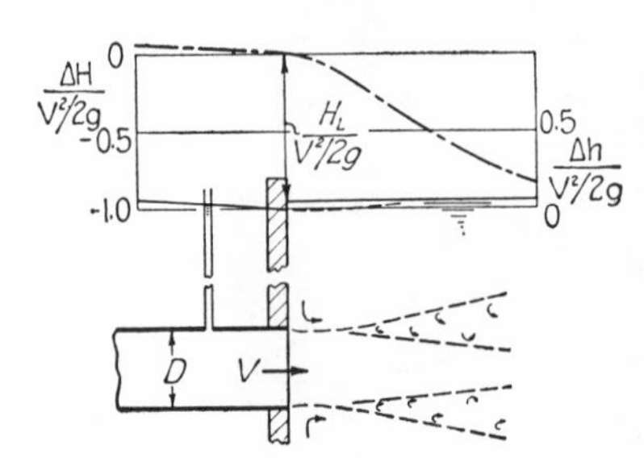

FIG. 133. Variation in head at a submerged outlet.

The converse of the inlet to a conduit from a reservoir is likewise found in the outlet from a conduit into another reservoir, as in Fig. 133. This represents the limiting condition of Eq. (183), which indicates that the rise in h as the jet expands will be almost nil. The extent to which this rise can be increased by flaring the outlet is subject to the same conditions as the streamlining of any abrupt enlargement.

Conduit bends. The conduit bend is perhaps the only type of conduit transition which is not analogous to any of the elementary forms of immersed body discussed in the foregoing sections, although it does possess certain characteristics common to the lifting vanes of the following chapter. Aside from a peculiar type of secondary motion which a bend produces, however, this type of transition embodies essentially the same combination of viscous and form effects as do the others. Thus a short-radius bend (Fig. 134) shows a tendency toward boundary-layer separation along the inner wall, which may be eliminated by increasing the bend radius. No matter how great the radius, however, the change in flow direction requires an increase in pressure intensity (i.e., piezometric head) along the outside of the bend and a decrease along the inside. Evidently, the longer the radius, the smaller the difference in h between the outside and the inside, since the total force required to produce the momentum change does not vary with the length of bend. It must nevertheless be recalled that the velocity of the approaching flow is zero along all portions of the boundary and hence cannot decrease further to provide the required rise in pressure

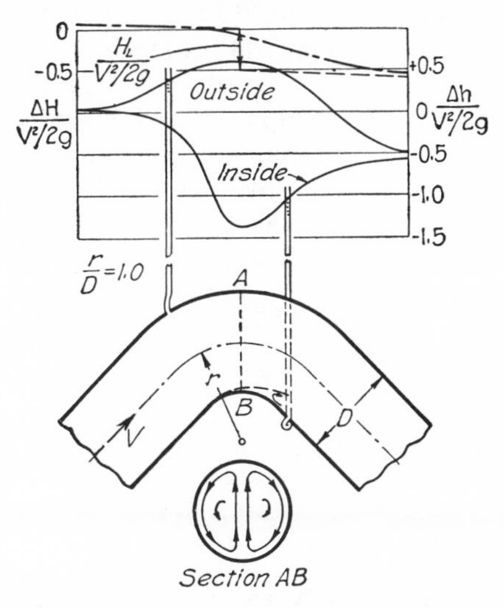

FIG. 134. Variation in head at a 90° short-radius bend.

along the outside. Just within this boundary zone, however, the velocity is finite, and the unbalanced centrifugal force of this portion of the fluid produces a secondary flow outward along the plane of the bend and inward around the boundary, as indicated in the figure. As a result, the fluid leaving the bend moves in a double spiral which diminishes in intensity only gradually as the fluid passes downstream. Noteworthy is the fact that such effects may be completely eliminated by replacing the standard bend with a miter elbow containing a series of vanes curved as shown in Fig. 135.

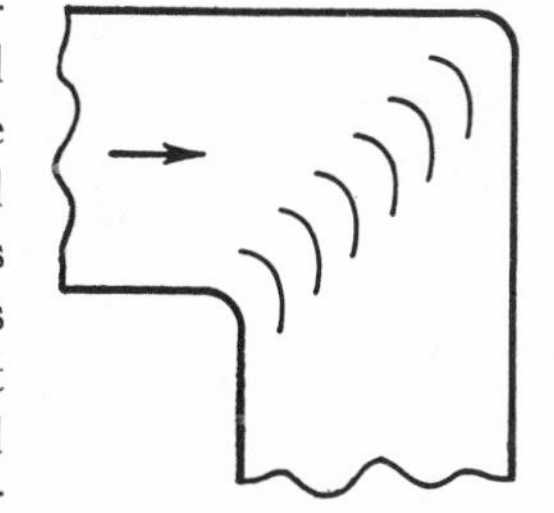

FIG. 135. Guide vanes at a miter bend.

Reasons for the study of head distribution. Knowledge of the various factors which influence the velocity and pressure variation at any boundary transition will generally be important for at least one of the following reasons: (1) Unless the dimensions of the conduit are so small that other considerations govern structural design (as is true in the case of most plumbing fittings), the pressure distribution in a transition may play a primary role in structural stability. (2) Any transition producing a known differential head between two points may be used to measure the rate of flow. (3) Transitions intended to yield particularly uniform flow conditions (such as the contraction preceding the test section of a wind tunnel) should be free from both separation and secondary flow. (4) Adjustable transitions such as valves and gates should be so designed as to cause a minimum disturbance when fully open and yet produce a steady reduction in recovery of head—and hence in rate of flow—during closure. (5) Unless transitions are well streamlined, the reduction or incomplete recovery of piezometric head due to form effects may amount to a considerable portion of the total resistance to flow.

Example 46. The abrupt entrance to a 5-foot conduit lies 20 feet (on center) below the free surface of a reservoir. Assuming a vapor pressure of 0.5 pound per square inch absolute, determine the rate of flow at which cavitation may be expected to commence.

At the point of greatest reduction in piezometric head, the magnitude of $\Delta h/(V^2/2g)$ is seen from Fig. 131*a* to be -2.35. Hence

$$\Delta h = -2.35 \frac{V^2}{2g}$$

Relative to the top of the inlet, the initial value of h is $20 - 2.5 = 17.5$ feet. Since the elevation z is thus eliminated, the pressure head at the point of lowest pressure intensity will be equal to the initial piezometric head plus the change as the flow enters the conduit:

$$\frac{p_{\min}}{\gamma} = 17.5 + \Delta h = 17.5 - 2.35 \frac{V^2}{2g}$$

Expressing the absolute intensity of the vapor pressure in relation to atmospheric pressure,

$$\frac{(0.5 - 14.7) \times 144}{62.4} = 17.5 - 2.35 \frac{V^2}{2 \times 32.2} \quad \text{whence} \quad V = 37.1 \text{ fps}$$

The rate of flow at which cavitation should take place is then

$$Q = VA = 37.1 \times \frac{\pi \times 5^2}{4} = 728 \text{ cfs}$$

PROBLEMS

295. Oil having a specific gravity of 0.85 and a dynamic viscosity of 2×10^{-4} pound-second per square foot is transmitted by an 8-inch pipe, the rate of flow being measured by a diaphragm orifice having a diameter of 6 inches. What rate of flow would correspond to a differential head of 5 inches of oil?

296. A Venturi meter having the characteristics shown in Figs. 128 and 129 is installed in a 2-foot water main to measure the rate of flow. When the main is carrying 35 cubic feet of water per second, what differential head would be read on a mercury manometer connected to the meter?

297. What percentage error in the computed rate of flow would result from using a discharge coefficient corresponding to piezometer openings located at the base of a

diaphragm orifice (Fig. 127) if the openings were actually one pipe diameter upstream and downstream from the orifice?

298. A pipe bend having the characteristics shown in Fig. 134 is used as a flow meter, piezometer openings located symmetrically at the inside and the outside of the bend being connected to a differential gage. Evaluate the corresponding coefficient C in the flow equation $Q = CA\sqrt{2g\,\Delta h}$.

299. In order to measure the rate of flow of air through an experimental duct 9 inches in diameter, a thin diaphragm containing a concentric hole 6 inches in diameter is inserted at one of the conduit joints. What rate of flow would be indicated by a differential head of 3 inches of water in a manometer connected to piezometers either side of the diaphragm?

300. Estimate from the distribution curves of Fig. 127 the longitudinal force exerted upon a similar orifice plate in a 12-inch pipe during a flow of 7 cubic feet of water per second.

301. A ventilation duct of circular cross section changes abruptly from 6 feet to 3 feet in diameter. If the pressure intensity in the 6-foot approach section corresponds to a 2-inch head of water during a flow of 20,000 cubic feet of air per minute, what minimum pressure intensity will prevail just beyond the transition?

302. If the metal lining of the 5-foot conduit of Example 46 had projected into the reservoir, what reduction in the maximum rate of flow would have resulted?

303. Compare the rates of efflux from three 12-inch pipes each 3 feet long and set horizontally into the side of a reservoir 15 feet below the water surface, pipe A projecting into the reservoir, pipe B being flush with the reservoir wall, and pipe C having a rounded entrance (see Fig. 131).

304. Two reservoirs 600 feet apart have a difference in surface level of 60 feet. The reservoirs are connected by a horizontal pipe line consisting of the following three sections: 200 feet of 4-inch pipe, 200 feet of 6-inch pipe, and 200 feet of 3-inch pipe, in the order given. Assuming each transition to be abrupt, sketch to approximate scale (10 vertical to 1 horizontal) the lines of total and piezometric head between the two reservoirs.

42. LOSS OF HEAD AT CONDUIT TRANSITIONS

Evaluation of the loss coefficient. As in the case of immersed bodies, the variation in normal and tangential stress at any conduit transition will produce a resultant longitudinal force upon the boundaries which varies with the boundary form and the Reynolds number of the flow. Judging from the discussion of the foregoing section, however, it would appear that tangential stresses usually play a very minor role in determining the resistance of a conduit transition, owing in part to the fact that commercial fittings are seldom well streamlined, and in part to the high Reynolds numbers of flows which are generally encountered in engineering practice. In general, therefore, the resistance to flow caused by a conduit transition will be equal to the integral of the longitudinal component of normal force over the transition boundary.

As in the case of immersed bodies, such resistance may be evaluated from curves of boundary pressure or piezometric head similar to those of the foregoing section. Unlike the immersed-body problem, nevertheless, the problem of conduit resistance is not so much the determination of the longitudinal force exerted upon the transition as the evaluation of the rate at which work must be done upon the flow by external forces to overcome such resistance—i.e., the power expenditure or rate of energy dissipation which the flow transition entails. Under the latter circumstances, recourse must be had to the variation in total head rather than piezometric head as the fluid passes through the transition, for the drop in piezometric head is a true measure of energy dissipation only if the inlet and outlet diameters of the transition are the same. As may be seen from the diagrams of the foregoing section, however, for any given boundary form the variation in total head and piezometric head will, at high Reynolds numbers, be interdependent. For instance, the overall change in total head to be expected at an abrupt enlargement may readily be evaluated from the corresponding changes in piezometric head [Eq. (183)] and velocity head, as follows:

$$\frac{\Delta h}{V_1^2/2g} + \frac{\Delta(V^2/2g)}{V_1^2/2g} = 2\frac{A_1}{A_2}\left(1 - \frac{A_1}{A_2}\right) + \frac{A_1^2 - A_2^2}{A_2^2}$$

whence

$$\frac{\Delta H}{V_1^2/2g} = -\left(1 - \frac{A_1}{A_2}\right)^2 \tag{184}$$

Graphical representation of conduit losses. However short a boundary transition may actually be, the fact remains that the eddies which it produces will have an influence upon the velocity distribution (and hence upon the surface resistance) for a considerable distance downstream. Evidently, not until the flow again becomes truly uniform can the full effect of the transition be evaluated. Owing to the resistance of the uniform conduit, on the other hand, the lines of total head and piezometric head will approach sloping rather than horizontal asymptotes beyond the transition (refer to Figs. 130–134), so that only by extrapolation of these asymptotes to the section of the transition can the loss in total head for the transition itself be determined. As indicated by the broken lines in the several diagrams, their intercept on the vertical scale invariably represents a characteristic decrement of the parameter $H/(V^2/2g)$, commonly known as the *loss coefficient* C_L of the transition. Letting H_L represent the net loss in total head, it follows that

$$H_L = C_L \frac{V^2}{2g} \tag{185}$$

TABLE IV

APPROXIMATE VALUES OF THE LOSS COEFFICIENT FOR VARIOUS PIPE TRANSITIONS

Abrupt contractions $H_L = C_L V_2^2/2g$		90° Bends $H_L = C_L V^2/2g$	
D_2/D_1	C_L	r/D	C_L
0.8	0.13	1.0	0.40
0.6	0.28	1.5	0.32
0.4	0.38	2.0	0.27
0.2	0.45	3.0	0.22
0.0 (Inlet from reservoir)	0.50	4.0	0.20
Abrupt enlargements $H_L = C_L V_1^2/2g$		Valves $H_L = C_L V^2/2g$	
$C_L = \left[1 - \left(\frac{D_1}{D_2}\right)^2\right]^2$			C_L
		Gate (open)	0.2
(Outlet into reservoir $C_L = 1.0$)		Globe (open)	10.0

Typical loss coefficients for common types of transition are given in Table IV. These, it must be realized, refer to transitions which are almost the complete antithesis of streamlined forms, corresponding in general to the abrupt changes in section discussed in the foregoing pages. As such, the coefficients represent maximum values, the magnitude of which can be reduced to a considerable degree by proper streamlining (under the latter circumstances, of course, the loss coeffi-

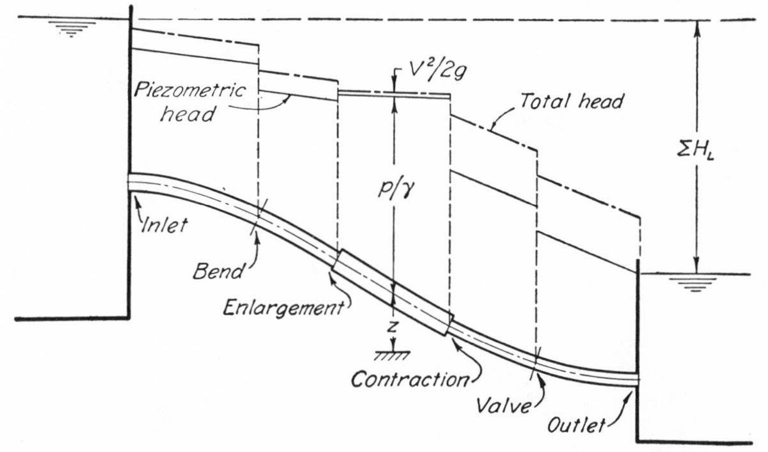

FIG. 136. Simplified representation of the change in head along a conduit of non uniform cross section.

cients would ultimately become dependent upon surface rather than form effects, and hence would vary to some extent with the Reynolds number). In making use of such coefficients, one is restricted perforce to the one-dimensional method of flow analysis; in other words, although any transition has been seen to have an effect upon the lines of total head and piezometric head for a considerable distance downstream, it must arbitrarily be assumed that the full effect is concentrated at the transition section. On a given conduit diagram such lines will therefore slope in proportion to the surface resistance in the uniform reaches and drop abruptly in proportion to the form resistance at the sections of rapid transition, as in Fig. 136. The Bernoulli equation must then include the summation of all losses which occur between any two sections:

$$\frac{V_1^2}{2g} + \frac{p_1}{\gamma} + z_1 = \frac{V_2^2}{2g} + \frac{p_2}{\gamma} + z_2 + \Sigma H_L \qquad (186)$$

Use of loss coefficients in problems of unsteady flow. Although the one-dimensional method of analyzing conduit resistance to steady flow must, at best, be regarded merely as a convenient simplification for purposes of rapid calculation, it should also be realized that the assumption of a resistance coefficient which is independent of the Reynolds number permits an approximate solution of problems in unsteady flow which would otherwise be hopelessly involved. Perhaps the simplest example of this circumstance is found in the establishment of flow in a long pipe, a case discussed in Chapter V without regard to boundary resistance. It is now possible to incorporate the effect of loss in head as a first approximation by introducing in the initial equation a coefficient bulk ΣC_L which embodies, in terms of the variable velocity head, the effect of surface and form resistance over the entire length of conduit. Thus,

$$\frac{L}{g}\frac{\partial V}{\partial t} = H_1 - H_2 - \Sigma H_L = h - (1 + \Sigma C_L)\frac{V^2}{2g}$$

and

$$h = (1 + \Sigma C_L)\frac{V_{max}^2}{2g}$$

The subsequent integration is affected by the presence of this factor only to the extent of modifying the numerical characteristics of the result, the counterpart of Eq. (96) being simply

$$t\frac{\sqrt{2gh}}{L}\sqrt{1 + \Sigma C_L} = \log_e \frac{1 + V/V_{max}}{1 - V/V_{max}} \qquad (187)$$

Evidently, the time required to establish a given percentage of the ultimate velocity in the case of resistance can be found from Fig. 61 by dividing the corresponding value of t for zero resistance by the radical $\sqrt{1 + \Sigma C_L}$.

Example 47. Water flows between two reservoirs through a 6-inch rough pipe for which $f = 0.03$. Assuming the difference in reservoir elevation to have the constant magnitude of 30 feet, determine the rate of flow and the percentage loss due to form resistance (abrupt entrance and exit) if the pipe is (*a*) 50 feet long, (*b*) 500 feet long, and (*c*) 5000 feet long.

A general expression for the rate of flow may be developed by equating the total change in head to the losses at the inlet, along the pipe, and at the outlet; thus,

$$\Sigma H_L = 30 = \left(0.5 + \frac{0.03L}{0.5} + 1.0\right)\frac{V^2}{2g}$$

whence

$$Q = AV = \frac{\pi \times 0.5^2}{4}\sqrt{\frac{30 \times 2 \times 32.2}{\frac{0.03L}{0.5} + 1.5}} = \frac{35.3}{\sqrt{L + 25}}$$

Similarly, the loss ratio will be

$$\frac{H_L\,(\text{form})}{\Sigma H_L} = \frac{(0.5 + 1.0)\frac{V^2}{2g}}{\left(0.5 + \frac{0.03L}{0.5} + 1.0\right)\frac{V^2}{2g}} = \frac{1}{1 + 0.04L}$$

(*a*)

$$Q = \frac{35.3}{\sqrt{50 + 25}} = 4.08 \text{ cfs}$$

$$\frac{H_L\,(\text{form})}{\Sigma H_L} = \frac{1}{1 + 0.04 \times 50} \times 100 = 33.3\%$$

(*b*)

$$Q = \frac{35.3}{\sqrt{500 + 25}} = 1.54 \text{ cfs}$$

$$\frac{H_L\,(\text{form})}{\Sigma H_L} = \frac{1}{1 + 0.04 \times 500} \times 100 = 4.8\%$$

(*c*)

$$Q = \frac{35.3}{\sqrt{5000 + 25}} = 0.498 \text{ cfs}$$

$$\frac{H_L\,(\text{form})}{\Sigma H_L} = \frac{1}{1 + 0.04 \times 5000} \times 100 = 0.5\%$$

Evidently, the relative importance of such form effects becomes insignificant in long reaches of uniform flow.

PROBLEMS

305. Two reservoirs are connected by 400 feet of horizontal 10-inch pipe for which $f = 0.025$. Midway between the reservoirs is a turbine, the controls of which regulate the rate of flow between the reservoirs. If the difference in reservoir elevation is 35 feet and the change in head at the turbine is 20 feet, what is the rate of flow? Sketch the lines of total and piezometric head.

306. Substituting a pump ($\Delta H = 50$ feet) for the turbine of Problem 305 and assuming that the direction of flow is reversed, determine the rate of flow and sketch the lines of total and piezometric head.

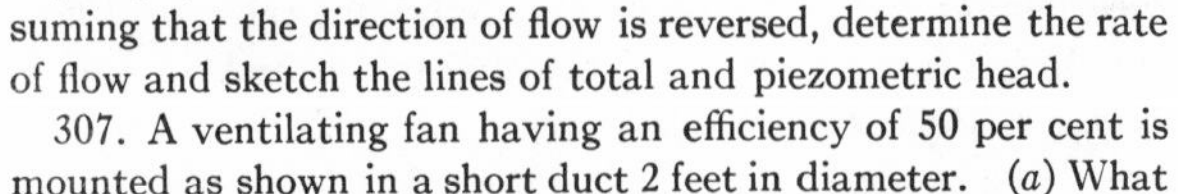

307. A ventilating fan having an efficiency of 50 per cent is mounted as shown in a short duct 2 feet in diameter. (*a*) What power must be supplied to the fan to deliver 5000 cubic feet of air per minute? (*b*) What percentage increase in the rate of flow would result from rounding the inlet as shown, the power input remaining the same?

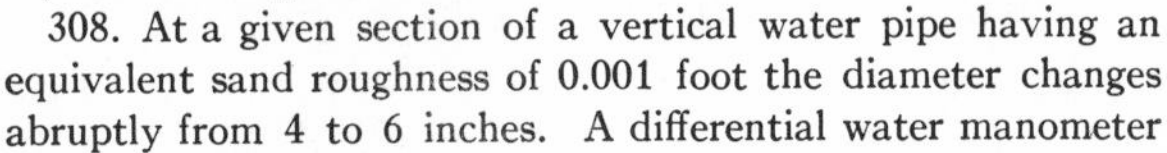

PROB. 307.

308. At a given section of a vertical water pipe having an equivalent sand roughness of 0.001 foot the diameter changes abruptly from 4 to 6 inches. A differential water manometer is connected between points (1) 5 feet above (i.e., upstream from) and (2) 5 feet below the enlargement. To what rate of flow would the gage reading of $h_1 - h_2 = -0.35$ foot correspond?

309. Natural gas, having a specific weight of 0.05 pound per cubic foot and a kinematic viscosity of 1.6×10^{-4} square foot per second, is discharged from a pressure tank through a 3-inch galvanized-iron pipe. If an open globe valve is located 10 feet beyond the abrupt inlet, what rate of flow should prevail when a differential pressure head of 1.5 inches of water exists between the tank and a point 50 feet down the pipe?

310. Reservoirs A and B are connected with a pipe line made up of the following consecutive elements: (1) a horizontal 6-inch pipe 50 feet long, having its inlet flush with the wall of reservoir A at a depth of 10 feet below the water surface; (2) a 6-inch 90° bend having a 12-inch radius and directed downward; (3) a vertical 6-inch pipe 15 feet long; (4) a 6-inch 90° bend having an 8-inch radius; and (5) 10 feet of horizontal 6-inch pipe with outlet in the wall of reservoir B. Assuming the pipe roughness to be such as to yield a coefficient of 0.025, determine the rate of flow when the water surface in reservoir B is 20 feet below that in reservoir A.

311. A water tank is generally emptied through a horizontal 5-foot length of smooth 3-inch pipe having an abrupt inlet and containing a gate valve at its midpoint. Would a greater increase in rate of efflux be obtained by rounding the pipe inlet or by changing to a 4-inch line?

312. If the nozzle of Fig. 54 delivers a 1.5-inch jet of water 35 feet below the reservoir level, and if the delivery line consists of 250 feet each of 4-inch smooth pipe and 3-inch smooth hose, evaluate the rate of flow and sketch the lines of total and piezometric head. Assume a nozzle loss of 5 per cent of the efflux velocity head.

313. Determine the rate of flow through the pipe system of Problem 304, assuming a value of $f = 0.02$ for the surface resistance.

314. If a ventilation duct of galvanized sheet metal changes abruptly from 4 feet to 3 feet in diameter, what differential head of water should exist between points 50 feet upstream and downstream from the contraction during a flow of 12,000 cubic feet of air per minute?

315. Water is pumped between two tanks through 300 feet of 2-inch galvanized pipe containing an open gate valve and 3 elbows of 4-inch radius. If the total lift is 50 feet, what power input will be required to maintain a flow of 15 cubic feet of water per minute?

316. Two storage tanks for oil are connected by 120 feet of steel pipe 4 inches in diameter with abrupt inlet and outlet, a gate valve midway between the tanks permitting regulation of the flow. What is the greatest rate at which oil having a specific gravity of 0.85 and a kinematic viscosity of 10^{-4} square foot per second could be transferred from one tank to the other under a head of 25 feet?

317. A 6-inch pipe 100 feet long, containing three 90° elbows of 8-inch radius and two open globe valves, is used to transmit water from a pump to a tank 30 feet above. If the pump operates 6 hours a day 300 days a year and pumps 1000 gallons per minute, determine the annual saving which could be effected by substituting gate valves and bends of 24-inch radius. Assume the overall efficiency of motor and pump as 70 per cent and the cost of power as 2 cents per kilowatt-hour.

318. Assuming a constant surface-resistance coefficient of 0.02 and equal tank diameters of 9 feet in Problem 316, estimate the time required to bring the free surfaces in the two tanks to the same level.

319. Compare the times computed for the establishment of flow under the conditions of Problem 159 (*a*) without regard to energy loss and (*b*) including the effect of an abrupt inlet, an open gate valve, and boundary roughness yielding a coefficient of 0.02.

QUESTIONS FOR CLASS DISCUSSION

1. The statement has been made that no drag can be exerted upon a body by steady, irrotational flow; why is this true? Why is it not true if the flow is unsteady?

2. Separation of flow from the upper surface of an airplane wing is sometimes diminished by slotting the wings. Explain the principle involved.

3. Distinguish between deformation drag, surface drag, and form drag.

4. Which type of drag predominates in the motion of the following bodies: a model-T Ford; a wing strut of an airplane; a bubble in honey; a baseball; a snowflake; an arrow; a parachute?

5. Compare from observation the orders of magnitude of the drag coefficients of an ordinary desk blotter when dropped in vertical and horizontal positions.

6. Why is the drag of a hollow hemisphere greater when held with the concave side upstream than when held with the convex side upstream?

7. Why is the drag coefficient of a square plate lower than that of a very long rectangular plate of the same width?

8. Why is the streamlining of automobiles of less practical importance than that of airplanes?

9. What is the cause of vibration in telephone wires in a high wind? Under what circumstances could such vibration easily lead to rupture?

10. Under what conditions will roughening the surface of a body (*a*) increase, and (*b*) decrease, its drag coefficient?

11. Why is considerable attention paid to the streamlining of large conduits in water-supply and hydroelectric installations, but very little to the streamlining of plumbing fixtures?

12. The discharge coefficient of a pipe orifice increases with increasing relative roughness of the pipe. Explain this tendency in terms of velocity distribution and its effect upon the contraction of the jet.

13. Explain the manner in which a valve controls the rate of flow through a conduit. When is it advantageous to use a gate valve instead of a globe valve?

14. Use of the momentum principle has been shown to yield the loss given in Table IV for an abrupt enlargement. Why is this method not applicable to abrupt contractions?

15. The resistance characteristics of Table IV are often classified as "minor" losses. Under what circumstances is such a term quite apt? When is it a misnomer?

SELECTED REFERENCES

GOLDSTEIN, S. *Modern Developments in Fluid Dynamics.* Vol. II, Oxford, 1938.

Fluid Meters, Their Theory and Application. American Society of Mechanical Engineers, 1937.

KING, H. W. *Handbook of Hydraulics.* McGraw-Hill, 1939.

CHAPTER IX

LIFT AND PROPULSION

43. CIRCULATION AND THE MAGNUS EFFECT

Significance of circulation. An important concept of fluid kinematics which was not discussed in Chapter II is that known as *circulation.* As the word implies, circulation refers to flow which follows a circuitous course back to the starting point. If there is no other motion of the fluid, the stream lines themselves must form the closed curves around which the circulation takes place. More generally, however, the circulatory motion is superposed upon the basic translatory motion of the fluid, and it therefore becomes necessary to express circulation in terms of an arbitrary curve passing through a system of stream lines as shown in Fig. 137.

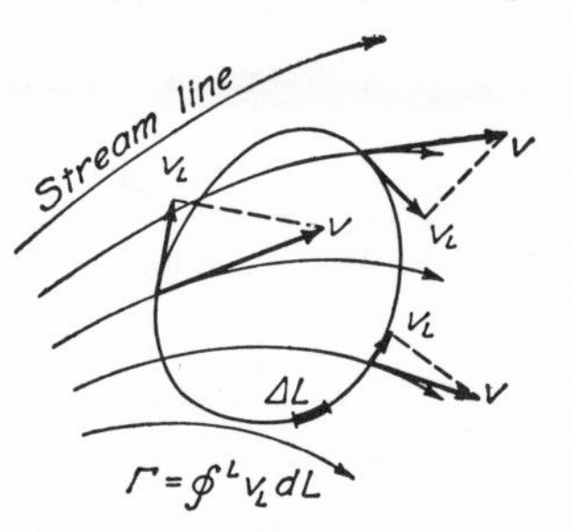

FIG. 137. Definition sketch for circulation around a closed curve.

At any point on this curved line the velocity vector will be seen to have a tangential component v_L, which varies in magnitude between 0 and v and generally changes sign as one proceeds around the circuit. If the product of v_L and the increment of length dL of the curved line is integrated completely around the circuit, the magnitude of the integral will indicate to what extent circulation exists. In other words, circulation is defined as the integral of the quantity $v_L\,dL$ completely around a closed curve. Thus, designating circulation by the symbol Γ (gamma),

$$\Gamma = \oint v_L\,dL \tag{188}$$

As a simple example, consider the case of flow in concentric circles according to the condition $v = \omega r$. Around any stream line (see Fig. 138) the magnitude of the circulation is evidently

$$\Gamma = \omega r \cdot 2\pi r = 2\pi r^2 \omega$$

which indicates that Γ increases in direct proportion to the square of the radius of the circuit. In the same manner one may determine the circulation around any arbitrary curve not coincident with a stream line—such as that composed of two circular arcs and two segments of radii as indicated in the figure; in the latter case,

$$\Gamma = \omega r_2 \cdot \theta r_2 - \omega r_1 \cdot \theta r_1 = \omega\theta(r_2{}^2 - r_1{}^2)$$

If, in either case, the resulting circulation is divided by the area of the surface enclosed by the closed curve, it will be found that the result will be simply 2ω. As a matter of fact, circulation and rotation are intimately associated, in that the ratio of the circulation to the enclosed surface area will, as the area approaches the limit zero, be equal to twice the rate of fluid rotation about an axis normal to the surface at the limiting point:

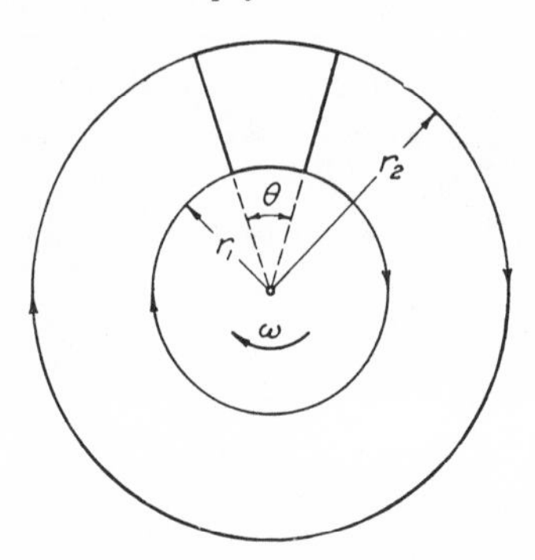

FIG. 138. Rotational flow in concentric paths.

$$\lim_{A \to 0} \frac{\Gamma}{A} = 2\omega_{\perp A} \tag{189}$$

Consider now the case of flow in concentric circles according to the condition $v = C/r$. Although this was previously stated to be irrotational flow, it will be found that circulation nevertheless exists around any stream line:

$$\Gamma = \frac{C}{r} \cdot 2\pi r = 2\pi C$$

The magnitude of Γ now, however, is independent of the radius, since $\Gamma/2\pi = C$. On the other hand, the circulation around any curve not including the center will invariably be found to equal zero; thus, for the same element used in the foregoing example (Fig. 138) it will be seen that

$$\Gamma = \frac{C}{r_2} \cdot \theta r_2 - \frac{C}{r_1} \cdot \theta r_1 = 0$$

In other words, there is zero rotation at all points except the center; there $\omega = \infty$, since, from Eq. (189), $2\omega = \Gamma/0$.

Side thrust due to circulation in irrotational flow. The concept of irrotational flow with circulation is of particular interest because the flow net for constant circulation around any two-dimensional body may be combined with the net for flow past the body to simulate a lateral displacement of stream lines often encountered in actual flow phenomena. For instance, if the stream-line pattern for constant circulation around a cylinder (Fig. 139*b*) is superposed upon the pattern for flow past a cylinder (Fig. 139*a*), and if intersections of stream lines

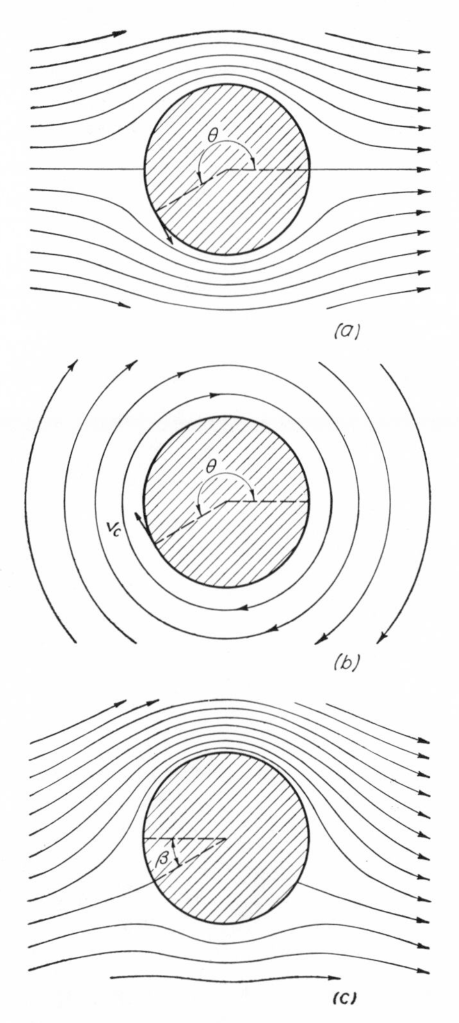

FIG. 139. Effect of circulation upon irrotational flow past a cylinder.

are connected by smooth curves as described in Section 8, the pattern shown in Fig. 139*c* will result. Evidently all stream lines are displaced systematically across the flow, with the result that the velocity is increased on one side and decreased on the other.

Just as this composite flow pattern is obtained by graphical addition of two stream-line systems, the resulting velocity at any point may be obtained by vector addition of the corresponding velocity components. Thus, mathematical analysis of irrotational flow past a cylinder yields the quantity $2v_0 \sin \theta$ for the tangential velocity at any point on the circumference; since the circumferential velocity due to the circulation is simply $\Gamma/2\pi r_0$, the velocity of the combined flow should vary around the cylinder according to the expression

$$v = 2v_0 \sin \theta + \frac{\Gamma}{2\pi r_0}$$

Evidently, the resulting tangential velocity at any point will differ from that of the symmetrical flow around the cylinder to an extent depending upon both Γ and the location of the point. Since Γ was taken as clockwise, it will be seen from Fig. 139 that the tangential velocity is, in general, increased around the upper portion of the cylinder and decreased around the lower portion. That is, expressing the tangential velocity due to circulation alone as $v_c = \Gamma/2\pi r_0$ and dividing by v_0,

$$\frac{v}{v_0} = 2 \sin \theta + \frac{v_c}{v_0}$$

from which it is seen that the change in velocity at any point on the cylinder will depend upon the ratio of v_c to the velocity v_0 of the approaching flow.

As indicated by Fig. 139, the change in stream-line spacing due to superposition of circulation upon the basic flow pattern will necessarily result in a displacement of the two points of stagnation around the circumference of the cylinder. The extent of this displacement may be determined in terms of the ratio v_c/v_0 by setting the ratio v/v_0 in the foregoing expression equal to zero (i.e., by locating the point of zero velocity); thus

$$0 = 2 \sin \theta + \frac{v_c}{v_0}$$

whence

$$\theta = \sin^{-1}\left(-\frac{1}{2}\frac{v_c}{v_0}\right) = \beta + 180°$$

Apparently, β will be 30° when the circulation is such that $v_c = v_0$, and 90° when $v_c = 2v_0$. Further increase in circulation will simply cause the point of stagnation to move away from the cylinder, which will then be entirely surrounded by fluid moving in the same angular direction.

Inasmuch as the pattern of stream lines (and hence the distribution of velocity and pressure) is still symmetrical about the vertical axis, it is obvious that in irrotational flow no amount of circulation can change the longitudinal force upon the cylinder from its initial magnitude of zero. On the other hand, the fact that the velocity increases above the cylinder and decreases below clearly indicates a rise in pressure intensity around the lower portion and a drop around the upper portion—in other words, a resultant force in the lateral direction is produced by circulation. The magnitude of the lateral force exerted upon the cylinder may be determined by integrating over the surface of the cylinder the lateral component of force due to the pressure upon an increment of surface area; thus, since

$$\Delta p = \frac{\rho}{2}(v_0^2 - v^2)$$

and

$$v = 2v_0 \sin\theta + \frac{\Gamma}{2\pi r_0}$$

it will be found that

$$F = L\int_0^{2\pi} (\Delta p \, r_0 \, d\theta) \sin\theta$$

$$= \frac{L\rho r_0}{2}\int_0^{2\pi}\left[v_0^2 - \left(2v_0 \sin\theta + \frac{\Gamma}{2\pi r_0}\right)^2\right]\sin\theta \, d\theta$$

which reduces to the very significant expression:

$$F = L\rho v_0 \Gamma \qquad (190)$$

Lift and drag of a rotating cylinder. Were it physically possible to produce such conditions of flow past a cylinder, the side thrust F could be expected to vary in direct proportion to the length of the cylinder, the density, the velocity of approach, and the circulation, in accordance with Eq. (190). As a matter of fact, flow of this nature may be approximated through rotation of the cylinder itself, thereby producing local circulation through surface drag. Although the resulting viscous shear eventually induces some degree of circulation even at a considerable distance from the cylinder, Γ necessarily decreases radially instead of remaining constant as in the irrotational case. Nevertheless, the local circulation actually does produce considerable side thrust—commonly known as *lift*—first noted by a German scientist named Magnus approximately a century ago.

The actual magnitude of this so-called *Magnus effect* must evidently depend upon the Reynolds number of the flow as well as the density and velocity of the fluid and the size and peripheral speed of the cylinder. At high Reynolds numbers, nevertheless, viscous effects upon the lift should approach an asymptotic limit, just as in the case of drag. Under such circumstances, the lateral force may be evaluated in terms of a *coefficient of lift*, according to the expression

$$F_L = C_L L D \frac{\rho v_0^2}{2} \tag{191}$$

similar to the equation for drag discussed in the foregoing chapter. C_L must then be evaluated from experiments as a function of the ratio v_c/v_0, v_c now representing the peripheral speed of the cylinder. In Fig. 140 will be found a plot of both C_L and C_D obtained in this manner. A similar coefficient of lift may be obtained for the irrotational case of constant circulation through the following modification of Eq. (190):

$$C_L = \frac{F/LD}{\rho v_0^2/2} = \frac{2\Gamma}{D v_0} = 2\pi \frac{v_c}{v_0}$$

This function is shown as a broken line in Fig. 140, from which may be judged the extent of the circulation which can be produced by boundary drag. As a matter of fact, visual studies indicate that in the case of the rotating cylinder the magnitude of the ratio v_c/v_0 at which the point of stagnation leaves the boundary is approximately 4, rather than 2 as in the irrotational case, corresponding in Fig. 140 to the zone in which the lift coefficient approaches a maximum value. It would appear from these studies that the local circulation produced by surface drag and viscous shear is about half as effective as the constant circulation required for truly irrotational flow.

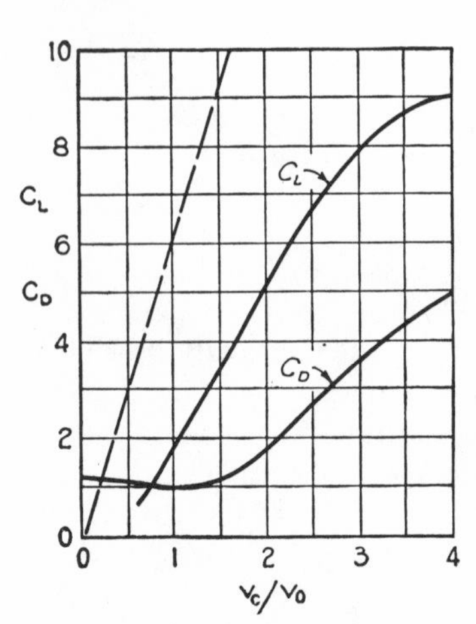

FIG. 140. Coefficients of lift and drag for a rotating cylinder.

Despite the considerable variation between the two conditions of motion, the fact remains that the phenomenon of side thrust or lift is fully explained by the circulation principle; this principle will hence have considerable bearing upon the following section in which characteristics of the airfoil are discussed. So far as the Magnus effect itself is concerned, the deviation of spinning balls and projectiles from their normal trajectories is simply a further illustration of this phenomenon. Indeed, the side thrust of a rotating cylinder has been utilized at large scale in the Flettner rotor as a possible (though not the most practicable) means of propelling ships and windmills.

Example 48. Two 20-foot rotors 5 feet in diameter are used to propel a ship. Estimate the total longitudinal force exerted upon the rotors when the relative wind velocity is 25 miles per hour, and the angular velocity of the rotors is 250 revolutions per minute, in the directions shown.

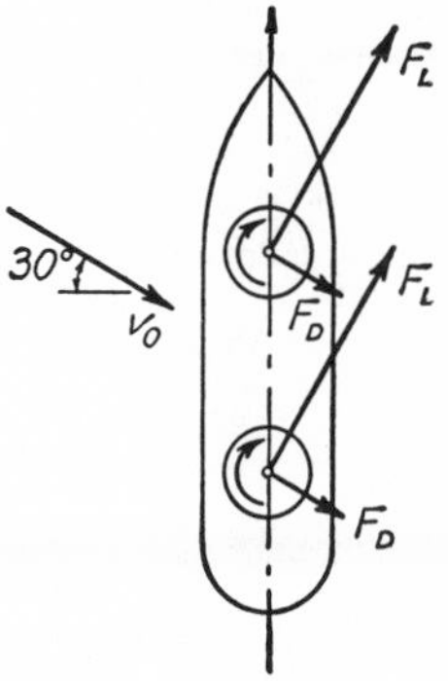

The tangential velocity of the rotors will be

$$v_c = \omega r_0 = \frac{2\pi \times 250}{60} \times 2.5 = 65.5 \text{ fps}$$

And the relative velocity of the wind

$$v_0 = \frac{25 \times 5280}{3600} = 36.7 \text{ fps}$$

whence

$$\frac{v_c}{v_0} = \frac{65.5}{36.7} = 1.78$$

From Fig. 140, $C_L = 4.4$ and $C_D = 1.5$, so that, per rotor,

$$F_L = C_L L D \frac{\rho v_0^2}{2} = 4.4 \times 20 \times 5 \times \frac{0.0025 \times \overline{36.7}^2}{2} = 741 \text{ lb}$$

and

$$F_D = C_D L D \frac{\rho v_0^2}{2} = 1.5 \times 20 \times 5 \times \frac{0.0025 \times \overline{36.7}^2}{2} = 253 \text{ lb}$$

In the direction of motion (see accompanying sketch) the total force will then be

$$F = 2(F_L \cos 30° - F_D \sin 30°) = 2(741 \times 0.866 - 253 \times 0.5) = 1030 \text{ lb}$$

PROBLEMS

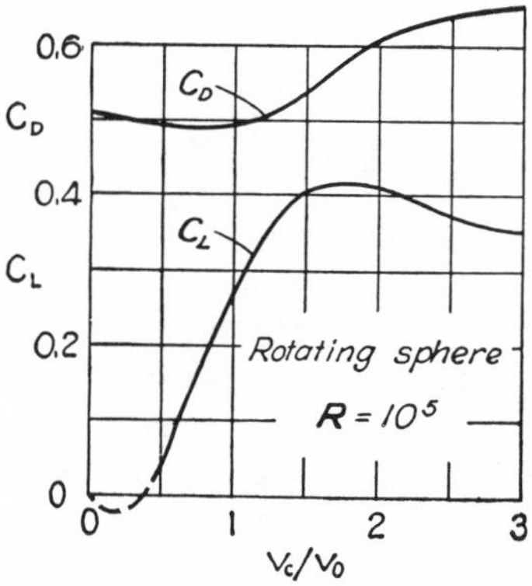

PROB. 320.

320. If a baseball having a circumference of 9 inches is pitched with a speed of 80 feet per second, what rate of backspin must it have in order to drop at the lowest possible rate? At what speed would it have to travel in order for the lift produced by backspin to balance its weight of 5 ounces?

321. A ping-pong ball weighing 0.085 ounce and having a diameter of 1.4 inches is served in such a manner that it passes over the net in a horizontal direction with a velocity of 40 feet per second and a forward spin of 90 revolutions per second. Determine, with reference to the plot accompanying Problem 320, the instantaneous magnitude and direction of its acceleration.

322. What orientation of the vector of relative wind velocity would yield the greatest propelling force upon the rotor ship of Example 48? Compute the magnitude of this force.

323. Assuming the optimum ratio of lift to drag, determine how nearly into the wind a rotor ship could sail; i.e., ignoring the wind effect on the ship itself, at what wind angle would the resultant propelling force on the rotors approach zero?

324. In investigating the possibility of using rotors in place of airplane wings, it is assumed that each of two rotating cylinders would have a diameter of 3 feet and a length of 15 feet. If the weight of the entire plane is estimated to be 9000 pounds, at what rate would the rotors have to be turned to support this load at a 150-mile-per-hour cruising speed? What power would then be required to overcome the rotor drag?

325. The six blades of an experimental windmill consist of cylinders having a diameter of 2 feet and a length of 10 feet, mounted radially so that the outer end of each cylinder is 15 feet from the windmill hub. If each cylinder is rotated at the rate of 180 revolutions per minute, what torque will have to be exerted upon the windmill shaft to prevent it from turning during a normal wind of 15 miles per hour?

44. LIFT AND DRAG OF THE AIRFOIL

Development of circulation around an inclined plate. One of the most noteworthy contributions of classical hydrodynamics is a mathematical method of determining the pattern of irrotational flow, either with or without circulation, around not only the cylinder but also a wide variety of other two-dimensional body profiles. While this process, known as conformal transformation, is well beyond the scope of the present text, the results obtained for a certain family of body forms bear so directly upon the theory of lift and propulsion as to warrant careful attention at this point.

If irrotational flow is assumed to take place past a very thin plate held parallel to the direction of motion, the flow pattern will consist simply of a series of equidistant stream lines. If, on the other hand, the plate is inclined to the direction of motion at a small angle, the pattern of irrotational flow will be similar to that shown in Fig. 141*a*, the two points of stagnation corresponding to those of Fig. 139*a* for the cylinder. In the latter case, it will be recalled, the symmetry of the pattern led to a zero resultant force upon the cylinder. Although the flow around the inclined plate is not truly symmetrical, the eccentricity of the stagnation points simply produces a couple tending to rotate the plate about its midpoint, the resultant force upon the plate still being zero. If a pattern of constant circulation (Fig. 141*b*) is superposed upon such motion, both stagnation points will be displaced toward the rear or trailing edge of the plate, corresponding to the displacement around the circumference of the cylinder shown in Fig. 139*c*. If, moreover, just the right degree of circulation is chosen, the upper stagnation point can be made to coincide with the trailing edge (Fig. 141*c*), with the result that the stream line leaving the plate is tangent to it. As in the case of the cylinder, the increase in pressure intensity below the plate, together with the decrease above, produces a resultant force. Indeed, the side thrust or lift may be again evaluated from

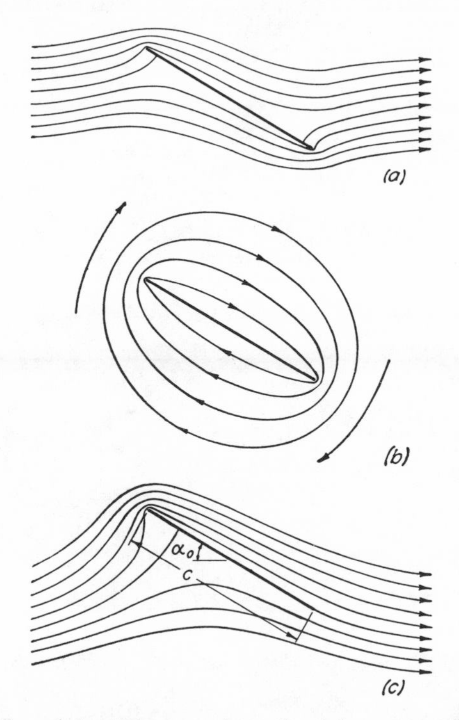

FIG. 141. Effect of circulation upon irrotational flow past an inclined plate.

Eq. (190), the circulation required to produce a tangential stream line at the trailing edge being found from the relationship

$$\Gamma = \pi c v_0 \sin \alpha_0 \tag{192}$$

in which c is the width or *chord* of the plate and α_0 is its inclination or *angle of attack.*

Flow with circulation approximately similar to the irrotational pattern has been shown to be obtainable in a viscous fluid through rotation of a cylinder, but it is obviously impossible to expect the analogous pattern required to result from rotation of a plate. However, viscous shear again provides the necessary effect, though in a somewhat different manner. In the present case, that is, the zero velocity at the base of the boundary layer causes separation to take place at the trailing edge of the plate, with the result that the stream line next to

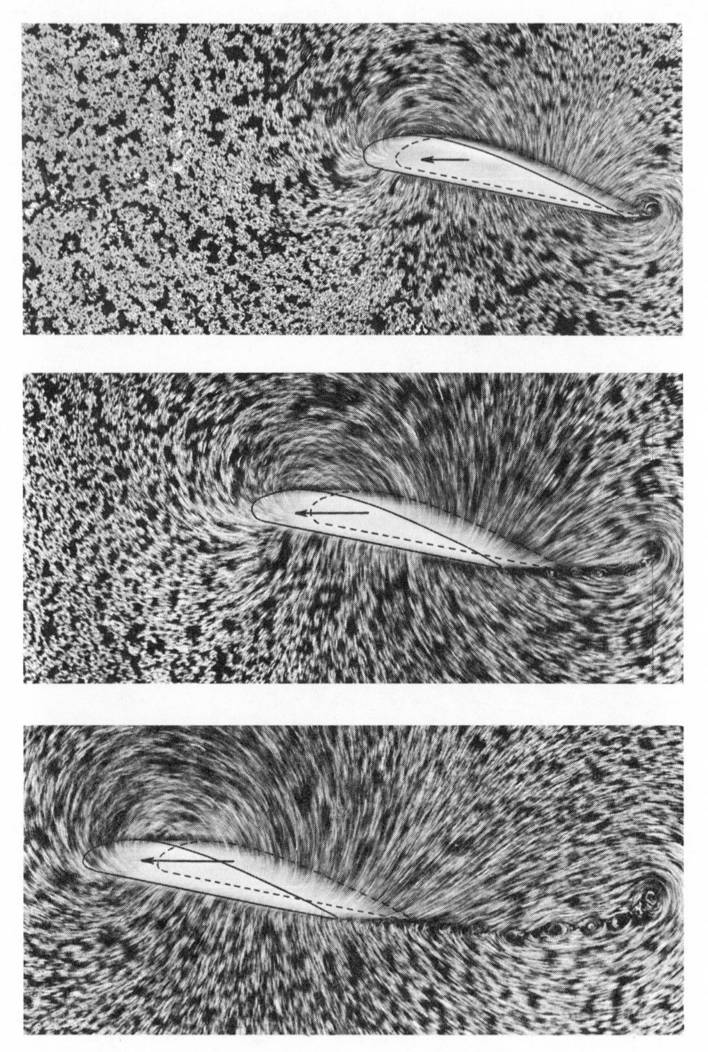

Plate XXI. Development of circulation around an airfoil as motion begins (camera is stationary relative to fluid).

the lower side actually leaves the trailing edge in the tangential direction, thereby producing in this zone a pattern of motion closely approaching that of Fig. 141*c* and resulting in a definite lifting force.

Lift characteristics of the Joukowsky profile. Although conditions at the trailing edge of an inclined plate thus approximate the characteristics of irrotational flow with circulation, the sharp leading edge also produces boundary-layer separation and a consequent departure from the potential pattern along the entire upper surface. This may be remedied to a considerable extent by rounding the leading edge and tapering the profile gradually toward the rear, as in Fig. 142*b*. Finally, to reduce the tendency toward separation even at moderately high values of α_0, the profile is given a slight *camber* as in Fig. 142*c*. The resulting pattern of irrotational flow is shown in Fig. 143*a*, the constant circulation in Fig. 143*b*, and the combined flow in Fig. 143*c*. Since the mathematical method of obtaining the patterns of flow for foils of this nature was developed, early in the present century, by a Russian scientist named Joukowsky,

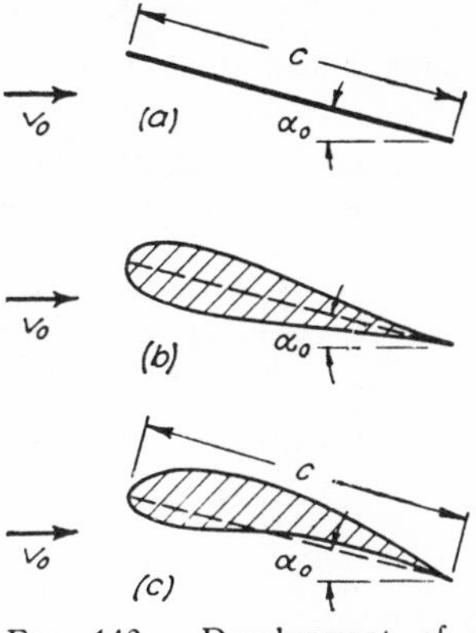

FIG. 142. Development of a cambered foil.

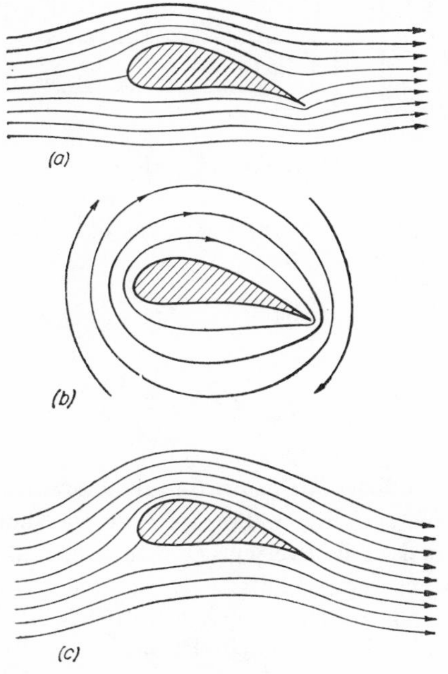

FIG. 143. Effect of circulation upon irrotational flow past a cambered foil.

such a section is generally known as the *Joukowsky profile*. Also due to Joukowsky is the original derivation of Eqs. (190) and (192); indeed, the inclined plate and the cylinder are in reality limiting forms of the Joukowsky profile.

The relative lift and drag of such an airfoil may readily be evaluated in terms of the coefficients C_L and C_D developed for the cylinder, if the length L and the diameter D of the cylinder are replaced by l

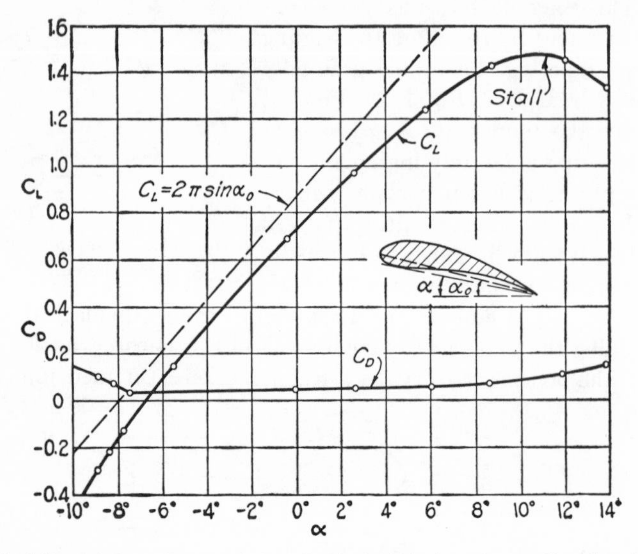

FIG. 144. Lift and drag characteristics of a typical Joukowsky profile, after Prandtl.

and c, the *span* and *chord* of the foil. The equations of lift and drag thus become

$$F_L = C_L l c \frac{\rho v_0^2}{2} \tag{193}$$

$$F_D = C_D l c \frac{\rho v_0^2}{2} \tag{194}$$

from which the coefficients C_L and C_D are determined experimentally for any foil geometry. Figure 144, for example, shows the results of tests on a Joukowsky foil of very great span; herein, it should be noted, the angle of attack for zero lift α_0 is replaced by the more readily measured inclination α of the tangent to the lower boundary of the profile (see inset sketch). It will be seen from this plot that the measured curve for C_L lies much closer to the theoretical than that for the cylinder in Fig. 140. Evidently, not only does the circulation resulting from separation at the trailing edge of the foil more nearly agree with the irrotational pattern, but the effect of separation in the

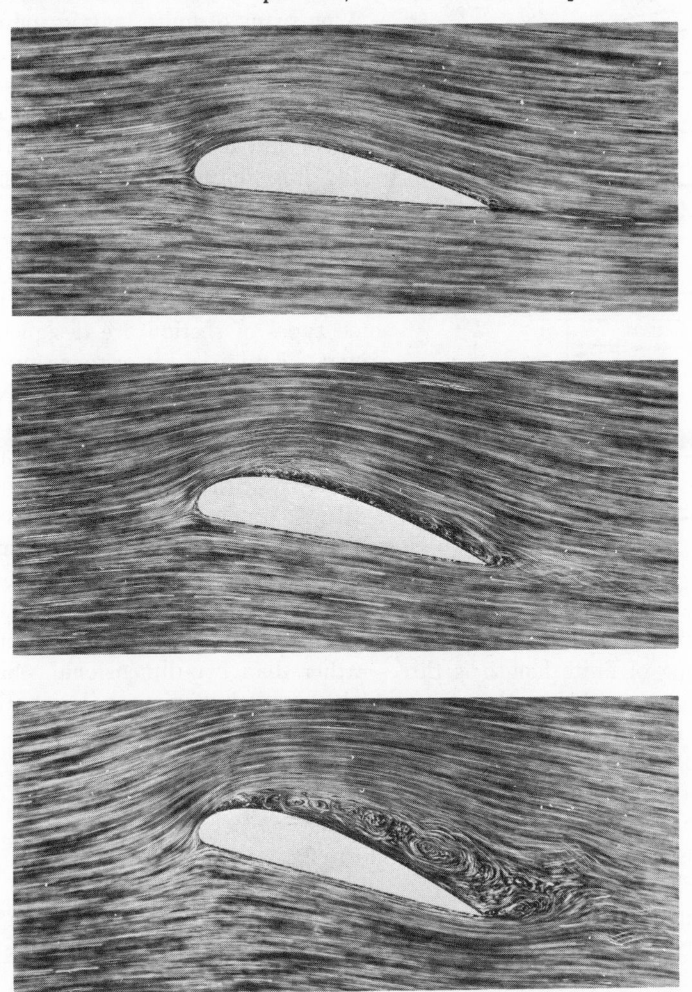

PLATE XXII. Patterns of flow past an airfoil at angles of attack of 5°, 10°, and 15° (camera is stationary relative to foil).

wake is of far less consequence. In other words, whereas the net lift of the rotating cylinder is much higher than that of the airfoil, the drag is still more pronounced; as a result, the efficiency of the foil as a lifting device is far superior. Nevertheless, even the best-formed foil is eventually subject to separation at the leading edge as the angle of attack is increased, the condition known as *stall* being followed by a rapid drop in the lift-drag ratio.

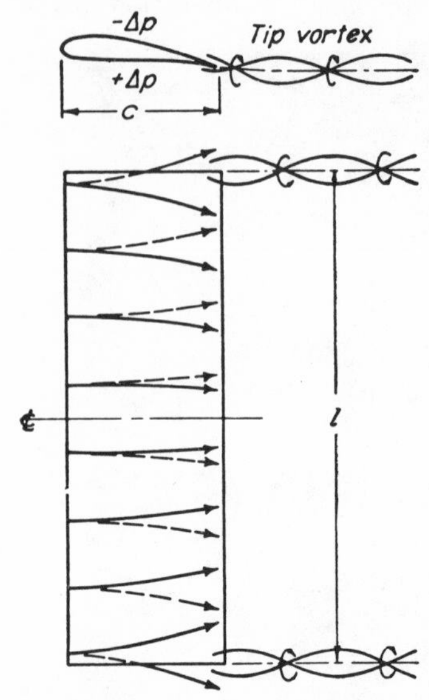

FIG. 145. Secondary flow produced by a wing of finite length.

Wing spans of finite length; induced drag. Although the Joukowsky-profile series includes a wide range of relative thickness and camber, and hence permits considerable systematic variation in lift-drag characteristics, the performance and structural requirements of the modern airplane have made it advisable to depart considerably from this series in actual wing profiles. Thus, some types of section are designed to provide high initial lift or to permit low landing speeds; others possess relatively high angles of stall; and still others display efficiency curves which change only slowly with angle of attack. Evidently, no profile can satisfy all requirements at once, and wing design must be guided by the particular needs of the given type of plane.

Further complexities in design result from the fact that flow around a wing of finite length is three- rather than two-dimensional. Since the pressure below the wing is higher, and that above is lower, than the intensity of the surrounding air, a lateral component of velocity in the direction of the pressure decrease must exist on both the lower and the upper surfaces of the wing, as indicated in Fig. 145. As a result, circulation occurs not only around the wing itself but about two vortex filaments extending into the wake from the wing tips. In the zone directly behind the wing, therefore, an appreciable *downwash* is produced by the tip vortices, so that the relative velocity between wing and air has a vertical as well as a horizontal component, and hence differs from the absolute velocity of the wing.

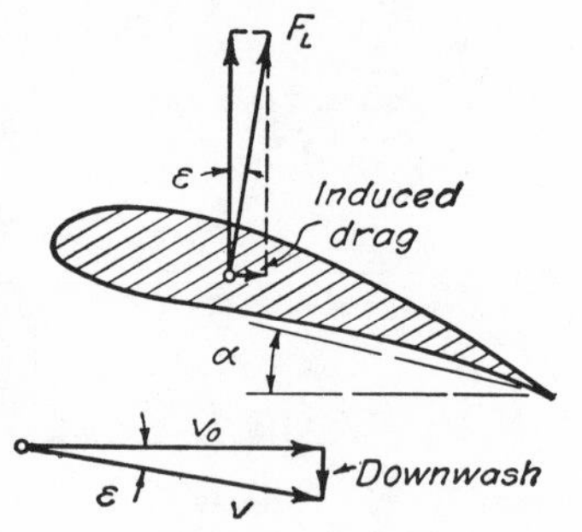

FIG. 146. Definition sketch for induced drag.

With reference to Fig. 146 it will be seen that this downwash necessarily reduces the effective angle of attack. In other words, since the angle α is measured from the line of motion (here assumed to be horizontal) of the wing itself, the vector sum of the horizontal velocity and that of the downwash yields a resultant which is inclined to the horizontal at an angle ϵ. Since the true lift is still defined as the force component at right angles to the mean relative velocity, it will now have a component in the horizontal direction. So long as ϵ is small, the horizontal and resultant velocities and the vertical and resultant lifts will be essentially the same; that is,

$$v_0/\cos\epsilon \approx v_0 \quad \text{and} \quad F_L \cos\epsilon \approx F_L$$

The horizontal component of the true lift will not be negligible, however, and hence may increase appreciably the actual drag. This component is commonly known as the *induced drag*, which evidently has the magnitude

$$F_{Di} = F_L \sin\epsilon \approx F_L\epsilon$$

By making an arbitrary assumption as to the distribution of the intensity of lift over a wing of finite length, a coefficient of induced drag may be expressed approximately as

$$C_{Di} \approx \frac{C_L{}^2}{\pi l/c} \tag{195}$$

in which l/c is the *aspect ratio* of the wing—the ratio of span to chord. From the foregoing equations the following approximate relationships for evaluating the angle ϵ may be obtained:

$$\epsilon \approx \frac{F_{Di}}{F_L} \approx \frac{C_L}{\pi l/c}$$

Interpretation of the polar diagram. By means of these expressions it is possible, as a close approximation, to evaluate from measurements on a wing having a particular aspect ratio the lift-drag characteristics of the given wing profile at any other aspect ratio—including those for the infinite span. In this process it is expedient to plot the measured values of lift and drag in the form of a so-called *polar diagram* (see Fig. 147), which shows C_L as a function of C_D for the tested aspect ratio and various angles of attack. If, for the same aspect ratio, Eq. (195) is plotted on this diagram, the horizontal intercept between the two curves will represent the drag coefficient for two-dimensional flow around a span of infinite length; since $F_L \cos\epsilon \approx F_L$, the lift coefficient is essentially the same for both cases. It should be noted in passing that the polar diagram permits ready determination of the *maximum lift-drag ratio;* that is, the line which is tangent to the curve and passes through the pole evidently has a slope which is equal to the optimum value of the ratio C_L/C_D, the point of tangency therefore indicating the most efficient angle of attack.

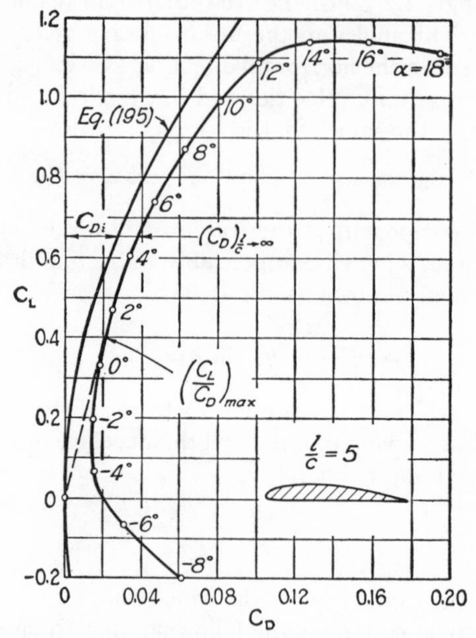

FIG. 147. Polar diagram for a wing of finite length, after Prandtl.

Example 49. Determine the lift and drag of an airplane wing having a span of 40 feet, a chord of 7 feet, and the profile characteristics shown in Fig. 147, when traveling at a velocity of 200 miles per hour at the angle of attack yielding a maximum lift-drag ratio. What percentage of the total drag is due to the finite span of the wing?

The aspect ratio of the wing will be

$$\frac{l}{c} = \frac{40}{7} = 5.71$$

Since this does not differ greatly from the aspect ratio for the curve plotted in Fig. 147, it may be assumed without replotting that the greatest lift-drag ratio occurs at the same angle of attack—i.e., $\alpha = 1°$.

For this value of α, $C_L = 0.4$, and

$$F_L = C_L l c \frac{\rho v_0^2}{2} = 0.4 \times 40 \times 7 \times \frac{0.0025 \times \left(\frac{200 \times 5280}{60 \times 60}\right)^2}{2} = 12{,}000 \text{ lb}$$

From Fig. 147 the drag coefficient 0.011 for a wing of infinite span is found as the intercept between the curves for C_D and C_{Di}. The induced drag for the given aspect ratio is, from Eq. (195),

$$C_{Di} \approx \frac{C_L^2}{\pi l/c} = \frac{\overline{0.4}^2}{\pi \times 5.71} \approx 0.009$$

whence

$$C_D = 0.011 + 0.009 = 0.020$$

Therefore

$$F_D = C_D l c \frac{\rho v_0^2}{2} = 0.020 \times 40 \times 7 \times \frac{0.0025 \times \left(\frac{200 \times 5280}{60 \times 60}\right)^2}{2} = 600 \text{ lb}$$

and the percentage of the total drag represented by C_{Di} is

$$\frac{C_{Di}}{C_D} \times 100 = \frac{0.009}{0.020} \times 100 = 45\%$$

PROBLEMS

326. A kite having an effective area of 7 square feet and a weight of ½ pound is so rigged that its surface is held at 120° to the anchor cord. If, during a wind of 20 miles per hour, the pull on the cord is 5 pounds and the cord is inclined at an angle of 45° to the horizontal, determine the corresponding coefficients of lift and drag.

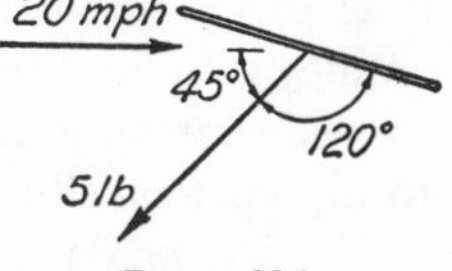

PROB. 326.

327. If a guide vane having the form of the symmetrical foil shown in Fig. 142 has a chord of 3 feet, estimate from the equations of the circulation theory the side thrust per foot of length which would be exerted upon the vane by water traveling with a velocity of 12 feet per second at an angle of 5° to the vane axis; assume a lift efficiency of 70 per cent.

328. An airfoil section is found to yield a zero lift coefficient at an angle of attack of −5° and a lift coefficient of 0.7 at an angle of 2°. What is the efficiency of the foil under the latter conditions, based upon the circulation theory?

329. When the line of descent of a glider is held at 5° to the horizontal, it is found to attain a maximum speed of 110 miles per hour. If its gross weight is 350 pounds, what are the magnitudes of the corresponding lift and drag?

330. The total wing area of a small monoplane having a weight of 1400 pounds is 110 square feet. What must be the lift coefficient of the wing if the cruising speed of the plane is 135 miles per hour? If the motor delivers 200 horsepower at this speed, and if 60 per cent of this power represents propeller loss and body resistance, what is the drag coefficient of the wing?

331. Plot the ratio of lift to drag against the angle of attack for the data of Fig. 144, indicating the angle of best performance.

332. What percentage reduction in (*a*) induced drag and (*b*) total drag could be obtained by increasing to 10 the aspect ratio of the wing in Example 49?

333. An airfoil section is tested in a wind tunnel under conditions simulating an infinite aspect ratio (i.e., through use of thin plates mounted parallel to the flow at each end of the section). At an angle of attack of 3° the lift coefficient is deter-

mined to be 0.7 and the drag coefficient 0.05. What would be the corresponding magnitude of the coefficient of induced drag for an aspect ratio of 7?

334. If each of two paravanes towed through water by a minesweeper has the lift-drag characteristics of Fig. 147 and is rigged as shown in the accompanying plan view, at what cable angle α will a stable position relative to the ship be attained?

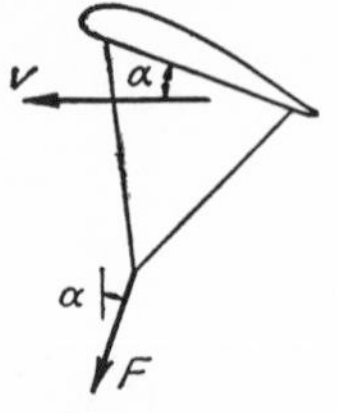

PROB. 334.

335. A wing having the infinite-span lift and drag coefficients of Problem 333 has a span of 50 feet and a chord of 8 feet. What thrust would the propeller of the plane have to develop in order to overcome the total drag of this wing at a speed of 250 miles per hour and an angle of attack of 3°?

336. As a means of eliminating the wave resistance of speed boats, it has been suggested that the hulls be supported well above the water surface on struts extending to cambered vanes below the surface. If such a vane has the characteristics plotted in Fig. 147, what must be its dimensions to support a load of 5 tons at the most favorable lift-drag ratio when the boat is cruising at a speed of 30 miles per hour?

45. CHARACTERISTICS OF PROPELLERS

Vector analysis of a blade element. Although the airfoil finds its simplest application in providing the supporting force required by all heavier-than-air craft, the principles of lift and drag discussed in the foregoing section are by no means limited to the design of airplane wings. Indeed, not only are these principles applicable to liquids as well as gases at comparable values of the Reynolds number, but they may readily be adapted to the non-rectilinear motion of vanes exemplified by windmills, fans, airplane propellers, and ship screws. While the detailed analysis of such phenomena is rendered somewhat more complex by secondary factors due to boundary rotation, the basic principles will be found identical in every instance.

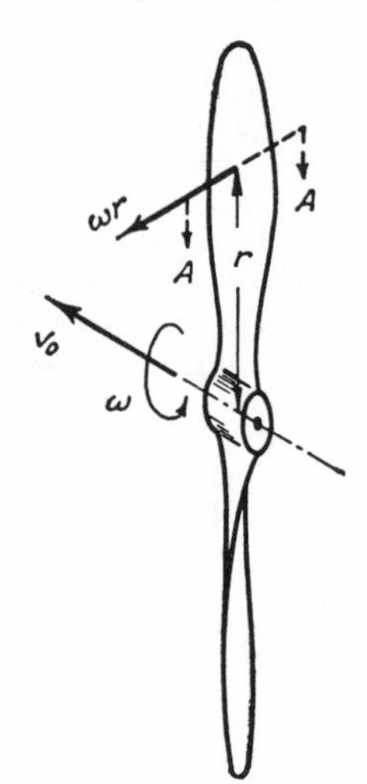

FIG. 148. Velocity vectors due to translation and rotation of a propeller.

For example, let Fig. 148 represent a typical propeller which rotates with an angular velocity ω at the same time that it advances as a whole through the fluid at the linear velocity v_0. Further, let Fig. 149 represent a tangential section A–A through the blade at the distance r from the axis of rotation. If the undisturbed fluid is assumed to be at rest, the actual velocity v of the section relative to the fluid as a whole will evidently be the vector sum of the longitudinal velocity v_0 and the tangential velocity ωr, as shown in the illustration. The local blade angle β thus bears the same relationship to the local angle of advance $\phi = \tan^{-1} v_0/\omega r$ as the *geometric pitch* of the section bears to the *effective pitch* of the propeller as a whole—i.e., its axial displacement $2\pi v_0/\omega$ during one complete revolution.

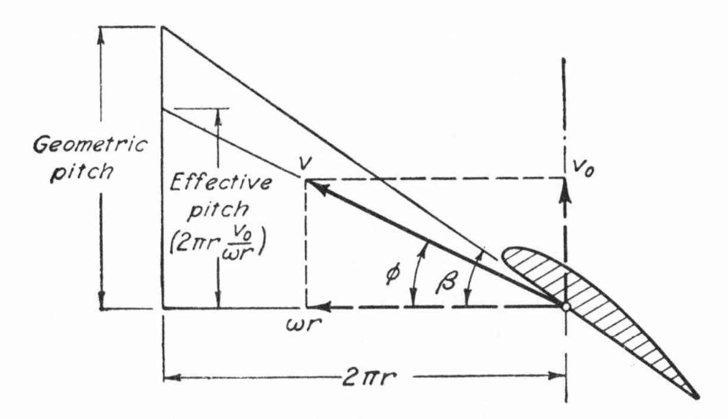

FIG. 149. Definition sketch for the pitch of a blade element.

The corresponding conditions of steady flow relative to the section may be obtained as usual by adding the velocity v in the opposite direction to both fluid and body, with the result shown in Fig. 150. The angle $\beta - \phi$ is now seen to represent the counterpart of the angle of attack for the elementary airfoil. It is therefore to be expected that the blade section will be subject to the same combination of fluid forces

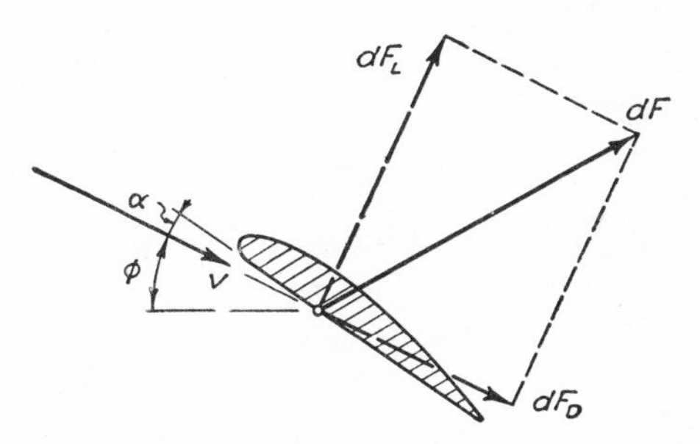

FIG. 150. Lift and drag components of force on a blade element.

as the airfoil. That is, owing to the circulation produced by separation at the trailing edge of the section, the moving fluid will exert, upon an incremental length of blade dr having the given section, the differential lift dF_L at right angles to the velocity v, and the differential drag dF_D parallel to the velocity v. For present purposes, however, it is more expedient to resolve their resultant dF into components in the axial and tangential directions as shown in Fig. 151; the axial component dF_a thus represents the elementary *thrust*, and the tangential component dF_t the elementary *torque per unit radius*, of the blade section.

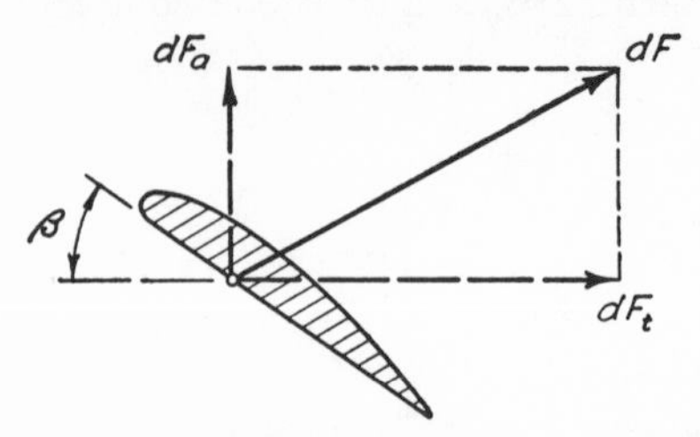

FIG. 151. Axial and tangential components of force on a blade element.

Propeller thrust, power, and efficiency. As in the case of the airfoil, the relative magnitudes of the elementary forces at any section of a propeller blade will vary with both the form of the section and the angle of attack. In the case of the propeller, however, the pitch of the section is also involved, since the lift component contributes not only to the thrust of the blade but also to the torque. Moreover, the angle of attack depends upon both the blade angle and the angle of advance of the section, and the latter (see Fig. 149) necessarily varies with radial distance from the axis of rotation. If, therefore, the blade is to attain high efficiency as a whole (i.e., a high thrust-torque ratio at all sections), the geometric pitch as well as the cross-sectional form of a well-designed propeller will have to vary continuously along the blade. It then becomes necessary to characterize the blade as a whole in terms of its *nominal pitch*—i.e., its geometric pitch at some arbitrary section, generally taken at two-thirds the blade radius.

The thrust-torque ratio, of course, becomes secondary to structural considerations in the proximity of the hub, but here the tangential velocities are so low, relatively speaking, that considerable departure from proper blade form in this zone is of little consequence. On the other hand, near the tip of the blade the same type of secondary motion is encountered which was found to produce downwash and induced drag in the wing of finite span. Moreover, tip speeds of airplane propellers approaching the velocity of sound lead to density disturbances which greatly lower the blade efficiency, while the reduction in pressure on the back of ship screws often produces severe cavitation. Further treatment of such details is obviously beyond the scope of the present text. Suffice it to say that in the refined design of a propeller the force characteristics of each elementary portion of the blade must be investigated in the manner shown, and the overall thrust and torque determined through integration of the elementary force components over the entire length of blade.

At the present time, it is needless to remark, propeller models are invariably tested in the wind or water tunnel prior to acceptance of the prototype design. Test results are evaluated in terms of dimensionless coefficients of thrust and power, comparable to the coefficients of lift and drag characterizing the performance of the airfoil, and plotted, together with the efficiency, as functions of the ratio of the axial to the tangential velocity of the tips of the blades. For convenience, it is standard practice to use the revolutions per second n

PLATE XXIII. Powered scale model of the Republic P-47 in the Langley atmospheric wind tunnel of the National Advisory Committee for Aeronautics.

instead of the angular velocity ω, the overall diameter D of the propeller instead of the radius of the blade tip, and the square of the diameter instead of the actual surface or projected area of the blades. The *thrust coefficient* thus becomes

$$C_T = \frac{F_T}{\rho n^2 D^4} \tag{196}$$

and the *power coefficient*

$$C_P = \frac{P}{\rho n^3 D^5} \tag{197}$$

Since $n = \omega/2\pi$, the quantity v_0/n represents the effective pitch or advance per revolution; when divided by the propeller diameter this becomes the basic parameter v_0/nD known as the *advance-diameter ratio.* The *efficiency* η (eta) is at once expressible in terms of the advance-diameter ratio and the thrust and power coefficients:

$$\eta = \frac{F_T v_0}{P} = \frac{C_T}{C_P} \frac{v_0}{nD} \tag{198}$$

A typical test diagram in terms of these parameters is shown in Fig. 152.

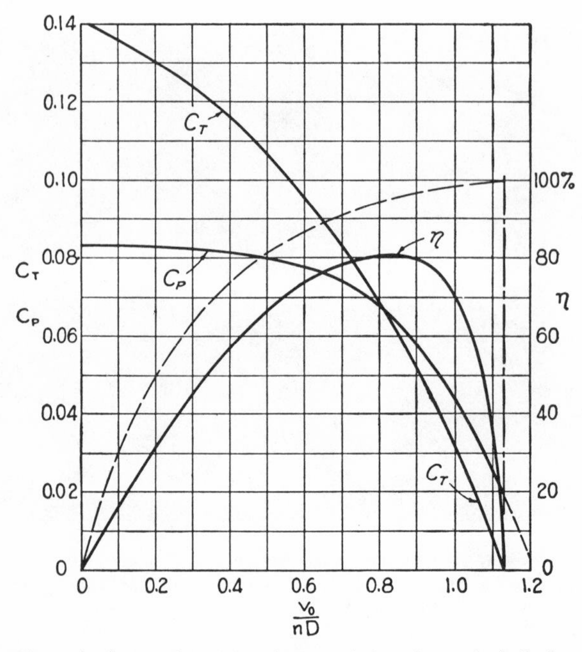

FIG. 152. Dimensionless performance characteristics of a typical airplane propeller.

The values of the several curves at their intercept with the vertical axis evidently correspond to starting conditions of a ship or plane, at which the relative velocity v_0 between the propeller and the fluid as a whole is zero; under such circumstances, the thrust coefficient is seen to attain its maximum value, the efficiency being zero owing to the use of v_0 as the characteristic velocity (the efficiency of a stationary fan, obviously, must be otherwise evaluated). At the other extreme of the abscissa scale the thrust coefficient attains the limit zero, which corresponds approximately to the position of zero lift for the airfoil; the efficiency is then zero, for appreciable torque is still required to overcome the fluid resistance to the relative motion of the blade.

Between these limits of the performance curves—but well toward the right of the diagram—the efficiency of the propeller is seen to reach a maximum value, corresponding to best operating conditions. At peak efficiency the effective pitch of the propeller is somewhat below the geometric pitch, which results in a so-called "slip" of the blades; indeed, the abscissa parameter v_0/nD is sometimes replaced by what is known as the *slip function*, which varies from 100 per cent to 0 per cent as the ratio of the effective to the geometric pitch varies from 0 to 1. It is of interest at this point to note that propeller blades

PLATE XXIV. High-speed photograph of a cavitating model ship propeller under test at the David Taylor Model Basin of the U. S. Navy.

of adjustable pitch in effect permit control of the slip function to maintain a high degree of efficiency over a wide range of operating conditions.

One-dimensional analysis of the slipstream. Although the foregoing discussion has dealt specifically with propellers used to produce motion of craft through fluids at rest, essentially the same analysis is adaptable to stationary fans and windmills. In this connection, the one-dimensional method once used exclusively in propeller analysis is of particular value for the general clarity which it lends to the problem as a whole. Before proceeding to a description of this method, however, it must again be emphasized that the motion in the *slipstream* of any propeller is quite complex, even when reduced to the schematic form of Fig. 153. Evidently, the circulation around the blades of finite length continues in the slipstream as a system of helical tip vortices, the tangential component of the resultant force upon the blades produces rotation of the slipstream as a whole, and the thrust results in a change in axial velocity—regardless of whether the rotating blades are those of a propeller, windmill, or fan.

In the case of a propeller, addition of the velocity v_0 to the right will result—so far as the axial components of flow are concerned—in a constant velocity of approach v_0 and a mean velocity $v_0 + \Delta V$ within the slipstream at the right. In accordance with the continuity principle, the slipstream diameter must be smaller than that of the propeller, and the diameter of the corresponding zone in the approaching flow must accordingly be larger, as indicated in Fig. 154. From the principle of momentum, the resulting change in velocity may be related to the thrust in the following manner:

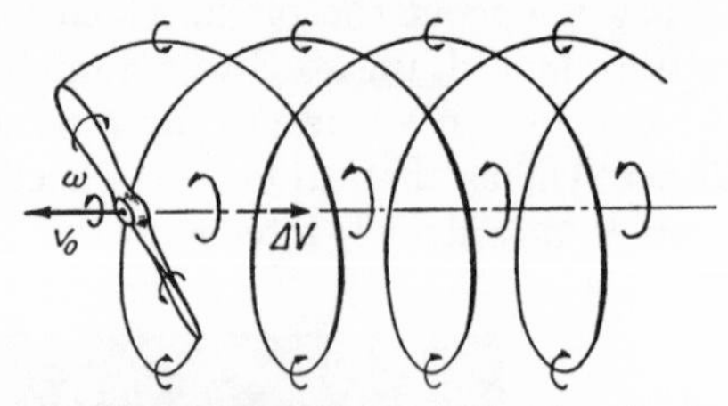

FIG. 153. Components of motion in the wake of a propeller.

$$F_T = Q\rho\,\Delta V \tag{199}$$

On the other hand, the propeller may be considered to produce an abrupt change in the pressure intensity of the flow as it passes the propeller section, the thrust necessarily being equal to the product of this mean pressure change and the cross-sectional area of the section:

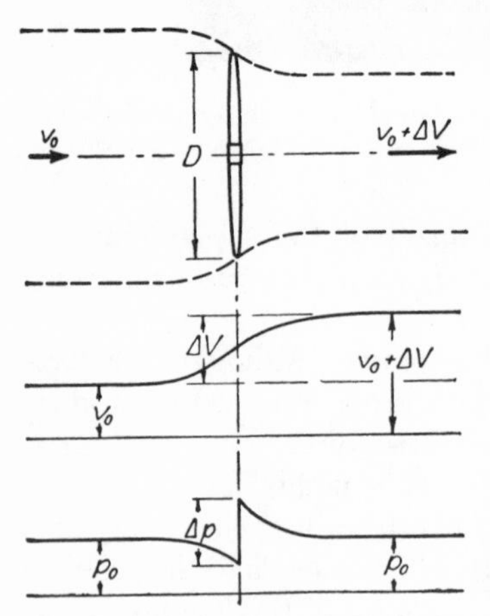

FIG. 154. One-dimensional analysis of flow past a propeller.

$$F_T = \Delta p \frac{\pi D^2}{4} \tag{200}$$

Assuming, in approximate accord with the accompanying change in mean velocity, that the pressure change consists of a reduction on the upstream side and an increase on the downstream side relative to the atmospheric intensity p_0, the mean velocity and pressure variation in the longitudinal direction will be essentially as shown in the figure.

Since the pressure intensity in the slipstream necessarily approaches that of the surrounding fluid, the power input of the propeller should, in accordance with the Bernoulli equation for one-dimensional flow, equal the rate of change in kinetic energy plus the power lost due to the surface drag of the blades. If, as a first approximation, both the power loss and that part of the kinetic energy due to circulation are ignored, the work done by the thrust of the propeller may be placed equal to the change in the mean kinetic energy of the axial flow:

$$\Delta p \approx \frac{\rho(v_0 + \Delta V)^2}{2} - \frac{\rho v_0^2}{2} \tag{201}$$

Elimination of F_T and Δp from Eqs. (199), (200), and (201) then yields for the rate of flow past the propeller section the quantity

$$Q \approx \frac{\pi D^2}{4}\left(v_0 + \frac{\Delta V}{2}\right) \tag{202}$$

which indicates that half the velocity change takes place in front of the propeller section and half behind. Introduction of this relationship for Q in Eq. (199) then permits evaluation of the velocity change (i.e., the actual velocity of a propeller slipstream) in terms of the thrust:

$$\Delta V \approx -v_0 + \sqrt{v_0^2 + \frac{8F_T}{\pi D^2 \rho}} \tag{203}$$

Since the stationary fan is equivalent to a propeller advancing with the velocity $v_0 = 0$, the corresponding one-dimensional representation would be essentially as shown in Fig. 155, the flow approaching

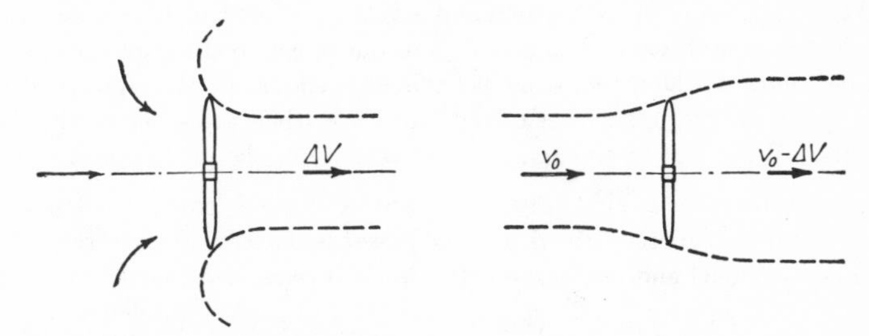

FIG. 155. Flow produced by a fan. FIG. 156. Flow past a windmill.

the fan section from all directions and again contracting as it passes into the slipstream. In the case of the windmill, on the contrary, the fact that the flow now does work upon the vanes requires a higher pressure on the upwind side and hence an increase in slipstream diameter beyond the vane section as shown in Fig. 156. In either case, nevertheless, the same equations may be used to evaluate the slipstream diameter and velocity, with essentially the same degree of approximation.

In making use of the one-dimensional approach, the fact must again be emphasized that neglect of the surface drag of the blades and the circulatory motion in the wake leads to considerable inexactness in the results. Indeed, if the efficiency of a propeller is computed by this method, the idealized curve shown as a broken line in Fig. 152 will result, from which the actual extent of the discrepancy may be judged. Although the magnitude of the energy of circulation in the slipstream may be evaluated by a more refined analysis, it still represents lost power. It is therefore interesting to note that loss due to rotation of the slipstream as a whole may be avoided to a great extent through use of stationary straightening vanes (see Fig. 157) on the downstream side of the blades. The tip vortices and the resulting induced drag may likewise largely be eliminated by placing a cylindrical sleeve or shroud around the blades; this also obviates the change in slipstream diameter, as shown in the figure, with a resulting increase in efficiency. However, the unit then becomes, in effect, a duct fan, propeller pump, or turbine, and as such will be discussed more fully in the following section.

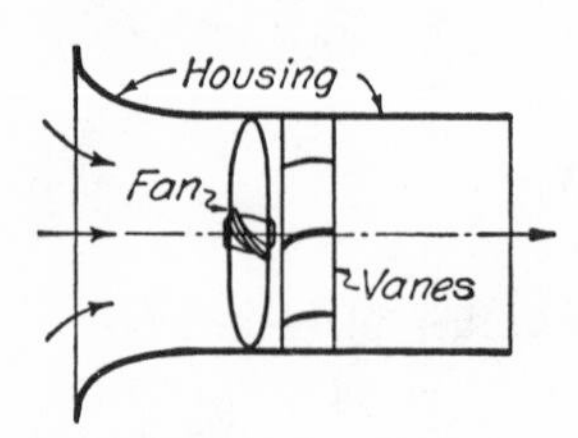

FIG. 157. Methods of improving fan performance.

Example 50. An 8-inch model screw, tested in a water tunnel at a rotational speed of 22.5 revolutions per second and a velocity of flow of 12 feet per second, yields a measured thrust of 28.5 pounds, a torque of 3.45 pounds, and an efficiency of 70 per cent. If the model scale is 1:25, what should be the equivalent rotational speed, thrust, and power of the prototype when the speed of the ship is 15 knots? What will be the approximate velocity of the prototype slipstream?

Since the problem is evidently one of approximate dynamic similarity (viscous effects being ignored), the thrust and power coefficients must be respectively the same for model and prototype. It is thus at once seen that

$$n_p = \left(\frac{v_0}{D}\right)_p \left(\frac{nD}{v_0}\right)_m = \frac{15 \times 1.69}{25 \times 8/12} \times \frac{22.5 \times 8/12}{12} = 1.90 \text{ rps}$$

Moreover,

$$(F_T)_p = (\rho n^2 D^4)_p \left(\frac{F_T}{\rho n^2 D^4}\right)_m = 1.99 \times \overline{1.90}^2 \times \overline{25 \times 8/12}^4 \times \frac{28.5}{1.94 \times \overline{22.5}^2 \times \overline{8/12}^4}$$
$$= 81{,}500 \text{ lb}$$

Likewise,

$$P_p = (\rho n^3 D^5)_p \left(\frac{P}{\rho n^3 D^5}\right)_m = 1.99 \times \overline{1.90}^3 \times \overline{25 \times 8/12}^5 \times \frac{2\pi \times 22.5 \times 3.45}{1.94 \times \overline{22.5}^3 \times \overline{8/12}^5}$$
$$= 2{,}940{,}000 \text{ ft-lb/sec}$$

$$\frac{2{,}940{,}000}{550} = 5340 \text{ hp}$$

The velocity of the slipstream is estimated from Eq. (203) as follows:

$$\Delta V \approx -v_0 + \sqrt{v_0^2 + \frac{8F_T}{\pi D^2 \rho}} = -15 \times 1.69 + \sqrt{\overline{15 \times 1.69}^2 + \frac{8 \times 81{,}500}{\pi \times \overline{25 \times 8/12}^2 \times 1.99}}$$

$$= 6.5 \text{ fps}$$

PROBLEMS

337. A 6-foot windmill with vanes having a constant blade angle of 30° is found to have an effective pitch of 7 feet in a 20-mile-per-hour wind. Determine the angles of attack of the vanes at radii of 1, 2, and 3 feet, and sketch to scale the vector diagram of velocities in each case.

338. The rotational speed of a 6-foot propeller is 1500 revolutions per minute. If the blade angle at two-thirds radius is 35°, determine the nominal pitch and effective pitch of the propeller and the geometric pitch and angle of attack of the section when the speed of advance is 130 miles per hour.

339. Assuming the blades of a 9-foot propeller to be based upon the airfoil section shown in Fig. 144, determine the blade angle at a 3-foot radius which would yield an angle of attack of 2° when the speed of advance is 250 miles per hour and the rotational speed is 1800 revolutions per minute. If the section has a 6-inch chord, determine and plot to scale the corresponding lift and drag per radial foot of blade, the resultant force per foot, and its axial and tangential components.

340. When turning at the rate of 24 revolutions per minute, a screw is found to propel a ship at a speed of 15 knots. If the rotational speed were increased to 30 revolutions per minute, at what ship speed would the efficiency of the screw be the same? What would be the percentage increase in thrust and horsepower?

341. A ski-sled driven by an airplane propeller is found to require a thrust of 100 pounds when traveling at a speed of 60 miles per hour. Assuming the propeller to be operating at maximum efficiency and to have the characteristics shown in Fig. 152, determine the corresponding diameter, rotational speed, and horsepower.

342. What thrust would the propeller of the ski-sled of Problem 341 develop as the sled starts from rest, assuming that the motor speed remains constant?

343. At what speed of the ski-sled of Problem 341 would the propeller develop a zero thrust? Assume a constant motor speed.

344. If a 6-foot propeller having the characteristic curves of Fig. 152 is structurally designed to withstand a maximum thrust of 250 pounds, determine the highest permissible rotational speeds at a zero speed of advance (*a*) in air and (*b*) in water. What will be the power requirements in the two cases?

345. An 8-foot airplane propeller having the characteristics of Fig. 152 is driven by a motor turning at the rate of 1800 revolutions per minute. If the measured thrust at zero speed of advance is found to be 1300 pounds, at what speed of advance will the propeller attain maximum efficiency of operation? What power will then be developed by the motor?

346. Estimate the air speed in the slipstream when the propeller of Problem 345 is operating at a zero speed of advance.

347. Tests on a 10-inch model ship screw in a towing tank indicate a 50 per cent increase in relative water speed behind the screw when the towing speed is 6 feet per second and the rotational speed of the screw is 20 revolutions per second. Estimate the corresponding thrust on the propeller shaft.

348. A 24-inch stationary fan having the characteristics of Fig. 152 is operated at 600 revolutions per minute. Determine the fan efficiency based upon the kinetic energy of the "slipstream" which it produces.

46. AXIAL-FLOW AND RADIAL-FLOW MACHINERY

Duct fans, propeller pumps, and propeller turbines. As noted at the end of the previous section, a *duct fan* or *propeller pump* (Fig. 158) differs from an open fan or propeller primarily in that the blades are enclosed by a cylindrical casing of sufficient length to provide uniform flow on both the upstream and downstream sides. Such units also generally contain a system of guide vanes to prevent rotation of the "slipstream," together with a bell-mouthed intake to avoid undue contraction of the approaching flow. A *propeller turbine* evidently bears the same relationship to a propeller pump as a stationary windmill bears to a stationary fan. In a word, the method of analysis of the foregoing section applies as well to the design of duct fans, propeller pumps, and propeller turbines—which are all *axial-flow* machines—and hence requires only the following amplification at this point.

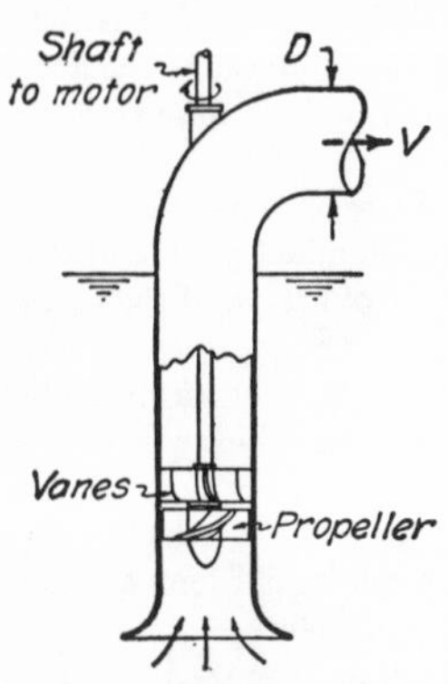

FIG. 158. Schematic diagram of a propeller pump.

So far as performance characteristics are concerned, it has long been customary to plot the head, horsepower, and efficiency for any given unit as functions of the capacity, as shown in Fig. 159 for a typical propeller pump. Duct fans, pumps, and turbines are, however, subject to the same generalized treatment as the airfoil and propeller, the performance curves for a given design thereby becoming independent of scale and fluid density. For greater convenience, however, in place of the thrust coefficient a *head coefficient* embodying the change in total head is used,

$$C_H = \frac{g\,\Delta H}{n^2 D^2} \tag{204}$$

and in place of the advance-diameter ratio a *capacity coefficient,*

$$C_Q = \frac{Q}{nD^3} \tag{205}$$

Since the quantities involved in evaluating the power requirements are identical in both instances, the *power coefficient* will be the same as before:

$$C_P = \frac{P}{\rho n^3 D^5} \tag{206}$$

The coefficients of head and power, together with the *efficiency*,

$$\eta = \frac{P_{\text{output}}}{P_{\text{input}}} \tag{207}$$

thus become functions of the capacity coefficient, and Fig. 159 reduces to the dimensionless performance diagram shown in Fig. 160. It is to be noted that two curves for power have now been included—the upper corresponding to the power input for a propeller pump,

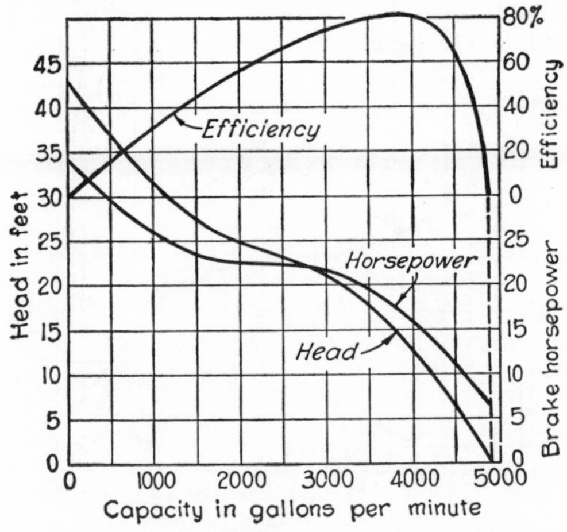

FIG. 159. Performance characteristics of a 12-inch propeller pump at 1500 revolutions per minute.

and the lower to the power output of a propeller turbine, for the same head-capacity characteristics. It is assumed, moreover, that the plotted characteristics apply to both liquids and gases, although structural considerations for the two cases must obviously be reflected in the performance diagram.

Curves such as those shown in this diagram are necessarily characteristic of one particular blade design, but the latter may be modified to a considerable degree to meet various operating requirements. Such units as propeller turbines, moreover, are frequently built with adjustable blades to permit maximum efficiency over a considerable range of capacity. On the other hand, all axial-flow machines have one characteristic in common: high flow capacity under relatively low head.

Blowers, centrifugal pumps, and Francis turbines. In many instances, of course, requirements quite opposite to those of axial-flow machines must be satisfied: high head at relatively low rates of flow. Such requirements can be fulfilled only by means of the so-called *centrifugal fan* or *blower*, the *centrifugal pump*, and the *Francis turbine*. In such units the flow is not wholly axial, as in propeller units, but almost entirely *radial*, as indicated schematically in Fig. 161. Although the action of each individual blade may again be analyzed in terms of

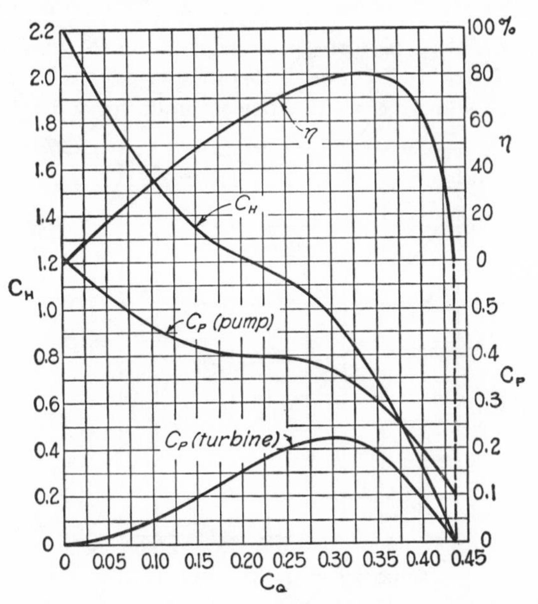

FIG. 160. Dimensionless performance characteristics of a typical axial-flow machine.

circulation, lift, and drag, it is perhaps more useful at present to adapt the one-dimensional approach to the case of axially symmetric motion, in the following manner.

Except for the necessary reversal in blade curvature and direction of flow (compare parts *a* and *b* of Fig. 161), there is little essential difference in the analysis of a centrifugal pump or Francis turbine by this method. Assume, therefore, that flow passing the guide vanes of a turbine (Fig. 162*a*) reaches the blades of the runner with the velocity v_1 at the angle α_1, and leaves the runner with the velocity v_2 at the angle α_2. The rate of flow through the runner evidently depends upon only the radial component of v_1 or v_2:

$$Q = 2\pi r_1 b_1 v_1 \sin \alpha_1 = 2\pi r_2 b_2 v_2 \sin \alpha_2 \tag{208}$$

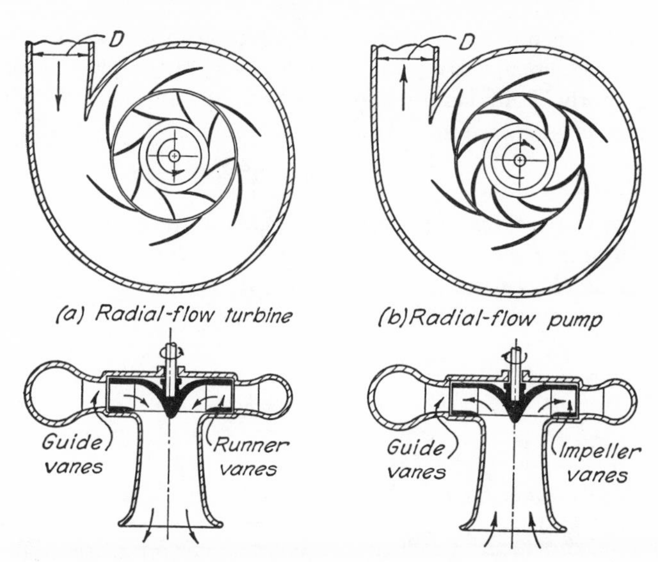

FIG. 161. Schematic diagrams of radial-flow machines.

The torque, on the other hand, depends upon only the tangential components of these velocities; that is, equating torque to rate of change

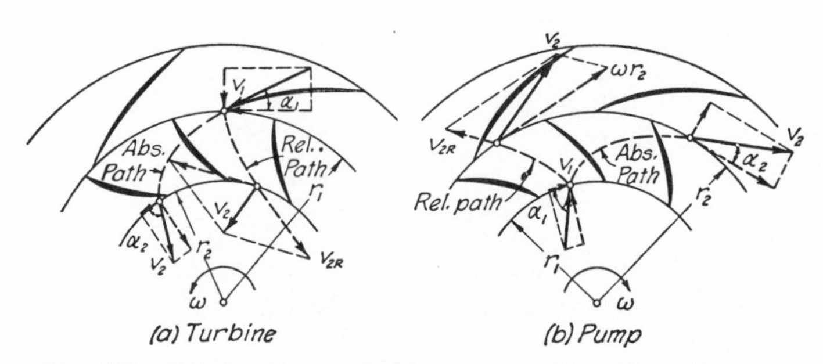

FIG. 162. Velocity diagrams for the runner and impeller of Fig. 161.

of angular momentum, there results an expression derived by Euler some two hundred years ago:

$$\text{Torque} = Q\rho(v_1 \cos \alpha_1 \, r_1 - v_2 \cos \alpha_2 \, r_2)$$

Multiplication of each term by the angular velocity ω then yields the power delivered to the runner:

$$P = Q\rho\omega(v_1 \cos \alpha_1 \, r_1 - v_2 \cos \alpha_2 \, r_2)$$

This relationship takes on particular significance when it is noted that the products $v_1 \cos \alpha_1 \, 2\pi r_1$ and $v_2 \cos \alpha_2 \, 2\pi r_2$ represent the *circulation* of the fluid before and after passage through the runner; thus, since $n = \omega/2\pi$, the power output becomes

$$P = Q\rho n(\Gamma_1 - \Gamma_2) \tag{209}$$

Evidently, the greater the change in circulation (i.e., the greater the change in both radius and tangential velocity), the greater the power delivered by the turbine. This power output corresponds, in effect, to a drop $\Delta H'$ in the total head of the flow, such that

$$P = Q\gamma \, \Delta H'$$

whence

$$\Delta H' = \frac{n(\Gamma_1 - \Gamma_2)}{g}$$

In addition, however, a certain amount of head is lost through surface drag and possible separation from the blades, not to mention losses of a similar nature in the scroll case, guide vanes, and draft tube. The total drop in head may therefore be written as the sum of that representing useful work upon the runner and that resulting from form and surface drag; thus, for the turbine,

$$\Delta H_t = \frac{n(\Gamma_1 - \Gamma_2)}{g} + \Sigma H_L \tag{210}$$

while for the centrifugal pump (or blower) it will be found that the head produced will be

$$\Delta H_p = \frac{n(\Gamma_2 - \Gamma_1)}{g} - \Sigma H_L \tag{211}$$

Since the efficiency is the ratio of the power input to the power output, the efficiency of the turbine will be simply

$$\eta_t = \frac{n(\Gamma_1 - \Gamma_2)}{g \, \Delta H_t} \tag{212}$$

and that for the pump (or blower)

$$\eta_p = \frac{g \, \Delta H_p}{n(\Gamma_2 - \Gamma_1)} \tag{213}$$

So far as the form of the runner or impeller blades is concerned, two salient facts will become clear from study of the velocity diagrams in Fig. 162. In order for the turbine to operate at peak efficiency, the total head of the flow which leaves a turbine runner must be a minimum; since the velocity head must then also be a minimum, optimum conditions—so far as direction of outflow is concerned—are reached when the tangential component of v_2, and hence the circulation Γ_2, is very nearly equal to zero; the trailing edges of the runner blades must therefore be at such an angle that the tangential velocity of the fluid relative to the blade is approximately equal and opposite to the tangential velocity of the blade itself. The same must be true, of course, at the leading edges of the blades of a blower or pump impeller, the fluid entering the runner under optimum conditions with practically zero circulation. At the leading edges of the turbine-runner blades, which correspond to the trailing edges of the blower or pump-impeller blades, for maximum efficiency there must also be a smooth transition of the

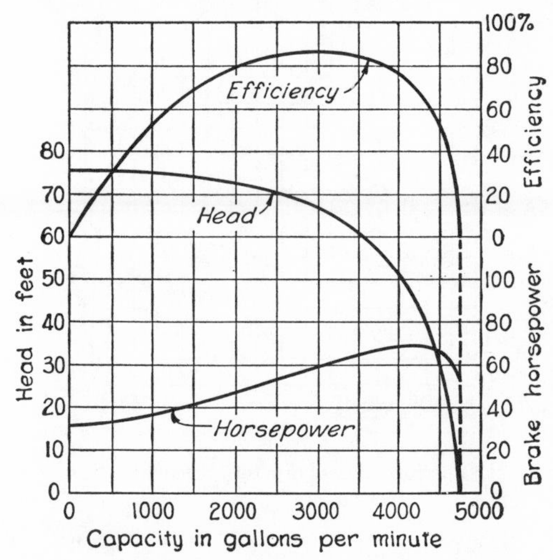

FIG. 163. Performance characteristics of a 12-inch centrifugal pump at 1000 revolutions per minute.

flow from or to the guide vanes; in other words, the blades in this zone must be so shaped as to be essentially tangent to the vector of relative velocity. Since the blades in this type of machinery are generally fixed in position, it is evident that a change in either speed or capacity from optimum conditions will result in fluid velocities which do not properly conform to the existing blade angles at entrance and exit. Separation then occurs, producing an increased loss in head and hence a reduction in operating efficiency.

Significance of the specific speed. Although the one-dimensional analysis of such radial-flow machinery differs in detail from that of the axial-flow type, the method of evaluating performance characteristics is essentially the same. The standard diagram of Fig. 163

for a particular centrifugal pump, thus corresponds to that of Fig. 159 for a particular propeller pump, while the dimensionless diagram shown in Fig. 164, like that of Fig. 160, may be used for any similar blower, pump, or turbine, regardless of size, speed, or fluid density. The essential difference between the radial-flow and axial-flow types, as already indicated, lies in the relative magnitudes of the coefficients of

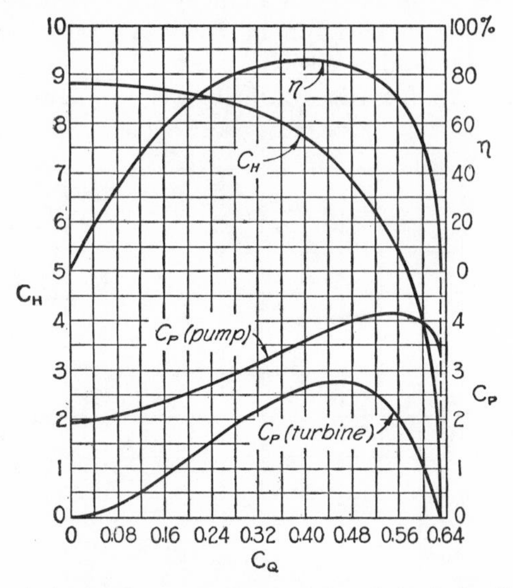

FIG. 164. Dimensionless performance characteristics of a typical radial-flow machine.

capacity and head at peak efficiency; the radial-flow type, in other words, is characterized by a low capacity-head ratio, and the axial-flow type by a high capacity-head ratio. If this ratio is so evaluated as to eliminate the factor D in both numerator and denominator, there results a very convenient index known as the *specific speed:*

$$N_s = \frac{C_Q^{1/2}}{C_H^{3/4}} = \frac{n\sqrt{Q}}{(g\,\Delta H)^{3/4}} \tag{214}$$

(This form of the index, it should be noted, is dimensionless and hence preferable to customary forms in which g is omitted and Q is expressed in gallons, or cubic feet, per minute.) The radial-flow turbine, pump, or blower is evidently characterized by a low specific speed, and axial-flow machinery by a high specific speed. There are, needless to say, many mixed-flow designs which combine axial and radial flow to varying degrees, so that a continuous trend exists from low to high specific speed, as indicated by the sequence of runner or impeller forms shown in Fig. 165. The performance characteristics of these units will, of course, range from the high specific-speed curves of Fig. 160 to the low specific-speed curves of Fig. 164.

Aside from their fundamental principles of operation, the several types of fluid machinery discussed herein have certain other features in common which warrant mention at this point. First, any enclosed unit is subject to the same cavitation or acoustic disturbances as the open propeller or fan, and should be so designed (and operated) as to reduce such effects to a minimum. Second, the dimensionless performance curves apply not only to one particular design but also, strictly speaking, to one particular Reynolds number; under normal conditions of operation, to be sure, the Reynolds number is usually so high that viscous influences may be ignored; nevertheless, tests on very small models or rates of flow, or with fluids of high kinematic viscosity, may invariably be expected to yield lower efficiencies. Third, the rated efficiency of any machine necessarily involves both fluid resistance and mechanical friction; the latter is generally higher on small units, which again will result in a lower model efficiency even though the Reynolds numbers are the same. Finally, although the units described herein were stripped to their barest essentials for the sake of clarity, any such device may be built in multiple-stage form—i.e., duplicate units arranged in series—in order to increase the overall change in head for the same rate of flow; such practice is, indeed, quite common in the case of centrifugal pumps.

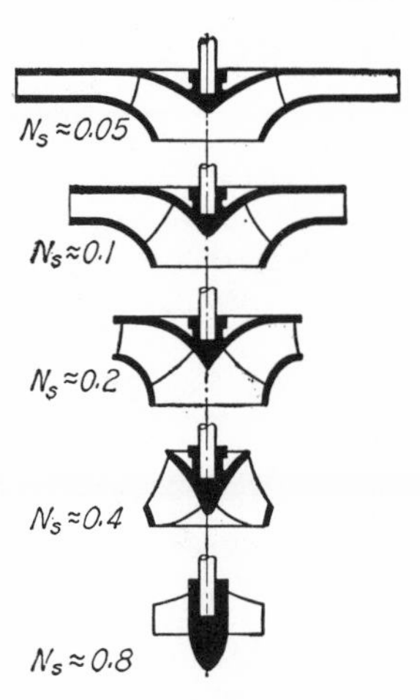

FIG. 165. Variation in specific speed with transition from radial to axial flow.

Example 51. When driven by a 20-horsepower motor at a speed of 1800 revolutions per minute, a ventilating fan delivers 35,000 cubic feet of air per minute under a pressure head of 1½ inches of water. What power and speed of motor would be required to increase the fan capacity by 25 per cent with the same pressure load? (Assume the fan to have the characteristics shown in Fig. 160.)

According to Eqs. (204) and (205), if ΔH remains constant and $Q_2 = 1.25\,Q_1$

$$\frac{n_2^2}{n_1^2} = \frac{(C_H)_1}{(C_H)_2} \qquad \text{and} \qquad \frac{n_2}{n_1} = 1.25\,\frac{(C_Q)_1}{(C_Q)_2}$$

Eliminating n_1/n_2,

$$\left(\frac{\sqrt{C_H}}{C_Q}\right)_2 = 0.8\left(\frac{\sqrt{C_H}}{C_Q}\right)_1$$

Presumably the original motor speed was chosen to obtain peak efficiency at the given pressure and rate of flow; therefore, according to Fig. 160, $(C_H)_1 = 0.80$, $(C_Q)_1 = 0.33$, and $(C_P)_1 = 0.33$, whence

$$\left(\frac{\sqrt{C_H}}{C_Q}\right)_2 = 0.80 \times \frac{\sqrt{0.8}}{0.33} = 2.17$$

By trial and error it is found from the diagram that $(C_H)_2 = 0.62$, and $(C_Q)_2 = 0.36$, which corresponds to a drop in efficiency of only 2 per cent.

Evaluation of n and P then proceeds as follows:

$$\frac{n_2}{n_1} = 1.25\frac{(C_Q)_1}{(C_Q)_2} = 1.25 \times \frac{0.33}{0.36} = 1.15$$

$$1800 \times 1.15 = 2070 \text{ rpm}$$

and for the value $(C_P)_2 = 0.28$,

$$\frac{P_2}{P_1} = \frac{n_1^3}{n_2^3}\frac{(C_P)_2}{(C_P)_1} = \overline{1.15}^3 \times \frac{0.28}{0.33} = 1.29$$

$$20 \times 1.29 = 25.8 \text{ hp}$$

PROBLEMS

349. Tests made on a $\frac{1}{10}$-scale model turbine indicate that under a head of 15 feet (corresponding to 150 feet in the prototype) the model will yield a shaft torque of 40 pound-feet when the gates pass a flow of 2 cubic feet of water per second. What is the model-prototype speed ratio? What are the corresponding values of the torque and discharge at full scale?

350. A duct fan used for ventilation yields a flow of 15,000 cubic feet of air per minute at 1200 revolutions per minute with a power input of $\frac{1}{2}$ horsepower. As it is eventually found that a 75 per cent increase in capacity will be necessary, it is decided to install a motor of higher speed and power. If the capacity coefficient is assumed to remain the same, what motor speed and horsepower will be required?

351. An axial-flow pump having the characteristics shown in Fig. 159 is increased in speed to 1800 revolutions per minute. What will be the capacity, horsepower, and efficiency when pumping water against a head of 30 feet?

352. Under a 9-foot head a small turbine having a diameter of 15 inches is found to develop 10 horsepower while discharging 12 cubic feet of water per second and turning at the rate of 180 revolutions per minute. If a geometrically similar turbine 4 feet in diameter is to operate at the same efficiency under a 25-foot head, what will be its speed, horsepower, and discharge?

353. The blades of a centrifugal fan or blower have a depth of 4 feet, their leading and trailing edges lying at radial distances of 3 and 4 feet from the axis of rotation. When the rate of flow of air is 75,000 cubic feet of air per minute and the rotational

speed is 360 revolutions per minute, it is found that the exit angle of the air is 20°. Assuming an entrance angle of 90°, determine the required power input.

354. If 20 per cent of the power input for the conditions of Problem 353 is dissipated in the inlet and volute of the blower, determine the efficiency of the machine and the pressure intensity in the 5-by-5-foot discharge duct to which it is connected.

355. Water passes from the guide vanes into the runner of a radial-flow turbine at an angle of 30° and a velocity of 8 feet per second at a radial distance of 6 feet from the vertical runner shaft, and leaves the trailing edges of the runner blades at a radial distance of 4 feet from the shaft. The height of the runner is 1.5 feet. If the torque on the shaft is 16,000 pound-feet when the rotational speed is 120 revolutions per minute, at what angle does the water leave the runner blades?

356. What is the brake horsepower of the turbine shaft of Problem 355? What change in average pressure head takes place as the water passes through the runner?

357. If a pump which has the characteristics of Fig. 163 when calibrated with water is used with gasoline (specific gravity = 0.7), what power would be required at the rated speed to pump against a head of 60 feet of gasoline? What rate of flow would be obtained under these conditions?

358. A water pump having the characteristics of Fig. 163 is to be driven at a 30 per cent greater rotational speed. Determine the factors by which the ordinate and abscissa scales of the given performance diagram should be multiplied in order to yield the corresponding values of head, power, efficiency, and capacity.

359. Compute the specific speeds of the pumps for which Figs. 159 and 163 represent the performance characteristics.

360. Air is to be supplied at the rate of 15,000 cubic feet per minute to a duct under a pressure head of 2 inches of water. Would an axial-flow or radial-flow type of blower be indicated by these requirements?

361. Carbon tetrachloride (specific gravity = 1.59) is to be pumped at the rate of 75 gallons per minute against a head of 60 feet. Determine whether the characteristic curves of Fig. 160 or those of Fig. 164 would require the lower power input at 6000 revolutions per minute. To what pump size would this correspond?

47. FLUID TRANSMISSION OF POWER

The fluid coupling. If a pump and a turbine of the same capacity characteristics were connected in series, it would be reasonable to expect that both the speed and the power output of the turbine could be controlled over a considerable range of operating conditions by variation in the speed and power input of the pump. This is, in effect, the principle of most machinery for the fluid transmission of power. However, certain practical requirements of such transmission devices result in two distinct types of design. One type, known as the *fluid coupling,* is intended only to absorb the shock caused in purely mechanical transmissions by abrupt changes in load. The other type, known as the *torque converter,* is the fluid equivalent of a series of mechanical speed-reduction gears.

In its primary details the fluid coupling is extremely simple. As indicated in Fig. 166, the blades of both impeller and runner lie entirely in radial planes and are housed in such manner that flow may take place radially outward through the impeller into the runner and then radially inward through the runner into the impeller. Although there is no mechanical means of preventing flow in the opposite direction, the centrifugal effect of a higher impeller speed will invariably produce flow in the direction indicated. Were the speeds of impeller and runner the same, to be sure, no flow would occur between the two units and hence no torque could be transmitted. The flow produced by a speed difference, however, will necessarily result in a continuous increase in the tangential velocity of the fluid as it passes through the impeller and a corresponding decrease in tangential velocity as it returns through the runner. Since the rate of flow, the radial displacement, and the change in tangential velocity are numerically the same in both impeller and runner, it follows that the torques of the input and output shafts of a fluid coupling will always be identical, regardless of their relative speeds. Owing to the necessary difference in angular velocity, of course, the power output will always be lower than the power input, the loss resulting in part from surface drag and in part from turbulence produced at the juncture of the two units. The peak efficiency of such a fluid coupling is very close to unity; in other words, at normal operating conditions the "slip" of the runner (and hence the power loss) is only a few per cent.

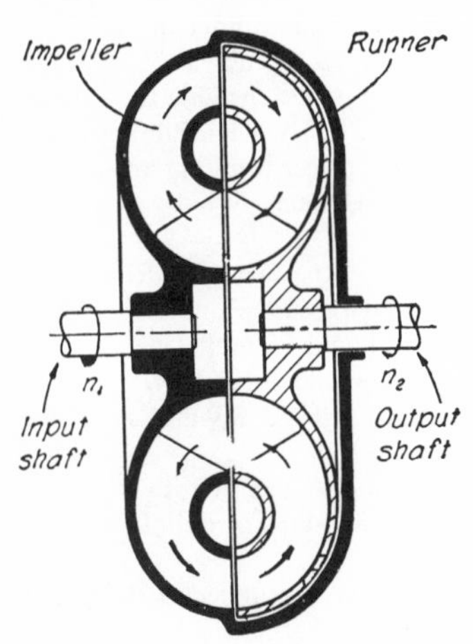

FIG. 166. Schematic diagram of a fluid coupling.

The torque converter. As its name implies, the torque converter differs from the fluid coupling in that not only is the torque of the output shaft greater than that of the input shaft but the torque ratio—instead of remaining constant—decreases as the speed ratio increases. In order to produce such torque conversion, a greater change must be made to occur in the tangential velocity of the fluid as it passes through the runner than takes place during passage through the impeller. This can be accomplished only by properly shaping the impeller and runner blades and by introducing one or more sets of stationary guide vanes (Fig. 167) curved to yield conditions of maximum efficiency. A change in speed ratio from such optimum conditions will, of course, result in a lowering of efficiency, owing to the accompanying change in the velocity diagrams at the transitions to and from the guide vanes. However, the velocity change through a torque converter will be such that a relative increase in runner torque will invariably accompany a relative decrease in runner speed, the torque output therefore reaching a maximum when the runner is completely stalled.

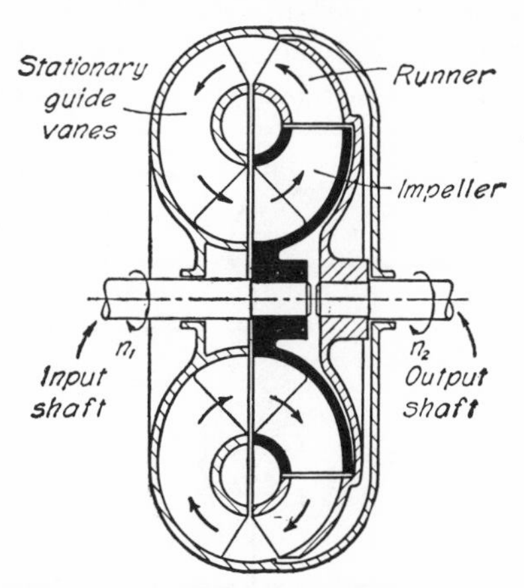

FIG. 167. Schematic diagram of a torque converter.

Of primary interest in the case of either the fluid coupling or the torque converter are the characteristic ratios of speed, torque, and power (i.e., the efficiency) of input and output shafts for given conditions of power input, fluid density, and unit speed and size. Since the latter dimensional variables may again be combined, as in the foregoing sections, to yield the dimensionless power factor $P/\rho n^3 D^5$, it follows that for any transmission design the several ratios should be func-

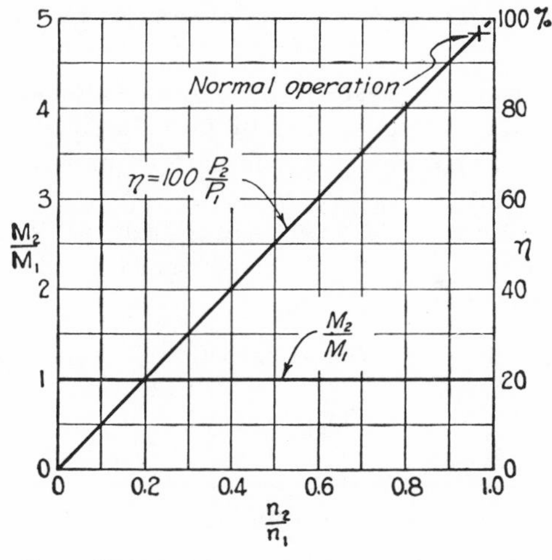

FIG. 168. Generalized performance characteristics of a fluid coupling.

tions of the power factor. The interrelationships of the ratios themselves, however, shown for the coupling and the converter in Figs. 168 and 169, permit ready comparison of the relative characteristics of the two types of unit. Evidently, the high peak efficiency of the coupling is offset by the high power input required under conditions of maximum slip, due to constancy of the torque ratio. Since the latter disadvantage is overcome by the torque-converter—which, however, is relatively inefficient even at optimum conditions—some

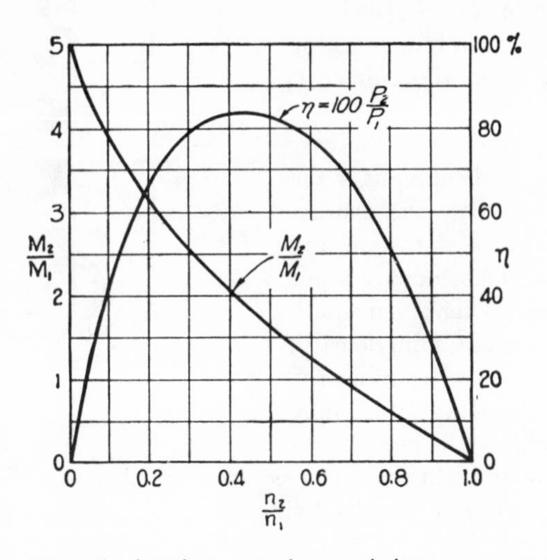

FIG. 169. Generalized performance characteristics ot a torque converter.

commercial transmission systems involve an automatically selective combination of the two units, thereby providing first a high starting torque and then a high operating efficiency entirely by fluid means.

Example 52. What reduction in the size of fluid power-transmission units would be permitted by the substitution of mercury for the oil customarily used? Why is this not practicable?

Since the power factor $P/\rho n^3 D^5$ may be considered constant as a basis of comparison,

$$\frac{D_{\mathrm{Hg}}}{D_{\mathrm{Oil}}} = \left(\frac{\rho_{\mathrm{Oil}}}{\rho_{\mathrm{Hg}}}\right)^{1/5}$$

whence, assuming that $\rho_{\mathrm{Oil}} = 0.9 \times 1.94$ slugs per cubic foot,

$$D_{\mathrm{Hg}} = D_{\mathrm{Oil}} \times \left(\frac{0.9 \times 1.94}{13.6 \times 1.94}\right)^{1/5} = 0.58 D_{\mathrm{Oil}}$$

Evidently, the slight saving in size would be more than offset by the increase in cost, weight, and corrosive effects if mercury were used.

PROBLEMS

362. How much larger would the transmission unit of Example 52 have to be to deliver the same power at the same speed and the same speed ratio if air were used instead of oil? How much faster would a unit of the same size have to run?

363. An automobile traveling at 50 miles per hour on level ground requires an engine output of 35 horsepower at 1100 revolutions per minute, the fluid coupling then having a slip of 4 per cent. To maintain speed at a slight increase in grade, a 15 per cent increase in power output of the coupling is required. If the coupling momentarily develops a 25 per cent slip, what will be the corresponding change in engine speed and power?

364. There is found to be a 4 per cent slip in a certain fluid coupling when the input shaft turns at the rate of 1500 revolutions per minute. To what power output would this correspond when the torque on the input shaft is 15 pound-feet?

365. The rotational speeds of the input and output shafts on a torque converter are 2400 and 900 revolutions per minute, and the input and output torques are 35 and 80 pound-feet, respectively. Determine the corresponding power input and efficiency.

366. A torque converter to be used on a Diesel locomotive unit is tested at a 1:15 scale reduction, the results indicating that at the peak efficiency a torque ratio of 2.5 and a speed ratio of 0.33 will prevail. If the prototype output shaft at its normal operating speed of 1200 revolutions per minute must maintain a torque of 30,000 pound-feet, what must be the power of the driving unit?

QUESTIONS FOR CLASS DISCUSSION

1. Why does a spinning tennis ball "curve" more markedly than a baseball?
2. What are some practical disadvantages of the Flettner rotor?
3. If a blotter is dropped in a slanting position, it will rotate rapidly as it falls. What is the cause of this phenomenon?
4. Explain the principle of the kite.
5. An airplane is considered to be supported by a lifting force exerted upon the wing by the air. With the aid of the principle of action and reaction, show that the weight of the plane is actually transmitted in full to the earth's surface.
6. Why do airplanes take off and land against the wind?
7. The efficiency of a wing may be measured in terms of its maximum lift-drag ratio. At what point in an airplane's flight is a high drag (and hence a low efficiency) of great importance?
8. Suggest a reason for reducing the chord of an airplane wing with distance from the base (i.e., tapering the wing toward the tip).
9. What practical consideration governs the difference in form between airplane propellers and ship screws?
10. Why does the efficiency of a fan or propeller eventually begin to drop as the number of blades is increased?
11. Suggest a suitable measure of efficiency for a fan, to replace Eq. (198).
12. If the power supply for a centrifugal pump fails, the flow sometimes changes direction before valves can be closed; the pump then behaves, in effect, like a turbine. Would its efficiency as such be high or low? Why?

13. Would turbines of high or low specific speed be selected for a hydroelectric plant at a Mississippi River dam?

14. Explain the necessity of using stationary guide vanes in a torque converter.

SELECTED REFERENCES

PRANDTL, L. "Applications of Modern Aerodynamics to Aeronautics." *Technical Report* 116, National Advisory Committee for Aeronautics, 1921.

GLAUERT, H. *The Elements of Aerofoil and Airscrew Theory.* Macmillan, 1943.

SPANNHAKE, W. *Centrifugal Pumps, Turbines, and Propellers.* Technology Press 1934.

KRISTAL, F. A., and ANNETT, F. A. *Pumps.* McGraw-Hill, 1940.

BAUMEISTER, T., JR. *Fans.* McGraw-Hill, 1935.

ALISON, N. L. "Fluid Transmission of Power." *S.A.E. Journal,* January, 1941.

CHAPTER X

SURFACE TENSION

48. MOLECULAR ATTRACTION AND SURFACE ENERGY

Work of cohesion and adhesion. All bodies of matter, whether fluid or solid, have been stated to exert upon each other an attractive force which is directly proportional to the product of their masses and inversely proportional to the square of the distance between their mass centers. Quite distinct from such mass attraction is an electrochemical force known as *molecular attraction*, which gives rise to the liquid properties of *cohesion* and *adhesion*. So far as present purposes are concerned, the chief difference between mass attraction and molecular attraction is the fact that the latter force is effective over only a very minute distance. Thus, as shown schematically in Fig. 170, a molecule at point A a short distance from the liquid surface will still be attracted equally in all directions by neighboring molecules, whereas molecule B very near the free surface will have exerted upon it a smaller force from the free-surface side where fewer molecules are present. It follows that molecules close to the surface of a liquid will be subject to a resultant force acting at right angles to the surface.

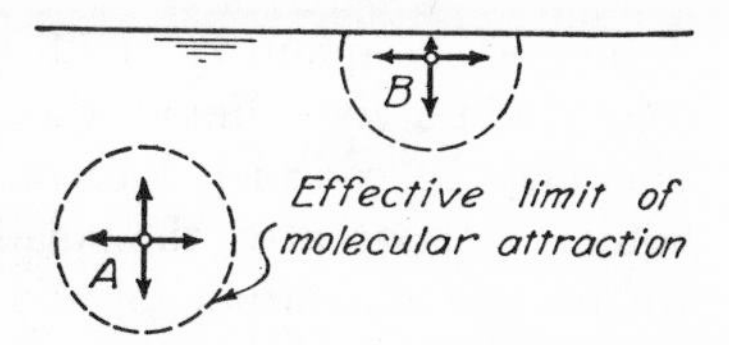

FIG. 170. Intermolecular forces near a liquid surface.

In order to form a free surface within a liquid body, it will be seen that work must be done against the mutual attraction of the molecules on the two sides of the zone of separation. The existence of a free surface therefore represents the presence of what is known as *surface energy*, which is equal in magnitude to the work that was done when the surface was formed. Designating such energy per unit surface area by the symbol σ (sigma), the work W which must be done, for instance, in producing separation of a liquid column at some arbitrary cross section will be simply $2A\sigma$, since two free surfaces of area A will thereby be formed. The quantity $W/A = 2\sigma$ is hence known as the liquid *work of cohesion.*

Owing to the molecular basis of such surface energy, it will be apparent that the magnitude of σ for a given liquid surface will depend upon the magnitude of the attractive force not only between the molecules of the liquid itself, but also between the liquid molecules and those of the medium on the other side of the surface. In the case of the liquid-vapor interface just discussed, σ represents simply the attractive force of the liquid molecules for one another. In the case of the interface between two different liquids, on the other hand, the attractive force between the unlike molecules may be either greater or less than that between the like molecules. In order to effect the separation of two liquids along such an interface of area A, the work required must be equal to the difference between the final surface energy of the two free surfaces $A(\sigma_a + \sigma_b)$ and the surface energy of the original interface $A\sigma_{ab}$; thus,

$$\frac{W_{ab}}{A} = \sigma_a + \sigma_b - \sigma_{ab} \tag{215}$$

Just as the quantity $W/A = 2\sigma$ was said to indicate the work of cohesion of a given liquid, the quantity W_{ab}/A of the foregoing equation may be considered to represent *work of adhesion* between two liquids. Apparently, the smaller the magnitude of W_{ab}/A (i.e., the greater the magnitude of the interfacial energy σ_{ab} in comparison with $\sigma_a + \sigma_b$), the less pronounced will be the adhesion of the two liquids; on the other hand, the larger the magnitude of W_{ab}/A (i.e., the smaller the relative magnitude of σ_{ab}), the more strongly will the two liquids adhere to one another. It then follows that if σ_{ab} becomes equal to zero—or even negative—the molecules of the two liquids will have at least as great an attraction for each other as those of either liquid have among themselves; a state of complete miscibility will then exist, and the resulting process of molecular mixing or diffusion will soon eliminate the original interface.

Evaluation of the pressure difference across a curved surface. Application of the simple concept of surface energy will serve to explain such phenomena as the spreading of certain oils on the surface of water, and the tendency of all liquids to form spherical droplets (i.e., bodies of minimum surface area) when sprayed into air. There are, however, two closely related aspects of the problem which have not yet been mentioned. In the first place, the formation of new surface entails not only mechanical energy but—if the temperature of the surface is to remain the same—thermal energy as well; since an analysis of the thermodynamics of surface phenomena is not within the scope of this text, the necessity for such considerations can merely be indicated. In the second place, and of particular importance to the present study, the existence of a resultant force normal to any surface must require an equal and opposite force within the fluid if static equilibrium is to prevail; within every liquid body, therefore, it must be realized that there will exist an internal pressure of such magnitude as to counteract the surface forces due to unbalanced molecular attraction.

So long as the surface in question is a plane, the magnitude of the pressure intensity due to such molecular forces will be not only very small but also impossible of direct measurement. Once the surface becomes curved, however, the resulting intensity of pressure will be both perceptible and measurable, for the presence of surface curvature is at once the cause of the additional pressure change and the means of its evaluation. Consider, for example, the spherical liquid-gas interface shown schematically in Fig. 171. The surface energy which it represents will be simply $\sigma(\Delta s)^2$. If, now, the surface is displaced the radial distance Δr, thereby increasing slightly in area, the surface energy at the end of the displacement will be approximately $\sigma(\Delta s + \Delta s\ \Delta r/r)^2$. If the work against the intermolecular forces which is entailed by this change in surface energy is done only by the radial force $\Delta p\ (\Delta s)^2$ due to the difference in pressure intensity between the concave and the convex side of the interface,

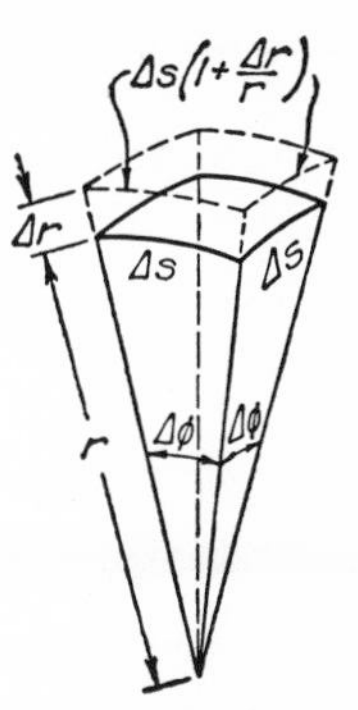

FIG. 171. Radial displacement of a curved interface.

$$\Delta p\ (\Delta s)^2\ \Delta r = \sigma\left(\Delta s + \frac{\Delta s}{r}\ \Delta r\right)^2 - \sigma(\Delta s)^2 = 2\sigma\ (\Delta s)^2\ \frac{\Delta r}{r} + \sigma\left(\Delta s\ \frac{\Delta r}{r}\right)^2$$

Upon dividing each term by $(\Delta s)^2\ \Delta r$ and letting Δr approach zero, the pressure difference at a spherical surface will be seen to be proportional to the ratio of the unit surface energy and the radius of curvature of the surface:

$$\Delta p = \frac{2\sigma}{r} \tag{216}$$

It is to be noted from the derivation that the pressure intensity is invariably higher on the concave side.

The bubble method of measuring surface energy. In order to use this relationship for the measurement of unit surface energy, it is necessary to restrict the surface curvature to a very small radius, so that gravitational effects upon the surface form—which were not taken into account in the derivation of Eq. (216)—will be relatively unimportant. Assume, then, that an open tube of very small bore is introduced vertically into a container of liquid, as shown in Fig. 172. As air is forced downward through the tube, a bubble will form at its tip, the size of the bubble being controlled by the pressure intensity of the air supply to the tube. As the bubble increases in volume with increasing air pressure, its radius of curvature will decrease until it is equal to the radius r_0 of the tube. Since further increase in volume would then cause the bubble to *increase* in diameter, this evidently marks the minimum radius of bubble curvature and hence, by Eq. (216), the maximum pressure intensity for equilibrium conditions. Thus, measurement of the pressure intensity p of the air supply just before the bubble becomes unstable permits the magnitude of σ for the given liquid to be determined from the following modification of Eq. (216):

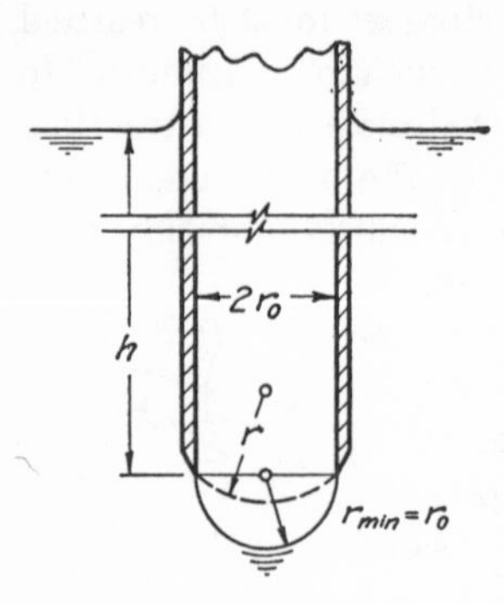

FIG. 172. Measurement of surface energy.

$$\sigma = \frac{r_0}{2}(p - \gamma h) \tag{217}$$

Example 53. A smoke generator used in military screening converts an oil having a unit surface energy of 0.0025 foot-pound per square foot into a mist of 10^{-4}-inch droplets at the rate of 50 gallons per hour. (*a*) What is the intensity of pressure within a typical liquid droplet? (*b*) Ignoring thermal effects, estimate the rate at which work is done by the generator.

The radius of curvature of a 10^{-4}-inch droplet is

$$r = \frac{10^{-4}}{2 \times 12} = 4.17 \times 10^{-6} \text{ ft}$$

Hence, from Eq. (216), the pressure intensity in the droplet will be

$$p = \frac{2\sigma}{r} = \frac{2 \times 0.0025}{4.17 \times 10^{-6}} = 1200 \text{ psf}$$

The number of droplets produced per second becomes

$$N = \frac{50 \times \frac{1}{7.48} \times \frac{1}{60 \times 60}}{\frac{4}{3} \times \pi \times (4.17 \times 10^{-6})^3} = 6.12 \times 10^{12}/\text{sec}$$

which corresponds to the following rate of surface-area production:

$$\frac{dA}{dt} = 6.12 \times 10^{12} \times 4 \times \pi \times (4.17 \times 10^{-6})^2 = 13{,}400 \text{ ft}^2/\text{sec}$$

Therefore, the rate at which work is done will be

$$P = \sigma \frac{dA}{dt} = 0.0025 \times 13{,}400 = 33.5 \text{ ft-lb/sec}$$

PROBLEMS

367. Determine the pressure intensity within a bubble of air 0.01 inch in diameter at a depth of 1 foot below the free surface of water at 60° Fahrenheit.

368. Ignoring thermal effects, compute the power required to convert 2 gallons of water per minute at a temperature of 50° Fahrenheit into a mist having an average droplet size of 10^{-5} inch.

369. What pressure intensity will exist within the mist droplets of Problem 368?

370. In measuring the unit surface energy of a mineral oil (specific gravity = 0.85) by the bubble method, a tube having a ¹⁄₁₆-inch bore is immersed to a depth of ½ inch in the oil. What magnitude of σ will a maximum bubble pressure intensity of 3.15 pounds per square foot indicate?

371. A soap bubble is blown at the end of a tube having a diameter of ¼ inch; if the soap solution has 75 per cent as great a unit surface energy as water at 60° Fahrenheit, what is the greatest pressure intensity required? (Note that a soap bubble has two surfaces.)

49. SURFACE "TENSION" AND CAPILLARITY

Force diagrams for adhesion and cohesion. For many years the surface phenomena discussed in the foregoing section were explained in terms of an apparent tension in an elastic skin or membrane which was thought to form at every liquid surface. It was, of course, never decided why the stress in such a skin remained constant no matter how much it was stretched, or how the skin could cling tenaciously to a solid boundary and at the same time slide freely along the boundary as the liquid surface was displaced. As a matter of fact, there are so many physical inconsistencies in the surface-tension concept that the continued designation of the quantity σ as the *coefficient of surface tension* is, to say the least, misleading. Oddly enough, however, the quantitative evaluation of surface phenomena by means of the surface-tension concept yields perfectly accurate results, despite the erroneous physical picture upon which it is based. Indeed, the use of the force diagrams involved in this concept still provides the only simple means of deriving many of the fundamental relationships for surface energy.

To illustrate this method, consider the force counterpart of the analysis leading to Eq. (216). Instead of investigating the work involved in increasing the area of the liquid surface shown in Fig. 171, it is merely necessary to assume a system of force vectors acting at the edges of the surface as indicated in Fig. 173. The components of these four vectors in the normal direction may then be equated to the increase in pressure on the concave side of the surface, as follows:

$$\Delta p\,(\Delta s)^2 = 4\,\sigma\,\Delta s \sin \Delta\phi/2$$

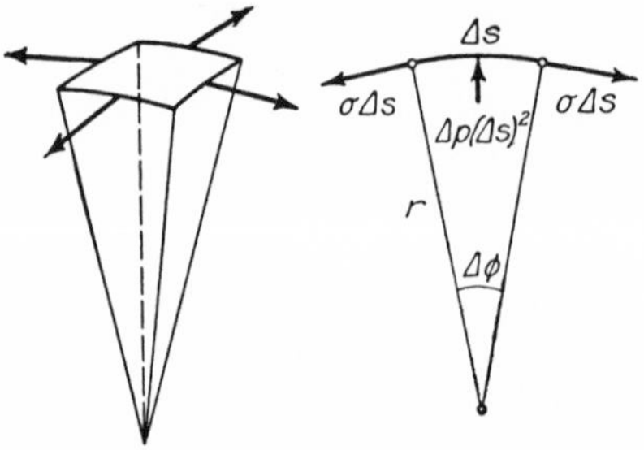

FIG. 173. Assumed vectors of surface "tension."

Since $\sin \Delta\phi/2 \approx \Delta\phi/2 \approx \Delta s/2r$, this at once reduces to the same expression as Eq. (216):

$$\Delta p = \frac{2\sigma}{r}$$

An example, on the other hand, of a derivation not so easily obtained by the surface-energy method is found in the problem of the *contact angle* at the juncture of a solid, a liquid, and a gas. While Eq. (215) is quite as applicable to a liquid-solid as to a liquid-liquid interface, its usefulness depends upon the elimination of the surface-energy term for the solid surface. If force vectors representing hypothetical tensions per unit length in the three surfaces are assumed as shown in Fig. 174, it follows that

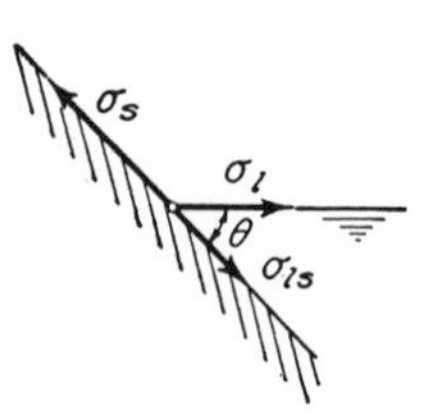

FIG. 174. Hypothetical surface forces at a boundary.

$$\sigma_s = \sigma_{ls} + \sigma_l \cos\theta$$

Introduction of this expression in the following counterpart of Eq. (215)

$$\frac{W_{ls}}{A} = \sigma_l + \sigma_s - \sigma_{ls}$$

then yields the basic relationship

$$\frac{W_{ls}}{A} = \sigma_l(1 + \cos\theta) \tag{218}$$

Inspection of the diagram will show that if the attraction between the liquid and solid molecules is equal to or greater than that between the liquid molecules themselves, the angle of contact θ will be zero—that is, $W_{ls}/A = 2\sigma_l$, which is the work of cohesion of the liquid; the liquid is then said to "wet" the solid. Accordingly, if the adhesive tendency between liquid and solid is only half as great as the cohesive tendency of the liquid, the contact angle will be 90°, and wetting will

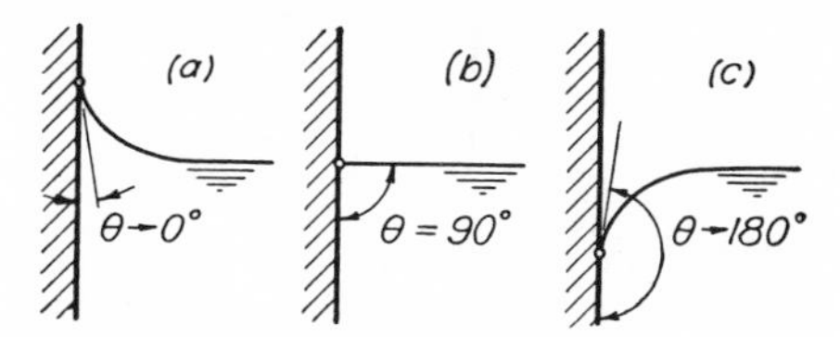

FIG. 175. Variation in contact angle with degree of wetting.

be incomplete. Although the adhesive stress between some liquids and solids is relatively small, in no case can it be equal to zero; hence a contact angle of 180° may be approached, but never attained; nevertheless, for all practical purposes the wetting tendency is then so small as to be negligible. For instance, a clean glass surface will be completely wetted by water but only imperceptibly by mercury, with the resulting contact angles shown in Fig. 175*a* and *c*.

Capillary rise of liquids. Perhaps the most important application of the principle of adhesion is in connection with the *capillary rise* of liquids in tubes and in the interstices of porous materials. If a tube is very small in diameter, and if the liquid completely wets its inner boundary, the curved surface or *meniscus* of the liquid will display the hemispherical form shown in Fig. 176. Since the pressure intensity within the liquid just below the free surface must be smaller in magnitude than that above, in accordance with Eq. (216), and since the liquid pressure intensity at a given elevation must be the same on the inside and the outside of the tube, in accordance with Eq. (45), it follows that the meniscus of the liquid column will lie above the free surface of the outer liquid. The extent of this difference in elevation may be evaluated by the following combination of Eq. (216) and the elementary principle of hydrostatics:

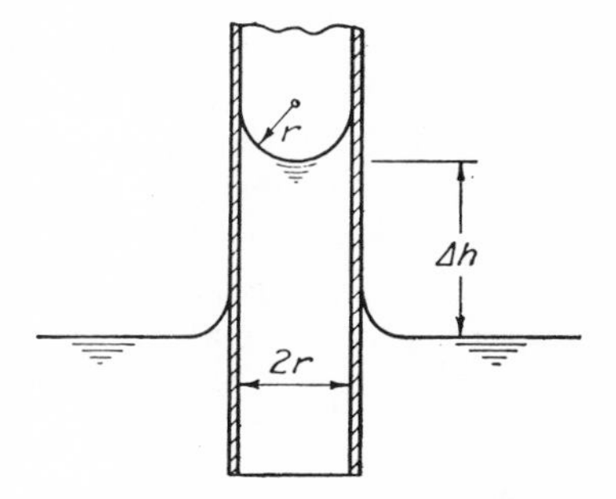

FIG. 176. Capillary rise of liquid in a tube.

$$\Delta h = \frac{1}{\gamma}\frac{2\sigma}{r} \qquad (219)$$

In applying this equation, the fact must be borne in mind that it—like Eq. (217)—becomes exact only as the radius of the tube becomes infinitesimal. Although the error involved will be inappreciable for

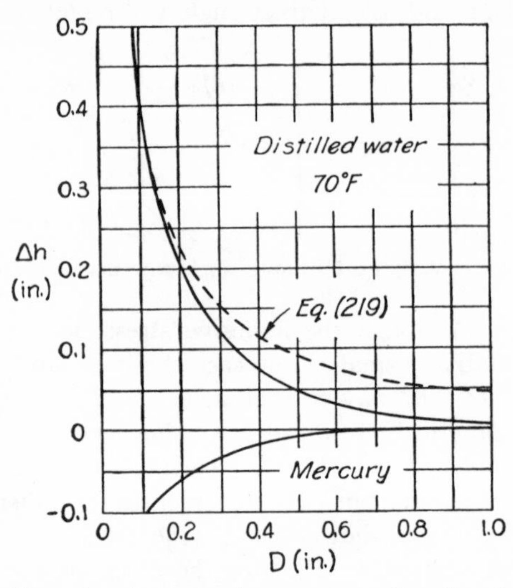

FIG. 177. Capillary rise of water and mercury as a function of tube diameter.

small but finite radii, continued increase in radius will eventually produce a marked departure from hemispherical surface form, the radius of curvature of the midpoint of the meniscus gradually increasing beyond that in the zone of boundary contact owing to the leveling effect of gravity upon any liquid surface. While the analysis of such a combination of mass attraction and molecular attraction is too complex to warrant further discussion, the extent of the gradually increasing error involved in Eq. (219) may be judged from Fig. 177. It must also be noted that this form of the relationship is restricted to liquids which completely wet the boundary and must be modified for angles of contact greater than zero; of particular importance is the fact that the slightest trace of uncleanliness of a liquid-solid interface may change the contact angle to an appreciable degree. As a rule, therefore, the measured capillary rise of liquids in tubes will be less than that computed for chemically clean conditions. Furthermore, in case the degree of wetting is negligible (i.e., for contact angles approaching 180°) there will result a capillary

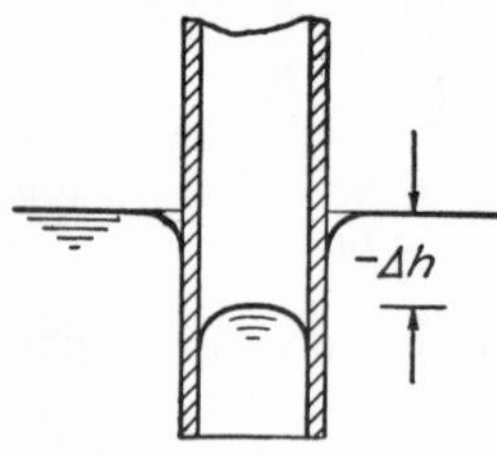

FIG. 178. Capillary depression of a liquid surface.

depression of the free surface, the meniscus being practically a mirror image of that for capillary rise shown in Fig. 176. As indicated by Fig. 178, evaluation of the magnitude of the depression will lead to a relationship which is identical to Eq. (219) except for a negative sign before the elevation term.

Example 54. Estimate the height to which water will rise in a clay soil having an average grain diameter of 0.0001 foot.

Since the size of the pores in the soil depends upon the compaction and grading as well as the average diameter of the particles, it is evident that only a rough estimate may be obtained with the given information. Assuming, as a first approximation, that the interstices are one-tenth the mean grain diameter, by Eq. (219)

$$\Delta h \approx \frac{1}{62.4} \times \frac{2 \times 0.005}{0.00001/2} \approx 32 \text{ ft}$$

PROBLEMS

372. Liquids A and B are used in a compound glass manometer, liquid A being lighter than liquid B. The surface tension between the two liquids is 0.0035 pound per foot, that between liquid A and glass is 0.0015 pound per foot, and that between liquid B and glass is 0.0040 pound per foot. Compute and show in a sketch the angle at which the liquid interface will meet the wall of the manometer.

373. The rise of sap in trees is sometimes erroneously explained as a purely capillary phenomenon. Assuming sap to have the same surface tension as water at 60° Fahrenheit, estimate the diameter of tube which would be necessary to carry it to a tree height of 100 feet.

374. It is found that water at 40° Fahrenheit rises through capillary action to a height of 7 feet in an earth dike. To what diameter of uniform tube would the interstices in the earth correspond?

375. Two glass plates 6 inches square are stood together in a shallow container of water in such manner that two vertical edges are 1/8 inch apart and the other two 1/64 inch apart. Determine and plot to scale the height of the meniscus over the 6-inch distance from edge to edge of either plate. (Note that the meniscus curves in one plane only.)

376. Glass tubing to be used in a differential water manometer is found to vary in diameter between 0.25 and 0.29 inch. Assuming that the gage is to be used for readings from 2 to 15 inches of water, estimate the maximum percentage of error that could occur.

377. Water at 75° Fahrenheit is contained between two concentric glass tubes. If the diameter of the inner tube is 1/2 inch and the radial distance between its outer wall and the inner wall of the outer tube is 1/100 inch, what capillary rise should take place? (Note that in the derivation of Eq. (219) surface curvature in two directions has been considered.)

50. SIGNIFICANCE OF THE WEBER NUMBER

Formulation of the Weber number. Like other fluid properties, the molecular attraction at the surface of a liquid may or may not play a primary role in a given problem of fluid motion, depending upon the magnitude of the surface energy in comparison with that of the flow as a whole. As in the case of fluid weight, of course, it is first necessary that a *free* surface or interface exist before the pattern of motion can be influenced by surface-energy phenomena. In other words, flow which is entirely confined by solid boundaries can no more be affected by molecular attraction than by gravitational attraction (for instance, the phenomenon of tangential shear and zero relative velocity at the surface of contact between a solid boundary and a moving liquid is in no way dependent upon either hydrostatic pressure or adhesion at the liquid-solid interface, for each of these stresses acts at right angles to the boundary). Once a free surface exists, on the other hand, any departure of this surface from a plane or any tendency of the flow to increase the surface area will bring molecular forces into play which may sensibly modify the basic distribution of velocity and pressure.

As a general measure of the relative magnitude of such surface effects, one may devise a dimensionless parameter analogous to the Euler, Froude, and Reynolds numbers already discussed. Thus, as in earlier pages, one may represent by the quantity $\rho V^2/L$ a typical unit inertial reaction to an accelerative force; expressing a typical unit surface force by σ/L^2, the ratio of these two quantities becomes simply $\rho V^2 L/\sigma$. More significantly, perhaps, this fraction may be regarded as twice the ratio of kinetic energy per unit volume, $\rho V^2/2$, to the surface energy per unit volume, σ/L. In any event, the radical form of this ratio

$$\mathbf{W} = \frac{V}{\sqrt{\sigma/\rho L}} \tag{220}$$

is known as the *Weber number*. The smaller the Weber number, evidently, the larger the relative influence of the molecular attraction should be, and vice versa.

Capillary effects in jets and waves. Since the magnitude of the Weber number should, for given boundary conditions, indicate the extent to which the flow pattern is modified by surface effects, one would expect the Euler number characterizing the flow pattern to vary therewith; that is,

$$\mathbf{E} = \phi(\mathbf{W}) \tag{221}$$

In the case of orifice flow, for example, the coefficient of discharge is known to be modified by both the adhesion of the liquid to the outer edge of the orifice and the tendency of the jet to attain a minimum surface area; in each case the effects become the more pronounced with decreasing Weber number, but of negligible magnitude as the Weber number becomes large. Unfortunately, the influence of both viscosity and gravity upon the jet characteristics at low Weber numbers makes the experimental analysis of surface effects so difficult that reliable quantitative information for orifice flow is not yet available.

The study of *capillary waves*, on the other hand, has not only been carried out successfully in the laboratory, but analytical methods have long since led to an accurate relationship for their velocity of propagation. If one assumes, for instance, that a liquid surface momentarily has the rippled form shown in Fig. 179, it will be evident from the

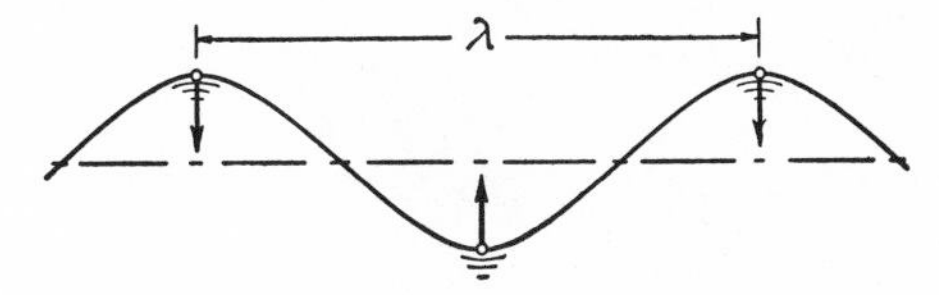

FIG. 179. Definition sketch for capillary waves.

foregoing discussion of surface energy that the surface will tend to contract until a minimum area is attained—that is, until the surface becomes a horizontal plane. In the process of contraction, however, the surface energy is transformed into kinetic energy, with the result that the motion does not cease at the minimum-area stage but continues as a series of standing waves oscillating about the mean surface line. By assuming the wave profile to be sinusoidal, and by superposing two standing-wave systems differing by a quarter period and a quarter wave length, the British physicist Lord Kelvin determined nearly a century ago that a train of capillary waves would travel over a liquid surface with the celerity

$$c = \sqrt{\frac{2\pi\sigma}{\lambda\rho}}$$

in which λ (lambda) is the distance between wave crests. Evidently, just as the Froude number in open-channel flow indicates the ratio of the velocity of flow to the celerity of a gravity wave, the Weber number may be considered to represent the ratio of the velocity of flow to the celerity of a capillary wave.

It would appear from the above expression that such a wave would move more and more rapidly as its size decreased. As is true of other capillary phenomena, nevertheless, only in the case of very small radii of curvature may gravitational effects safely be ignored. For instance, it has been mentioned at an earlier point in this text that a gravity wave in deep water will travel with the celerity

$$c = \sqrt{\frac{g\lambda}{2\pi}}$$

which at some intermediate stage will be of the same order of magnitude as that due to capillarity. As a matter of fact, the celerity of any

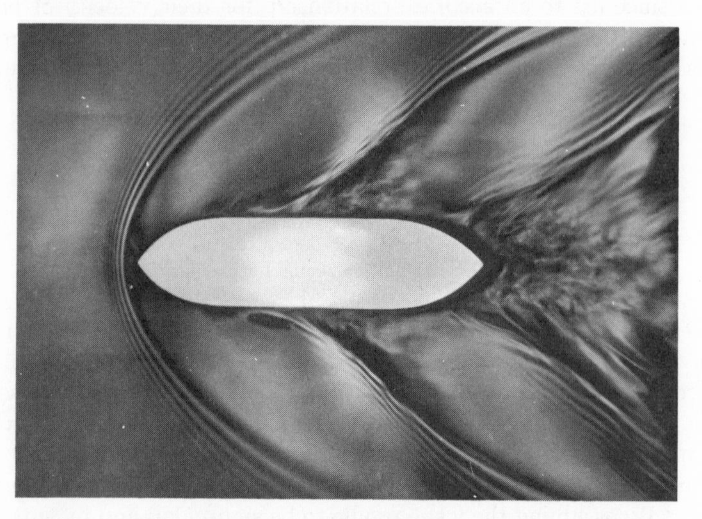

PLATE XXV. Capillary ripples superposed upon gravity waves in flow past a model bridge pier.

surface wave is governed by both gravity and capillarity in accordance with the expression

$$c = \sqrt{\frac{\lambda}{2\pi}\frac{\gamma}{\rho} + \frac{2\pi}{\lambda}\frac{\sigma}{\rho}} \tag{222}$$

Short waves (i.e., ripples) are evidently controlled primarily by capillarity and long waves primarily by gravity, intermediate wave lengths involving both influences. Moreover, since capillary waves travel faster with decreasing wave length and gravity waves with increasing wave length, there must exist a minimum celerity of wave propagation in the intermediate zone, as seen from Fig. 180. This explains the fact that wind has to exceed a certain velocity before it can ripple the surface of a river or lake—a phenomenon which evidently depends upon the Weber as well as the Froude number.

The Weber number, indeed, is a key parameter in the analysis of many flow phenomena, only a few of which can even be mentioned in these pages. Perhaps the most intriguing of these is the so-called

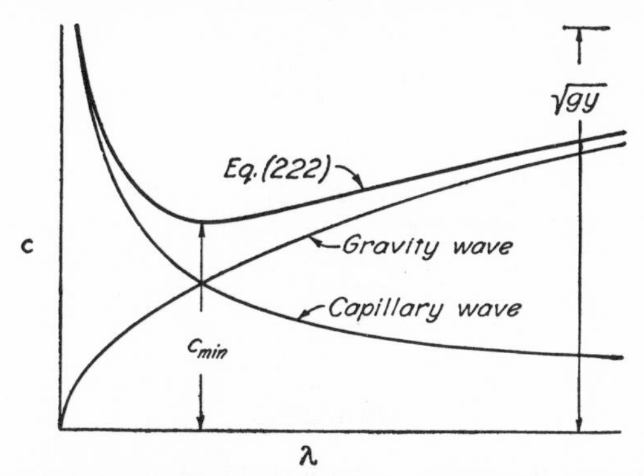

FIG. 180. Celerity of surface waves as a function of wave length.

inversion of jets from non-circular orifices, which is simply a modification of the foregoing problem. As indicated in Fig. 181, the tendency of a jet from an ellipsoidal opening to assume a circular (i.e., minimum-perimeter) cross section will give rise to a series of standing waves along the surface, the geometry of which will vary as the Weber number $V/\sqrt{\sigma/\rho d}$ changes. A phenomenon also peculiar to capillary effects in jets is the formation of drops or spray; the latter, in turn, is important in a number of industries which involve the production and size control of liquid droplets, whether molten shot or paint or agricultural insecticide. The application of surface-energy analysis to the field of chemistry, needless to say, is a subject in itself.

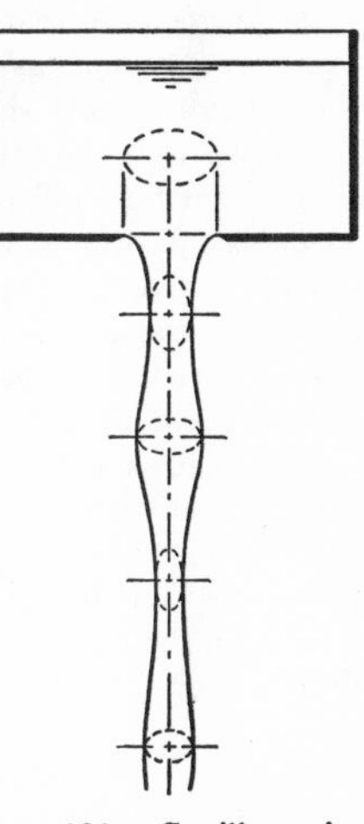

FIG. 181. Capillary inversion of a liquid jet.

Example 55. Determine the minimum celerity of waves on the surface of mercury ($\sigma = 0.035$ pound per foot).

A general expression for c_{min} may be obtained by placing the derivative of Eq. (222) with respect to λ equal to zero; thus,

$$\frac{dc}{d\lambda} = 0 = \frac{1}{2}\,\frac{\dfrac{\gamma}{2\pi\rho} - \dfrac{2\pi\sigma}{\lambda^2\rho}}{\sqrt{\dfrac{\lambda\gamma}{2\pi\rho} + \dfrac{2\pi\sigma}{\lambda\rho}}}$$

whence

$$\lambda_{c_{\min}} = 2\pi \sqrt{\frac{\sigma}{\gamma}}$$

and

$$c_{\mathrm{mm}} = \sqrt[4]{\frac{4\,\sigma\gamma}{\rho^2}}$$

Therefore, for mercury

$$\lambda_{c_{\min}} = 2\pi \sqrt{\frac{0.035}{62.4 \times 13.6}} = 0.0403 \text{ ft}$$

and

$$c_{\min} = \sqrt[4]{4 \times \frac{0.035 \times 62.4 \times 13.6}{(1.94 \times 13.6)^2}} = 0.642 \text{ fps}$$

PROBLEMS

378. Water is found to penetrate a 6-inch sand filter, through capillary action, in 15 seconds. What period would be required if the grain diameter of the filter sand were reduced by 75 per cent? (Note: a similarity parameter for such conditions of negligible acceleration may be based upon the ratio of a unit capillary force to a unit viscous resistance.)

379. It is known that the admixture of an aerosol will reduce the surface tension of water by as much as 50 per cent. If the Weber number only is used as the similarity criterion, what should be the ratio of heads on two identical orifices, the one discharging water and the other a water-aerosol mixture of the given characteristics, in order that the discharge coefficients of the two orifices will be the same?

380. The jet from a small triangular orifice is observed to undergo three inversion cycles in a given distance when the efflux velocity is 8 feet per second. Were the orifice dimensions to be doubled, at what velocity should the same fluid be discharged in order to yield a geometrically similar jet form (*a*) according to the Froude criterion and (*b*) according to the Weber criterion?

381. What should be the wave length of ripples produced just upstream from a 1-inch anchor cable when the relative velocity of the water is 8 feet per second?

382. Deep-water waves having a length of 8 inches are produced in a model of a harbor in which breakwater effects are under study. What discrepancy should there be between the actual wave celerity and that computed on the basis of gravitational effects alone?

QUESTIONS FOR CLASS DISCUSSION

1. Explain how certain insects are able to walk on the surface of water.

2. Contrast the trends of pressure variation in (*a*) blowing up a toy balloon and (*b*) blowing a soap bubble.

3. Noting that a bubble has two free surfaces, derive an expression for the internal pressure in terms of the bubble diameter.

4. If a tiny boat is rubbed at one end with camphor and then placed in water, it will move rapidly across the surface. In which direction will it travel, and why?

5. Why will a mold of damp sand retain its form, whereas either wet or dry sand will slump?

6. Under what conditions is the meniscus between two liquids in a glass tube (*a*) concave upwards and (*b*) concave downwards? (Explain in terms of the relative adhesions between the liquids and the boundary.)

7. Sketch the successive changes in the cross section of a liquid jet issuing (*a*) from a square orifice and (*b*) from a triangular orifice.

8. Indicate the role of capillary action in (*a*) an oil lamp; (*b*) a corn field; (*c*) a blotter; (*d*) an atomizer; (*e*) a fountain pen; (*f*) the process of blowing glass.

9. Why is it necessary to control the size of droplets formed in spraying (*a*) paint and (*b*) insecticide?

SELECTED REFERENCE

ADAM, N. K. *The Physics and Chemistry of Surfaces.* Oxford University Press, 2nd edition, 1938.

CHAPTER XI

THE ROLE OF COMPRESSIBILITY IN FLUID MOTION

51. PROPAGATION OF ELASTIC WAVES

Celerity of the elementary elastic wave. So much emphasis is usually laid upon the "incompressibility" of liquids in comparison with gases that one loses sight of the fact that liquids, solids, and gases alike are elastic media and hence subject to at least some change in volume with every change in compressive stress. Although the shear of solids and fluids results in the basically different phenomena discussed in Section 26, all matter may be characterized by a *bulk modulus of elasticity* equal to the ratio of a differential unit compressive stress to the relative reduction in volume which the stress produces. Since a decrease in the volume of any body must be accompanied by a proportional increase in its density, the bulk modulus may for present purposes be expressed most conveniently as follows:

$$E = \frac{dp}{d\rho/\rho} \tag{223}$$

If a finite compressive stress is applied to a body of matter gradually, it may be assumed for all practical purposes that each increment of stress is distributed instantaneously through the entire body. Actually, however, any such incremental change in pressure can be transmitted through the body only as an *elastic wave* traveling with a celerity which is far from infinite. A stress which is rapidly applied, therefore, may be of such short duration that quite different states of compression exist at various points in the body at a given instant. Since the terms "gradual" and "rapid" are by no means absolute, but are governed by the size of the body and the rate at which the given medium will transmit a compressive stress, the celerity of an elastic wave takes on immediate significance.

If a solitary wave of compression (or expansion) of differential magnitude is assumed to travel linearly through a given fluid medium which is otherwise at rest, the course of the wave might be followed (see Fig. 182*a*) by the local change in pressure intensity which it produces, just as the course of a gravity wave (Fig. 68*a*) is indicated by the local change in surface elevation. Like a gravity wave, an elastic wave must be accompanied by a local displacement of fluid, the resulting momentary change in velocity indicating conditions of unsteady flow. If, however, a velocity equal to the wave celerity is added in the opposite direction as in Fig. 182*b*, a state of steady flow will result, and the differential increase in pressure intensity within the standing wave will indicate a corresponding differential decrease in velocity. The interrelationship of the changes in velocity and pressure intensity may still be represented by the differential expression for convective acceleration

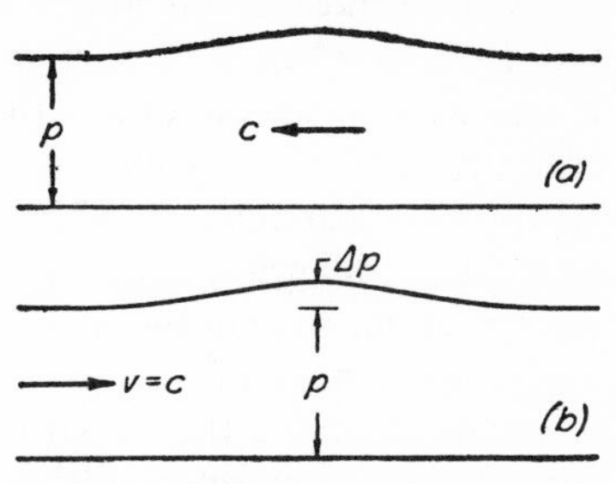

FIG. 182. Definition sketch for analysis of an elastic wave.

$$-dp = \rho\, v\, dv$$

but the continuity equation must now indicate a constant rate of *mass* flow past the two successive sections; that is, $d(\rho v) = 0$, whence

$$\rho\, dv + v\, d\rho = 0$$

If the quantity dv is eliminated from these two expressions, it will follow that

$$v = \sqrt{\frac{dp}{d\rho}}$$

Since the steady velocity v is equal to the celerity c, and since, by definition, $E = \rho\, dp/d\rho$, the celerity of a differential elastic wave will be simply

$$c = \sqrt{\frac{E}{\rho}} \tag{224}$$

Noteworthy is the fact that the celerity of an elastic wave is dependent upon the ratio of the elastic modulus to the density rather than upon the elastic modulus alone. Thus, although the elastic moduli of liquids and gases are vastly different, the corresponding values of E/ρ are much more nearly the same. In water and air, for example, the celerities of elastic waves are roughly 4700 and 1100 feet per second, respectively, which will probably be recognized as the *velocities of sound* in the two media. A sound wave is, of course, exactly the same as an elastic wave—but, as will soon be shown, elastic waves in fluids are sometimes responsible for far more than merely acoustic effects.

Pressure fluctuations due to rapid throttling of flow. Certain characteristics of wave motion in general may well be noted at this point. If one wave passes another, the effects of both must be added algebraically; two equal waves of compression will thus produce a momentary pressure change which is twice as great as that due to either wave alone; likewise, equal waves of compression and expansion will, upon passing, so counterbalance one another as to yield momentarily a pressure change of zero magnitude. A wave which reaches a solid boundary (such as the closed end of a conduit) will be positively reflected, the combination of the oncoming wave and its reflection doubling the local pressure change at the instant it reaches the boundary; the converse of a solid boundary (for instance, a large reservoir at the end of a conduit) will produce just the opposite effect—i.e., negative reflection—the combination of the positive wave and its inverted image resulting in a zero pressure change at this section. The cause of such reflection phenomena will become apparent if one notes that at the entrance to a large reservoir it is the pressure intensity rather than the velocity which must remain constant, while at the closed end of a conduit it is the velocity rather than the pressure intensity which cannot change.

From the foregoing discussion it becomes evident that the rate of travel of an elastic wave will have direct bearing upon the problems of unsteady flow in conduits discussed in Chapter V. Were the fluid and the conduit completely inelastic, as then assumed, the fluid column could change velocity only as a unit; in other words, the infinite magnitude of E required by this assumption would correspond to an infinite celerity of wave propagation, the change in pressure intensity due to valve adjustment therefore affecting instantaneously the entire fluid column. In the actual case, of course, a wave of compression caused by abrupt valve closure would travel upstream at a speed well below infinity; if the conduit were very long, moreover, the time required for the wave to reach the reservoir end might be relatively great—and not until it reached the end could the initial stoppage of flow be considered complete.

Presuming, for the moment, that the resistance to flow is negligible (see Fig. 183*a*), sudden closure of a valve at the end of the conduit would at once cause the local velocity head to become equal to zero, with an accompanying rise in pressure intensity. Reasoning that the change in kinetic energy per unit volume $\rho V^2/2$ must be equal to the work per unit volume $(\Delta p)^2/2E$ done by the change in pressure intensity Δp (i.e., the product of the average unit stress $\Delta p/2$ and the

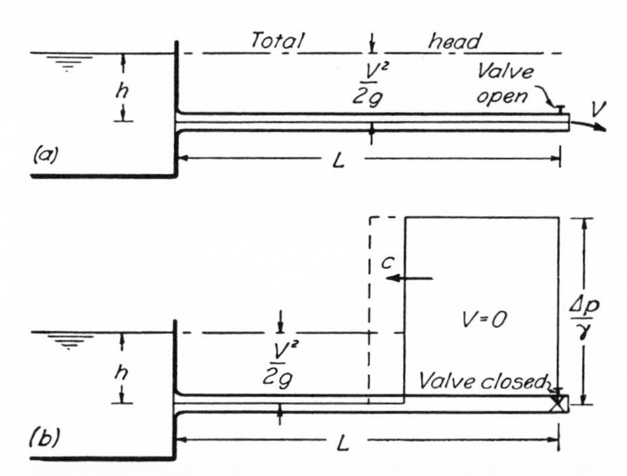

FIG. 183. Compression wave produced by abrupt throttling of flow, assuming resistance to be negligible.

resulting unit strain $\Delta\rho/\rho = \Delta p/E$), it follows that the rise in pressure within the wave may be evaluated from the expression

$$\Delta p = V\sqrt{E\rho} = \rho Vc \tag{225}$$

This then marks the extent of pressure rise within the conduit, which progresses upstream at the rate $c = \sqrt{E/\rho}$ as indicated in Fig. 183*b*. Upon reaching the open end of the conduit at time $t = L/c$, however, this compression wave will be reflected negatively, the resulting wave of expansion reducing the pressure to that in the reservoir as it travels back toward the valve; positive reflection of the expansion wave by the closed valve at time $t = 2L/c$ will then double the pressure drop, with the result that, under the assumed conditions, the original impulse will continue to travel back and forth over the length of the pipe, alternating between equally high and low pressure intensities about the line of zero flow, as shown by Fig. 184.

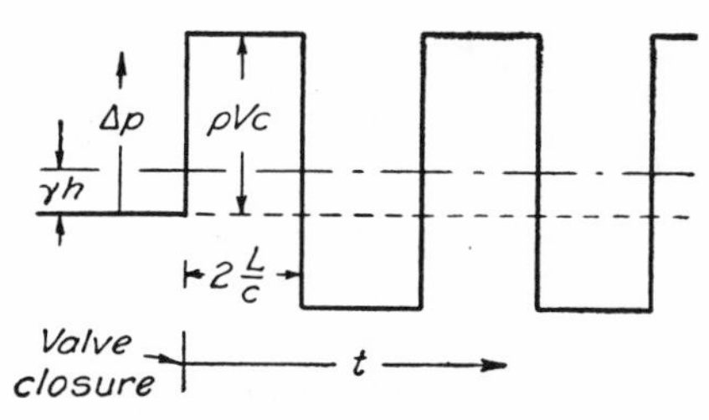

FIG. 184. Variation in pressure intensity with time for conditions of Fig. 183.

The matter of fluid resistance, of course, cannot wholly be ignored, since the influence of viscosity not only reduces the initial pressure rise but leads eventually to the complete dissipation of the wave energy. Thus, as may be seen from Fig. 185, the initial pressure rise will again

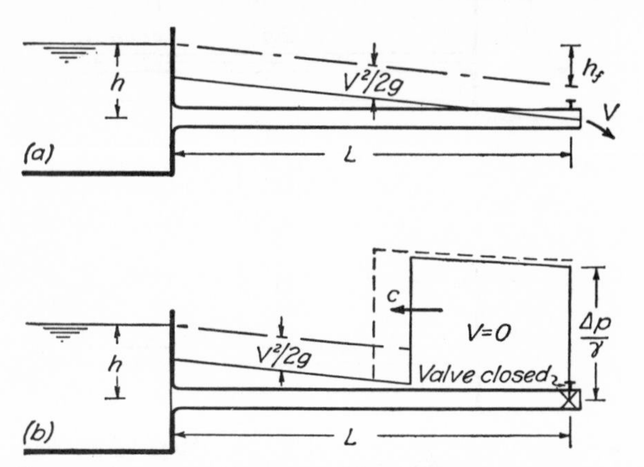

FIG. 185. Compression wave produced by abrupt throttling of flow, assuming resistance to be appreciable.

depend upon the original velocity head at the end of the conduit; but, since the piezometric head originally increases in the upstream direction as the result of the resistance to established flow, the pressure head at the valve must continue to rise as the wave travels along the pipe. From the same reasoning as led to Eq. (225) it will be found that the magnitude of this increase will be approximately $h_f/\sqrt{2}$. The resulting plot of variation in valve pressure with time is shown in Fig. 186, the successive pressure fluctuations gradually diminishing in magnitude as the result of energy dissipation.

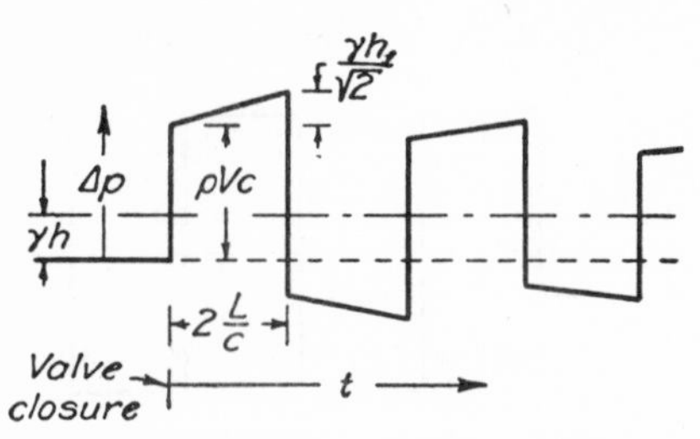

FIG. 186. Variation in pressure intensity with time for conditions of Fig. 185.

Although space does not permit their analysis, two additional features of such motion should be noted: First, the elasticity of the conduit itself further diminishes the magnitude of the pressure rise. Second, it is physically impossible to close a valve instantaneously, although the time of closure may still be small in comparison with the time required for the compression wave to reach the reservoir end. Actual measurements therefore result in pressure-time diagrams similar to that of Fig. 187. So far as time of closure is concerned, it should further be noted that the maximum pressure intensity at the valve will not be affected so long as the closure becomes complete before the wave reaches the valve end of the conduit. The longer the relative time of closure, therefore, the more important becomes the inertia of the fluid in comparison with its elasticity; in other words, the method of analysis of Chapter V becomes increasingly more exact as the rapidity of closure is reduced.

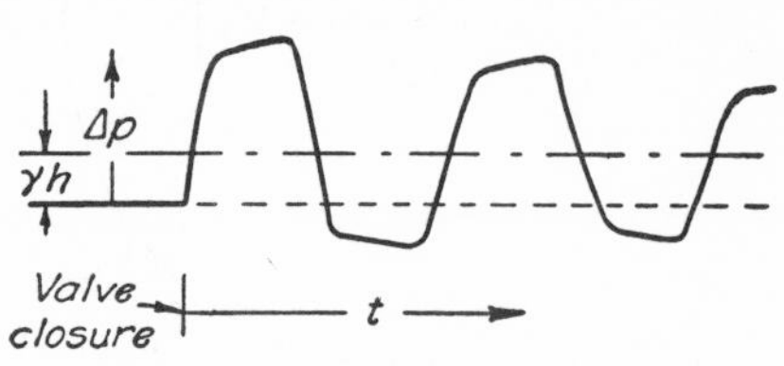

FIG. 187. Actual diagram of pressure wave produced by rapid closure of valve.

The phenomenon just described is commonly known as *water hammer*, owing to its frequent occurrence in plumbing and other hydraulic lines and to the violent nature of the pressure fluctuations which may result. Although the accompanying sound effects may not be quite as startling, it should nevertheless be realized that ventilation systems are subject to the same type of elastic disturbance on sudden throttling.

Submarine signaling and detection. The close similarity between the elastic behavior of gases and that of liquids becomes particularly striking when one considers the application of acoustics—normally associated with sound waves in air—to such problems as underwater signaling and ship detection. These problems are, in fact, merely useful versions of the water-hammer principle just discussed.

Assume, for instance, that elastic or sound waves are produced by the controlled vibration of a diaphragm at point 0 (Fig. 188) beneath the surface of water. The motion of the diaphragm relative to the otherwise still water thus produces the same sort of pressure impulse as the abrupt obstruction of moving water by the conduit valve, except that the continued vibration of the diaphragm results in a succession of positive and negative waves of controlled frequency. Unlike the conduit wave, however, the latter impulses will be propagated in all directions through the body of water. Upon reaching a boundary (whether the free surface, a ship, or the earth below), the waves will be reflected, and a small portion of each initial impulse will return to the point of propagation. An instrument to record the returning impulse against time should then permit evaluation of the distance to the reflecting surface in terms of the known velocity of sound, the controlled frequency of vibration of the diaphragm, and the measured displacement in time of the returning waves relative to those transmitted. Two such receiving instruments in slightly different positions, moreover, should yield an approximate directional indication, just as

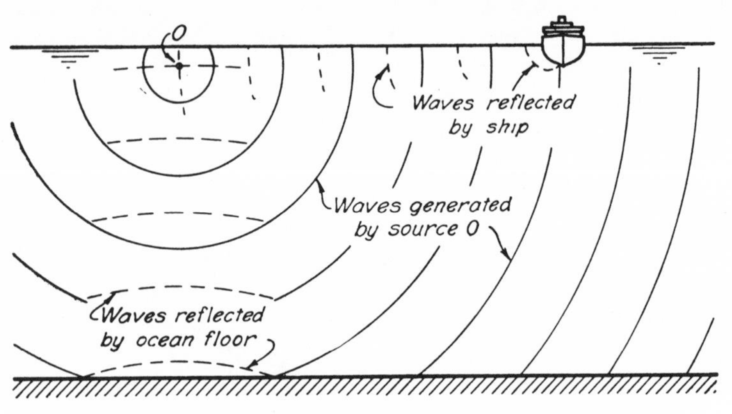

FIG. 188. Propagation and reflection of underwater sound waves.

turning one's head permits one to judge the location of a source of sound through a very sensitive perception of the differential time at which the sound reaches the two different eardrums. This is, basically, the principle of acoustic signaling and detection in both air and water.

PLATE XXVI. Silhouette of an underwater sound wave produced by the explosion of a aetonator cap at the David Taylor Model Basin of the U. S. Navy.

In present-day practice, of course, electrical methods of producing and receiving high-frequency sound waves in water have largely replaced the simpler acoustic instruments of the first world war, but the mechanics of elastic-wave propagation remains the same. Moreover, although the ultra-high-frequency electronic waves of radar possess so many practical advantages over the use of sound waves for detection purposes in air as to have superseded acoustic methods entirely, the analogy between these methods is very close.

Example 56. An electrical sounding device on an ocean vessel produces 2 pressure impulses per second. If the waves reflected from the ocean bottom are

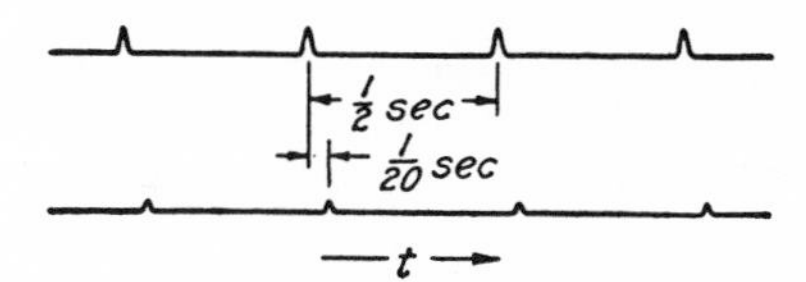

0.05 second out of phase with the initial signals, what depth (or depths) is indicated?

The celerity of the elastic wave in sea water is

$$c = \sqrt{\frac{E}{\rho}} = \sqrt{\frac{330{,}000 \times 144}{1.99}} = 4890 \text{ ft}$$

In 0.05 second the wave will travel

$$s = ct = 4890 \times 0.05 = 245 \text{ ft}$$

whence the indicated depth will be

$$h = \tfrac{1}{2}s = \frac{245}{2} = 122 \text{ ft}$$

However, in case the frequency is too great for the local depth conditions, the same indication would be obtained for the depth

$$h = \frac{c}{2}(0.5 + 0.05) = \frac{4890}{2} \times 0.55 = 1340 \text{ ft}$$

PROBLEMS

383. Assuming the density of salt water at atmospheric pressure to be 1.99 slugs per cubic foot, determine its change due to pressure increase at a depth of 5 miles below the ocean surface.

384. Compute the relative speed of sound in salt water, oil, and mercury, using the speed of propagation in fresh water as a reference.

385. Water flowing at the rate of 0.5 cubic foot per second through 2000 feet of horizontal 3-inch pipe undergoes a loss in head of 180 feet between the supply reservoir and the outlet. If the outlet valve is abruptly closed, determine: (*a*) the initial change in pressure head, (*b*) the time required for the pressure wave to travel to the reservoir and back to the valve, and (*c*) the maximum pressure head which will occur in the pipe, ignoring elastic effects of the pipe itself.

386. Assuming a linear rate of change in water velocity, what must be the minimum length of valve closure for the conditions of Problem 385 if the pressure head at the valve is not to exceed 200 feet?

387. Plot to scale against time after the instant of valve closure the variation in pressure head at the midpoint of the pipe line of Problem 385.

388. Flow through a ventilation duct having a 2-by-2-foot cross section is controlled at the outlet by a 2-by-2-foot plate pivoted about its mid-section. If the control is abruptly closed when the rate of flow is 7500 cubic feet of air per minute, what initial force will be exerted upon it by the air? Assume a density of 0.003 slug per cubic foot.

389. If sound-ranging equipment is used for submarine detection over a maximum radius of 5 miles, what should be the minimum time between sound impulses in order to receive all returning signals within the first half of the signal period?

52. EFFECTS OF COMPRESSIBILITY ON THE STEADY FLOW OF GASES

Limitations of the assumption of constant density. Now that emphasis has been laid upon the fundamental similarity between liquids and gases as elastic media, due heed must be given to the practical limitations of treating them invariably in the same category. First of all, the very fact that the ratio E/ρ is of the same order of magnitude in both cases is at once a warning that the elastic moduli must be just as different in order of magnitude as are the densities—the density of the average gas, in fact, being roughly one-thousandth that of the average liquid. No matter how great the compressive stress to which it is subjected, the value of E for every liquid happens to be so large that the resulting density change will still be very small in comparison with the density itself. The elastic modulus of any gas, on the other hand, is so small in comparison that even moderate compressive stresses may produce density changes of the same order as the original density. Thus, although comfortably audible sound waves in air still correspond to almost imperceptible density variations, waves of explosive violence may cause local density changes having magnitudes several times as great as the normal density of the atmosphere; under such circumstances, the elementary wave theory evidently cannot be expected to apply. Moreover, as will be recalled from earlier chapters, even in the case of steady flow both boundary form and boundary drag may entail considerable variation in pressure intensity; the effect of such variation upon the density of a gas was not taken into account, however, and it can only be concluded that application of the resulting equations to liquids and gases alike must involve at least some error in the latter case.

Since the detailed study of gases as compressible fluids is embodied in the science of thermodynamics, it should suffice for present purposes to determine the extent to which the density changes of a flowing gas may safely be ignored. To this end use will be made of the *ideal gas law* of elementary physics, which may be written most significantly in the form

$$\rho = \frac{p}{g_s RT} \tag{226}$$

Herein g_s is the standard gravitational acceleration (i.e., the magnitude of g at sea level and at latitude 45°), R is a so-called gas constant, and p and T are the *absolute* pressure intensity and temperature of the gas. The gist of this relationship is that the density of a gas having certain simplified characteristics will be directly proportional to the absolute intensity of pressure to which it is subjected, and inversely proportional to its absolute temperature.

Variation of atmospheric pressure with elevation. At once apparent from Eq. (226) is the fact that, if the familiar hydrostatic relationship $p + \gamma z =$ constant is applied over a considerable vertical distance, appreciable change in the pressure intensity will result in a change in the density—and hence in the specific weight—which was not taken into account in deriving this relationship. Thus, while the hydrostatic relationship for constant density would indicate a linear variation in atmospheric pressure with elevation above the earth, as shown by the straight line in Fig. 189, the actual pressure-elevation function must be a curve which approaches the vertical axis asymptotically. A closer approximation to this function than is permitted by the hydrostatic relationship may be determined quite simply from Eq. (226) by assuming that both the temperature and the gravitational force per unit mass do not change with elevation. Since this is equivalent to the assumption that p/γ remains constant,

$$\frac{p}{\gamma} \approx \frac{p_0}{\gamma_0}$$

in which p_0 and γ_0 correspond to the elevation $z = 0$; solution of this equation for γ and substitution in the differential expression for pressure variation

$$dp = -\gamma\, dz$$

will then yield the readily integrable form

$$\frac{dp}{p} \approx -\frac{dz}{p_0/\gamma_0}$$

Integration at once leads to the approximate expression

$$p \approx p_0\, e^{-\frac{z}{p_0/\gamma_0}}$$

in which the hydrostatic parameter p_0/γ_0 is seen to correspond to the height to which the atmosphere would extend if the density actually remained constant.

From the plot of this function on Fig. 189 it will be found that the assumption of constant density will lead to an error in the computed

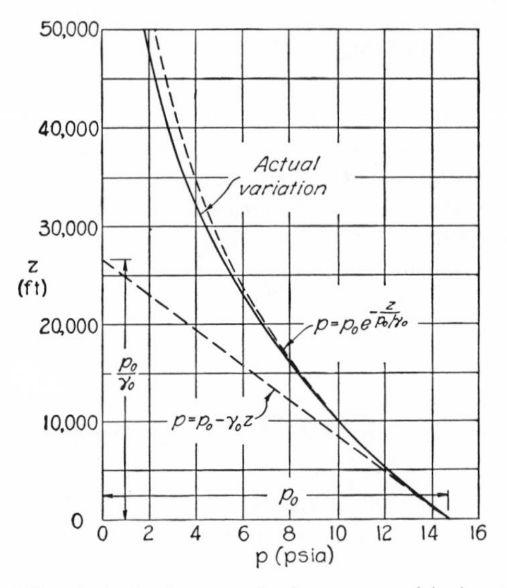

FIG. 189. Variation in atmospheric pressure with elevation.

pressure intensity of only 2 per cent in a vertical distance of 5000 feet, a difference in elevation which would seem very large in most engineering problems, but which is relatively small in certain phases of aeronautics and meteorology. Under the latter circumstances use must be made of the following empirical relationship for the variation of the absolute temperature of the atmosphere with elevation

$$T = 519 - 0.00357\,h$$

which applies with good approximation from $h = 15{,}000$ feet to the isothermal stratosphere. The corresponding variation in pressure intensity is shown as a full line in the figure.

Isothermal flow in conduits. Application of the same simplification of Eq. (226) to the case of the flow of a gas in a uniform conduit likewise permits evaluation of the error introduced by the assumption of flow at constant density. In the event that the conduit is not thermally insulated, the temperature of the gas—and hence the magnitude of the ratio $p/\rho = g_s RT$—will be the same at successive sections. Therefore, in terms of the pressure intensity and density at some reference section, for *isothermal* flow

$$\frac{p}{\rho} = \frac{p_1}{\rho_1}$$

Since the rate of mass flow must also be constant from section to section, in the case of a uniform conduit

$$\rho V = \rho_1 V_1$$

The decrease in pressure intensity along the conduit should depend only upon the boundary resistance and the acceleration of the fluid as the density decreases; therefore,

$$-dp = \frac{f}{D}\frac{\rho V^2}{2}\,dx + \rho V\,dV$$

However much the velocity may change as the result of the expansion of the fluid, the Reynolds number $\mathbf{R} = VD/\nu$ will be unaffected owing to the constancy of the product ρV; therefore f will be the same at all sections. Although the term $\rho V\,dV$ will obviously not be zero, as a first approximation it may be regarded as negligible in comparison with the resistance term. Thus, expressing ρ and V in terms of p, p_1, ρ_1, and V_1 by means of the foregoing expressions,

$$-\frac{p}{p_1}\,dp \approx \frac{f}{D}\frac{\rho_1 V_1^2}{2}\,dx$$

integration of which over the distance $L = x_2 - x_1$ then yields

$$\frac{p_1^2 - p_2^2}{2p_1} \approx f\frac{L}{D}\frac{\rho_1 V_1^2}{2}$$

Since

$$\frac{p_1^2 - p_2^2}{2p_1} = \frac{(p_1 - p_2)(p_1 + p_2)}{2p_1} = \frac{1}{2}(p_1 - p_2)\left(1 + \frac{p_2}{p_1}\right)$$

the foregoing relationship may be written in the form

$$p_1 - p_2 = \frac{2}{1 + p_2/p_1} f\frac{L}{D}\frac{\rho_1 V_1^2}{2} \tag{227}$$

This is seen to differ from the corresponding resistance equation of Chapter VI by a factor depending only upon the ratio $p_2/p_1 = 1 - \Delta p/p$. Evidently, so long as the pressure drop Δp is less than 5 per cent of the initial *absolute* intensity, neglecting this factor will involve an error of only 2½ per cent—which is permissible in many engineering calculations.

Acceleration under adiabatic conditions. Quite distinct from isothermal flow of this nature is motion which involves such rapid change in velocity that the expansion or compression of the gas is *adiabatic*—i.e., involving no change in the heat content per unit mass. Under such circumstances the isothermal relationship $\rho/\rho_1 = p/p_1$ is replaced by the adiabatic form of Eq. (226),

$$\frac{\rho}{\rho_1} = \left(\frac{p}{p_1}\right)^{1/k}$$

in which k is essentially constant for any given gas. If, as a first approximation, viscous resistance to motion is ignored, the differential equation of acceleration becomes simply

$$-dp = \rho V\,dV$$

Substitution of the value $\rho_1(p/p_1)^{1/k}$ for ρ in this equation then yields

$$-p_1^{1/k}\frac{dp}{p^{1/k}} = \rho_1 V\,dV$$

integration of which between sections 1 and 2 of a stream filament leads at once to the expression

$$\frac{p_1^{1/k}}{1 - 1/k}\left(p_1^{1-1/k} - p_2^{1-I/k}\right) = \frac{\rho_1}{2}\left(V_2^2 - V_1^2\right)$$

which is the adiabatic counterpart of the energy equation for flow at constant density. The error involved by assuming constant density may be estimated in the following manner: The equation is solved for p_2, the right side is expanded in a series, and p_1 is then subtracted from both sides, whereupon it will be found that

$$p_2 - p_1 = \frac{\rho_1}{2}\left(V_1^2 - V_2^2\right)\left[1 + \frac{\rho_1(V_i^2 - V_2^2)}{4kp_1} + \cdots\right] \tag{228}$$

Evidently, if adiabatic flow is treated as flow at constant density, the resulting error is represented as a first approximation by the term

$$-\frac{\rho_1(V_1^2 - V_2^2)}{4kp_1}$$

For instance, in the flow of air under normal atmospheric conditions, a change in velocity from 50 to 350 feet per second would involve a discrepancy in the computed pressure change of only +2½ per cent.

It would appear from the foregoing discussion that in very many instances gases may truly be treated as fluids of constant density without introducing errors of serious magnitude. Even when the resulting errors may no longer be neglected, ample correction may often be made by means of the approximate factors developed herein, thereby avoiding tedious evaluation of the more exact relationships. Needless to say, however, there is a limit to the applicability of such approximate methods, for certain problems of gas flow involve velocities approaching—and even exceeding—the speed of sound. If the complexities of thermodynamic analysis are added to those due to boundary curvature and viscous shear, the general problem of gas flow at high velocity obviously becomes extremely involved. However, as will be shown in the final section of this volume, there exists a fundamental parameter for variable-density flow which permits the conversion of test results on scale models to any desired prototype conditions.

Example 57. Natural gas is pumped at the rate of 1000 pounds per hour through a 6-inch wrought-iron conduit. At a given section the relative pressure intensity is 15 pounds per square inch, the corresponding density and kinematic viscosity being 0.0028 slug per cubic foot and 8.5×10^{-5} square foot per second. (*a*) If the density variation is ignored, what is the maximum length of line for which the resulting error in computed pressure drop will not exceed 2 per cent? (*b*) Show that the change in kinetic energy per unit volume is small compared with the energy dissipation per unit volume in the same distance.

For the stated conditions,

$$V = \frac{1000}{60 \times 60 \times 0.0028 \times 32.2 \times \frac{\pi}{4} \times \overline{0.5^2}} = 15.7 \text{ fps}$$

$$\mathbf{R} = \frac{15.7 \times 0.5}{8.5 \times 10^{-5}} = 92{,}300$$

$$\frac{D}{k} = \frac{0.5}{0.00015} = 3330$$

$$f \approx 0.02$$

(*a*) If $2/(1 + p_2/p_1) = 1.02$ (i.e., for 2 per cent error),

$$p_2 = \left(\frac{2}{1.02} - 1\right) p_1 = 0.96(14.7 + 15) = 28.5 \text{ psia} = 13.8 \text{ psi}$$

Therefore, from Eq. (227),

$$(15 - 13.8)144 = \frac{2}{1 + 0.96} \times 0.02 \times \frac{L}{0.5} \times \frac{0.0028 \times \overline{15.7^2}}{2}$$

and

$$L = \frac{1.2 \times 144 \times 1.96 \times 0.5 \times 2}{2 \times 0.02 \times 0.0028 \times \overline{15.7^2}} = 12{,}300 \text{ ft}$$

(b) At the distance L from the reference section

$$\rho_2 = \rho_1 \frac{p_2}{p_1} = 0.0028 \times 0.96 = 0.00269 \text{ slug/ft}^3$$

and

$$V_2 = V_1 \frac{\rho_1}{\rho_2} = 15.7 \times \frac{1}{0.96} = 16.4 \text{ fps}$$

Therefore, the change in kinetic energy per unit volume will be

$$\frac{\rho_2 V_2^2}{2} - \frac{\rho_1 V_1^2}{2} = \frac{0.00269 \times \overline{16.4^2}}{2} - \frac{0.0028 \times \overline{15.7^2}}{2} = 0.017 \text{ ft-lb/ft}^3$$

while the energy dissipation per unit volume may be approximated as

$$f \frac{L}{D} \frac{\rho_1 V_1^2}{2} = 0.02 \times \frac{12{,}300}{0.5} \times \frac{0.0028 \times \overline{15.7^2}}{2} = 170 \text{ ft-lb/ft}^3$$

The ratio of these values, obviously, will be exceedingly small:

$$\frac{0.017}{170} = 0.0001$$

PROBLEMS

390. Estimate from the data given in Fig. 189 the temperature and density of the atmosphere at an elevation of 6 miles.

391. What error would be involved in ignoring the effect of density variation when computing the pressure intensity of the atmosphere at an elevation of 1000 feet?

392. Air flows at the rate of 2000 cubic feet per minute in a sheet-metal duct having a diameter of 15 inches. If the pressure intensity at gage A is 5 pounds per square inch, and $f = 0.015$, what will be the reading of gage B 750 feet down the duct (a) assuming constant density, and (b) taking the density variation into account?

393. How far apart could the two gages of Problem 392 be without involving an error of more than 1 per cent between the given rate of flow and that computed from the measured pressure drop on the assumption of constant density?

394. What error would be involved in the constant-density indication of a stagnation tube mounted on the wing of an airplane traveling at a speed of 300 miles per hour?

395. Air under a pressure intensity of 15 pounds per square inch is discharged into the atmosphere through a 1-inch well-rounded orifice in the side of a tank. Estimate the error involved in computing the rate of flow without regard to the density change.

53. SIGNIFICANCE OF THE MACH NUMBER

Importance of c as a reference velocity. Considerable attention was paid in earlier pages of this chapter to the celerity of an elastic wave of small magnitude,

$$c = \sqrt{\frac{E}{\rho}}$$

which was shown to be synonymous with the speed of sound in any given medium. So far as liquids are concerned, E and ρ—and hence c—were found to be very nearly independent of the pressure intensity. Gases, on the contrary, vary with pressure to an extreme degree in both elasticity and density, but the net effect of such variation upon c happens to be nil provided that the temperature does not change. For instance, from the adiabatic relationship p/ρ^k = constant it follows that

$$\log p - k \log \rho = \text{constant}'$$

whence

$$\frac{dp}{p} - k\frac{d\rho}{\rho} = 0$$

and

$$E = \frac{dp}{d\rho/\rho} = kp$$

The celerity of a gaseous sound wave of small magnitude (which is propagated without appreciable loss in heat) is therefore

$$c = \sqrt{\frac{kp}{\rho}} = \sqrt{kg_sRT} \tag{229}$$

and hence ultimately dependent upon only the absolute temperature.

So long as the velocity of flow of either a liquid or a gas is very small in comparison with the corresponding speed of sound, elastic effects will generally be negligible in comparison with other factors influencing the motion. While it is unlikely that velocities of liquid flow even remotely approaching the sound-wave celerity will ever be attained, the tips of propellers have long since exceeded such velocities in air, and many high-speed projectiles likewise travel in advance of the sound waves which they produce. Obviously, elastic effects can by no means be ignored under such circumstances.

The importance of the wave celerity in problems of aeronautics and ballistics becomes the more immediate as c diminishes in magnitude with increasing elevation above the earth, owing to the rapid drop in atmospheric temperature. Such use of c as a reference velocity is, of course, by no means limited to the motion of bodies through air. For instance, since the mass rate of steady gas flow through a conduit must be the same at successive sections, continued decrease in pressure—and hence in density—requires a continued increase in velocity; the closer V approaches the limit c, the greater becomes the relative error in treating such flow as one of constant density. Similarly, in the adiabatic acceleration of gas through orifices or nozzles, the velocity in the zone of greatest contraction may easily approach—or even attain—the speed of sound, the error involved in ignoring the resulting density change again increasing with the ratio of V to c.

The Mach number. According to statements made in earlier chapters of this text, in the event that fluid properties other than density would have appreciable influence upon the flow, the Euler number could be expected to vary accordingly. If, as in the case of weight, viscosity, and surface tension, an elasticity parameter were formed of

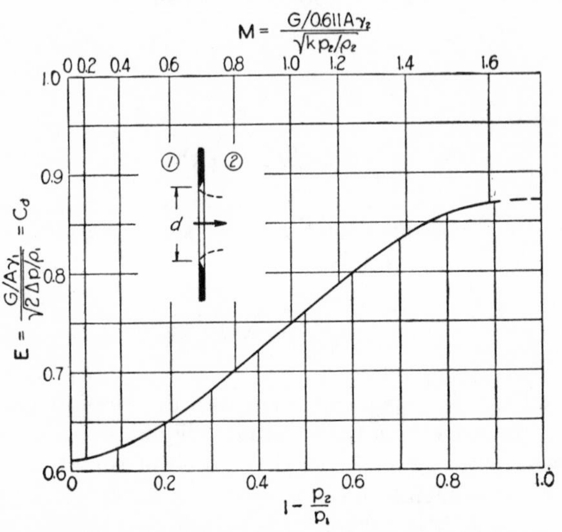

FIG. 190. Coefficient of orifice discharge as a function of the Mach number.

the ratio between a typical unit inertial reaction $(\rho V^2/L)$ and a typical unit elastic force (E/L), the result $(\rho V^2/E)$ should characterize the relative importance of elastic effects in a given state of flow. Inspection of this parameter will show that it is simply the square of the ratio V/c—the latter commonly being known as the *Mach number:*

$$\mathbf{M} = \frac{V}{\sqrt{E/\rho}} = \frac{V}{c} \tag{230}$$

It then follows, for given boundary geometry, that

$$\mathbf{E} = \phi(\mathbf{M}) \tag{231}$$

Since the discharge coefficient of an orifice is a particular form of the Euler number, it is therefore to be expected that C_d will vary with **M**, just as it has been found to depend, in turn, upon the parameters **F**, **R**, and **W**. That such is the case may be seen from the C_d : **M** plot for the efflux of gas from an orifice in a large pressure tank, shown in Fig. 190, G being the weight rate of flow. Noteworthy is the fact that as the contracted jet attains the speed of sound the efflux rate becomes independent of the external pressure; once this critical velocity is reached, in other words, outside disturbances can no longer be propagated in the upstream direction.

Variation in wave pattern with the Mach number. Interestingly enough, the formation of sound waves due to such high relative velocities between a solid and a gas is closely analogous to the formation of gravity waves on the surface of water by a ship or a channel constriction. In the latter case the form of the visible wave pattern has already been found to depend on a characteristic Froude number—the ratio of the relative velocity of the liquid to the celerity of a gravity wave. In the case of gases, the invisible sound-wave pattern may be photographed only through the variable refraction of light from a point source, but the similarity of the resulting pictures to those of gravitywaves produced by similar boundaries is in itself enough to convince one that the Mach number must have the same significance in elastic-wave phenomena as the Froude number has in gravity-wave phenomena. Such a conclusion is further justified by the following schematic analysis of wave propagation by a small body moving through an otherwise quiet fluid.

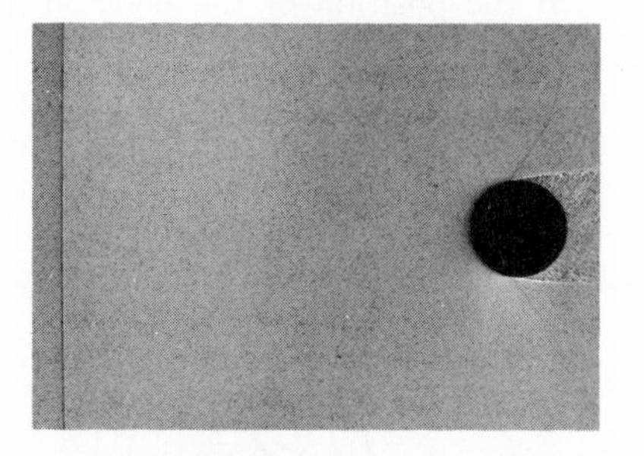

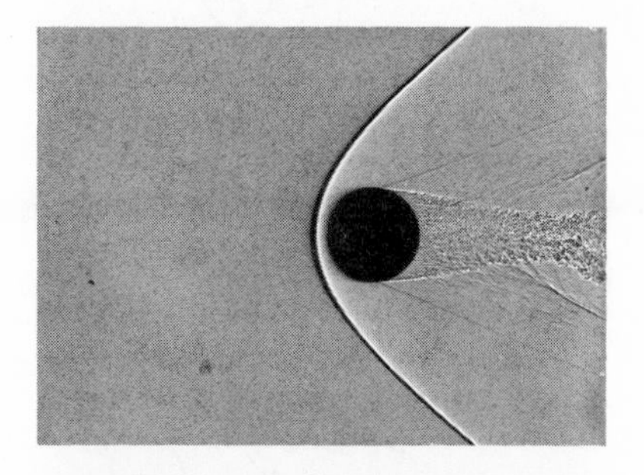

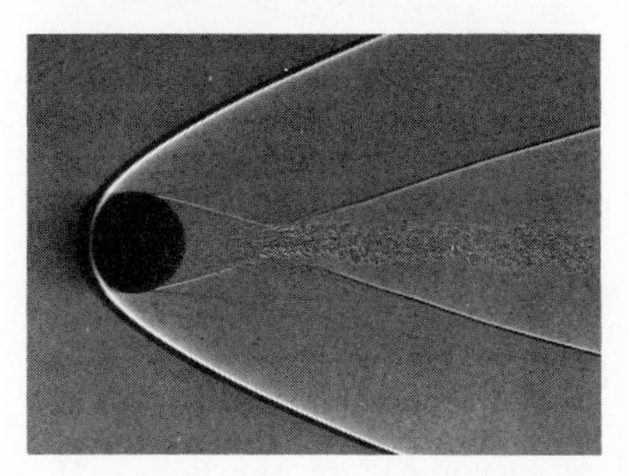

PLATE XXVII. Sound-wave silhouettes for a $9/16$-inch spherical projectile at Mach numbers of 1, 2, and 4, photographed at the Aberdeen Proving Ground.

If one assumes, for the sake of ease in visualization, that the motion of a body is transmitted to the surrounding fluid as a series of finite—rather than infinitesimal—impulses, each such impulse should give rise to a wave which travels radially outward at the celerity c from the position of the body at the instant of generation. If the speed of the body is very small in comparison with c, the two-dimensional pattern of the waves at any instant can be represented by a system of practically concentric circles, as in Fig. 191*a*. As v becomes larger, however, the centers of successive waves will be a series of points representing the location of the body as each wave in turn was

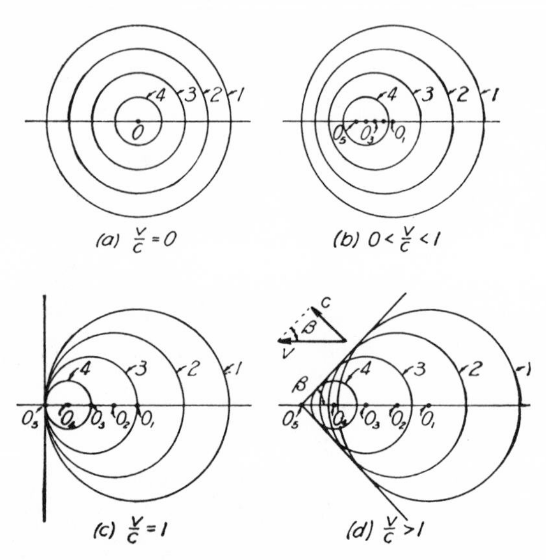

FIG. 191. Variation in wave pattern with relative speed of moving body.

generated—with the resulting pattern shown in Fig. 191*b*. Evidently, as v becomes exactly equal to c the body will be advancing at exactly the same speed as that part of each successive wave lying in its path; as a result (Fig. 191*c*) the waves will all be tangent to a single normal line just ahead of the body. If, then, v exceeds c, each successive wave will be left behind directly after generation, but through subsequent growth will remain tangent to a pair of lines intersecting at the body. As may be seen from Fig. 191*d*, the sine of the angle between either line and the path of the body (known as the *Mach angle*) will depend upon the ratio of c to v. If the forces transmitted to the fluid by the body are now visualized as continuous rather than a schematic series of successive impulses, it will be seen that the velocity conditions of Fig. 191*a* correspond to those treated in earlier chapters, wherein it was tacitly assumed that boundary forces would be transmitted instantaneously through any fluid; in other words, such flows correspond to Mach numbers which are very close to zero. Figure 191*b*, on the contrary, represents a phase of motion in which the flow pattern is already influenced by elastic effects to an appreciable degree, while Figs. 191*c* and 191*d* simply indicate intermediate and advanced stages of such influence; the corresponding Mach numbers are, respectively, <1, 1, and >1.

Form drag at supersonic speeds. Once **M** exceeds unity, of course, no elementary force can be transmitted ahead of the body; there is, however, a pronounced force concentration along the tangents shown in Figs. 191*c* and 191*d*, resulting in a very abrupt change in the fluid density—i.e., a so-called *shock wave* analogous to the breaking wave at the bow of a high-speed ship. Just as in the case of a liquid wave or surge, the celerity of an elastic wave increases with amplitude, so

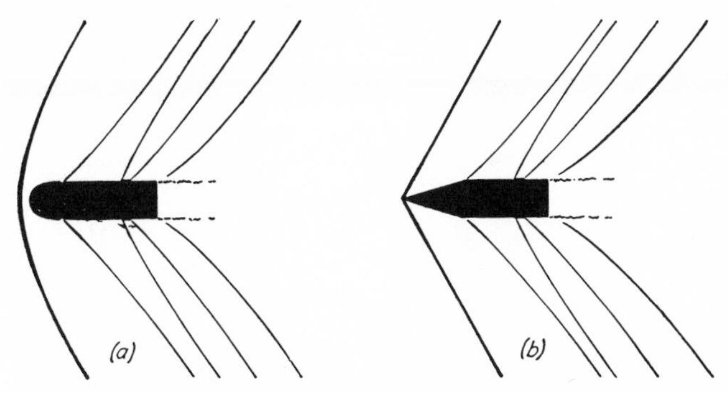

FIG. 192. Variation in wave pattern with form of projectile nose.

that shock waves travel more rapidly than sound waves of barely audible magnitude. That this must necessarily be true will be seen from Fig. 192*a*, which represents the wave formation produced by a blunt-nosed body traveling at a speed considerably in excess of the reference parameter $c = \sqrt{E/\rho}$. The faster such a body moves, the higher must be the pressure and density directly ahead—and hence the more rapidly this portion of the wave pattern can advance; such local conditions correspond, in effect, to the schematic pattern of Fig. 191*c*. If, on the other hand, the nose of the body is sufficiently tapered (as in Fig. 192*b*), the shock wave which it produces will form a conic surface with apex at the very tip; evidently, this will reduce the local wave intensity, since the pressure at the nose need no longer be sufficiently great to produce a local wave celerity equal to the speed of flight.

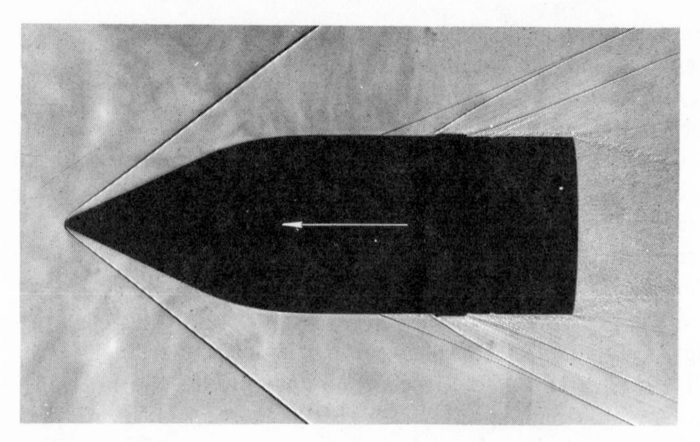

PLATE XXVIII. Wave pattern produced by a 75-millimeter shell traveling 2800 feet per second at the Aberdeen Proving Ground.

At velocities below that of sound, the process of streamlining is confined to the elimination of separation and the resulting formation of eddies at the rear of a body. Above the velocity of sound, on the

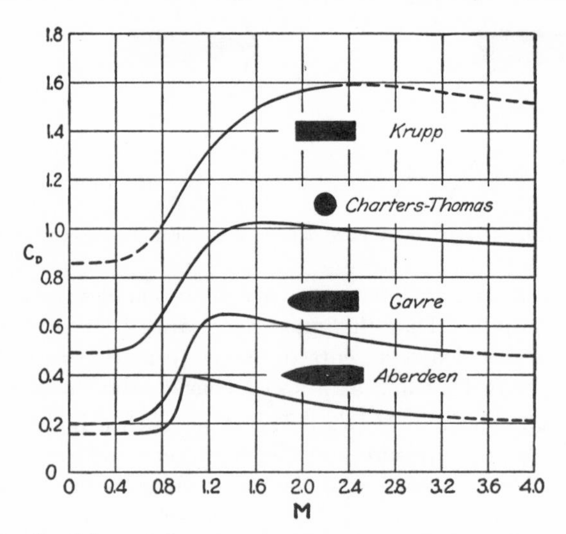

FIG. 193. Coefficient of drag as a function of the Mach number for typical projectile shapes.

contrary, it is the front of the body rather than the rear which is the source of form effects upon the resistance. Thus, if the Euler number is written as a coefficient of drag, as in Chapter VIII, the extent to which it varies with the Mach number will depend largely upon the shape of the front part of the body. Such curves of $C_D : \mathbf{M}$ for typical forms of projectile may be seen in Fig. 193, indicating both the extent to which elastic effects will increase the form drag of any body and the extent to which such effects may be reduced by streamlining. Much the same considerations apply to the design of high-speed propellers and will very likely have considerable influence upon the form of the airplane itself as ever greater velocities of flight are attained. Needless to say, the parameter $\mathbf{M}$ will have as much significance in model tests at supersonic speeds as the Reynolds number has at subsonic speeds.

Example 58. Assuming that the Gavre curve of Fig. 193 represents the drag function of a 1000-pound bomb having a diameter of 18 inches, determine the velocity at which such a bomb will strike the earth after release from a plane at a height of 20,000 feet.

If, as a first approximation, compressibility effects are ignored, it will be found from Eq. (180) that

$$v_0 = \sqrt{\frac{2F}{C_d A \rho}} = \sqrt{\frac{2 \times 1000}{0.2 \times \frac{\pi}{4} \times \overline{1.5}^2 \times 0.0025}} = 1500 \text{ fps}$$

Since this exceeds the speed of sound, it is evident that the drag coefficient will depend upon the Mach number—which, in turn, must depend for direct evaluation upon knowledge of the terminal velocity of descent. However, the velocity may be eliminated from this phase of the problem by simultaneous solution of the definition equations for C_d and $\mathbf{M}$, as follows:

$$C_d = \frac{F}{A\rho v_0^2/2} \qquad \mathbf{M} = \frac{v_0}{\sqrt{kp/\rho}} \qquad \text{whence} \quad C_d = \frac{2F}{Akp}\frac{1}{\mathbf{M}^2}$$

Evidently, for the given values of F, A, k, and p the intersection of the curve of this relationship with that of the Gavre function should correspond to the desired values of C_d and $\mathbf{M}$, as discussed in Sections 40 and 41. Taking arbitrarily the values $C_d = 0.4$ and $C_d = 0.5$, it will be found that the corresponding values $\mathbf{M} = 0.98$ and $\mathbf{M} = 0.87$, respectively, will be obtained for the condition that

$$\frac{2F}{Akp} = \frac{2 \times 1000}{\frac{\pi}{4} \times \overline{1.5}^2 \times 1.40 \times 14.7 \times 144} = 0.382$$

A second-degree hyperbola (i.e., $C_d \sim 1/\mathbf{M}^2$) passing through these points on Fig. 193 will be found to intersect the Gavre curve at $C_d = 0.42$, whence

$$v_0 = \sqrt{\frac{2F}{C_d A \rho}} = \sqrt{\frac{2 \times 1000}{0.42 \times \frac{\pi}{4} \times \overline{1.5}^2 \times 0.0025}} = 1040 \text{ fps}$$

PROBLEMS

396. Air is pumped at the rate of 200 pounds per minute through a 6-inch duct. If the gage readings at stations A and B, a considerable distance apart, are 60 and 5 pounds per square inch, respectively, what will be the corresponding Mach numbers of the flow?

397. When the pressure gage on a tank of compressed air reads 30 pounds per square inch, what will be the rate of outflow through a ½-inch orifice in the side of the tank? What would be the corresponding Mach number? What error would result if the rate of flow were computed without regard to elastic effects?

398. Estimate, from the wave angle formed by the projectile of Plate XXVIII, the Mach number and the speed of flight.

399. Assuming the sphere of Plate XXVII to be steel (specific gravity = 7.85), determine for each Mach number the ratio of the drag to the weight of the sphere.

400. Assuming the blunt (Krupp) and the pointed (Aberdeen) projectiles of Fig. 193 to have a diameter of 2 inches and a weight of 7 pounds, determine and sketch to scale the force components acting upon each when traveling horizontally at a speed of 1800 feet per second.

QUESTIONS FOR CLASS DISCUSSION

1. How would the pressure diagram of Fig. 184 vary if the section in question were midway between the ends of the pipe instead of directly upstream from the valve?
2. Why is it unlikely that an underwater body will ever attain a velocity as great as that of an elastic wave?
3. Should the failure to take into account the density change for flow through a thermally insulated pipe cause a greater or smaller error than for isothermal flow?
4. Why does the pressure intensity of a gas—but not of a liquid—represent potential energy?
5. Show the analogy between the gravity surges of Fig. 69 and sound waves of different intensities.
6. At what speed in miles per hour will an airplane travel as fast as sound? Why will the wings generate sound waves well before this speed is attained?
7. Compare the shapes of barges, freighters, and destroyers with those of low-speed and high-speed projectiles, and explain the reasons for the variation.
8. The high-density (pressure) wind tunnel has been shown useful in attaining high Reynolds numbers for model tests. What effect should this have upon the Mach number? May **R** and **M** be varied independently in such a tunnel?

SELECTED REFERENCES

GIBSON, A. H. "Phenomena Due to the Elasticity of a Fluid." Chapter VI, *The Mechanical Properties of Fluids*, Blackie, 1925.

PRANDTL, L. "The Flow of Liquids and Gases, Part III." Chapter VII, *The Physics of Solids and Fluids*, Blackie, 1936.

TAYLOR, G. I., and MACCOLL, J. W. "The Mechanics of Compressible Fluids." *Aerodynamic Theory*, Vol. III, Springer, 1935.

CHARTERS, A. C., and THOMAS, R. N. "The Aerodynamic Performance of Small Spheres." *Journal of the Aeronautical Sciences*, Vol. 12, No. 4, 1945.

APPENDIX

MECHANICAL PROPERTIES OF FLUID MATTER

54. APPLICATION OF DIMENSIONAL PRINCIPLES

Dimensional homogeneity. A fundamental requisite of any correct mathematical expression is that the quantities between which an equality is stated to exist be dimensionally as well as numerically equivalent. An equation must, in other words, be *dimensionally homogeneous* if the physical condition which it expresses is to be generally true.

Since all phenomena of mechanics in effect represent an application of Newton's laws of motion, which in turn are dimensionally describable in terms of length $[L]$, time $[T]$, force $[F]$, and mass $[M]$, these dimensions should be the only ones appearing in any equation of mechanics. If, moreover, the proportionality factor of Newton's second law is defined as a pure number, then any one of these four dimensions may be replaced by a particular combination of the other three. That is, by writing Newton's second law in the equation form

$$F = Ma$$

the force dimension is specifically related to the length, time, and mass dimensions:

$$[F] = \left[\frac{ML}{T^2}\right]$$

It follows therefrom that

$$[M] = \left[\frac{FT^2}{L}\right], \qquad [L] = \left[\frac{FT^2}{M}\right], \qquad \text{and} \qquad [T] = \left[\left(\frac{ML}{F}\right)^{1/2}\right]$$

or, more generally,

$$\left[\frac{ML}{FT^2}\right] = 1 \tag{232}$$

It is quite arbitrary which three of the four dimensions are chosen for general use; indeed, it is often convenient to follow different policies at different times, depending upon the problem under consideration. For example, it is more significant to think of density as having the dimension $[M/L^3]$ than $[FT^2/L^4]$ or $[M^4/F^3T^6]$, although all three are equally correct. On the other hand, it would be less significant—though more consistent with the foregoing example—to consider the dimension of pressure intensity as $[M/LT^2]$ than simply $[F/L^2]$. There is, of course, a practical limit to such freedom of selection, the more general use of length and time effectively restricting the alternative categories to force and mass.

As a typical example of the homogeneity test for the dimensional correctness of an equation, the pressure relationship frequently used in the foregoing pages may be considered:

$$\Delta p = \frac{\rho v^2}{2}$$

In dimensional terms this equation would have the form

$$\left[\frac{F}{L^2}\right] = \left[\frac{M}{L^3}\left(\frac{L}{T}\right)^2\right]$$

If force (or mass) is now replaced by its equivalent combination of the other three dimensions, the two sides of the equation will be found to be identical:

$$\left[\frac{ML/T^2}{L^2}\right] = \left[\frac{M}{L^3}\left(\frac{L}{T}\right)^2\right]$$

Evidently, therefore, the ratio $\Delta p/(\rho v^2/2)$ is truly a dimensionless quantity. Consider, on the other hand, the Chézy relationship:

$$V = C\sqrt{RS}$$

If C is assumed to be a dimensionless factor (as any general coefficient should be), it will be found that

$$\left[\frac{L}{T}\right] \neq [L^{1/2}]$$

Obviously, if the equation is to be dimensionally homogeneous, then C must have the dimension $[L^{1/2}/T]$ (i.e., $C = \sqrt{8g/f}$, as shown in Chapter VII).

Although dimensional homogeneity is thus a primary requisite for the general validity of an equation, it must not be assumed that any equation which is dimensionally homogeneous is perforce correct. There are, moreover, many empirical formulas in existence which are not dimensionally homogeneous [see, for example, Eq. (233)] but which nevertheless serve a useful purpose in faithfully reproducing experimental data over a limited range for a particular dimensional system. To be truly general, however, an expression must be dimensionally sound—a fact which in itself provides a far more significant basis for interpreting experimental results than simple curve fitting without regard to dimensional homogeneity. Even so far as the routine solution of problems is concerned, a rapid check of dimensional as well as numerical accuracy in all equations will prevent many an error of omission from occurring.

Dimensions of quantities describing boundary, flow, and fluid characteristics. In Table V are listed the most important of the quantities used in the study of fluid motion, together with their customary algebraic symbols, opposite each of which is given the corresponding dimension in both L-T-F and L-T-M terms. These are arranged, for maximum significance, in the following four groups: first, those describing the boundary geometry and shape of the flow pattern and hence involving only the length dimension; second, those describing the kinematics of the flow pattern and hence involving only the length and time dimensions (included in this group, because of their dimensional nature, are the ratios of two fluid properties to the fluid density); third, those terms required in describing the dynamics of the flow pattern and hence involving either force or mass in addition to length and time; fourth, the series of dimensionless parameters commonly used in analyzing general flow conditions. This tabulation will be found of constant reference value, both in checking the dimensional accuracy of flow equations and in converting units of measurement from one dimensional system to another.

Conversion of dimensional units. There is, unfortunately, no universal system of dimensional units in general use throughout the world, and even within any one country different systems are preferred by different professions. The second (1/86,400 of the mean solar day) is, to be sure, the standard unit in the measurement of time; but common use of the foot by English-speaking peoples and of the centimeter by others as the unit of length leads at once to obvious inconvenience in engineering intercourse. The situation becomes really involved, however, when one considers the units of force and mass. The gram is recognized as the unit of mass in most physical sciences, but continental engineers consider the kilogram (1000 grams) the unit of force. In England, similarly, the pound has long since been adopted as the unit of mass, but in America the pound is generally associated with the measurement of force.

If the metric gram, the British pound, and the American slug are all considered basic units of mass, these units will obviously be inde-

TABLE V

Dimensions of Quantities Describing Boundary, Flow, and Fluid Characteristics

Quantity	Symbol	Dimensions in terms of	
		L-T-F	L-T-M
Geometric			
Length (any linear measurement)	L	L	L
Area	A	L^2	L^2
Volume	V	L^3	L^3
Slope	S		
Kinematic			
Time	t	T	T
Velocity, linear	v	L/T	L/T
angular	ω	$1/T$	$1/T$
Acceleration, linear	a	L/T^2	L/T^2
angular	α	$1/T^2$	$1/T^2$
Volume rate of flow, total	Q	L^3/T	L^3/T
per unit width	q	L^2/T	L^2/T
Circulation	Γ	L^2/T	L^2/T
Gravitational acceleration (γ/ρ)	g	L/T^2	L/T^2
Kinematic viscosity (μ/ρ)	ν	L^2/T	L^2/T
Dynamic			
Mass	M	FT^2/L	M
Force	F	F	ML/T^2
Mass density	ρ	FT^2/L^4	M/L^3
Specific weight	γ	F/L^3	M/L^2T^2
Dynamic viscosity	μ	FT/L^2	M/LT
Surface tension	σ	F/L	M/T^2
Elastic modulus	E	F/L^2	M/LT^2
Pressure intensity	p	F/L^2	M/LT^2
Shear intensity	τ	F/L^2	M/LT^2
Impulse, momentum	I, M	FT	ML/T
Work, energy	W, E	LF	ML^2/T^2
Power	P	LF/T	ML^2/T^3
Dimensionless			
Euler number	$\mathbf{E}$		
Froude number	$\mathbf{F}$		
Reynolds number	$\mathbf{R}$		
Weber number	$\mathbf{W}$		
Mach number	$\mathbf{M}$		

pendent of any gravitational system, for they are measures purely of the quantity of matter within a given body. Since, according to the universally accepted significance of the Newtonian equation, 1 unit of force is that which will accelerate 1 unit of mass at a rate equal to 1 unit of length per unit of time per unit of time, then the corresponding force unit in each of these systems must be of such magnitude as to produce a unit acceleration upon the given unit of mass. That is, a force of 1 dyne will by definition accelerate a 1-gram body 1 centimeter per second per second; a force of 1 poundal will accelerate a 1-pound body 1 foot per second per second; and a force of 1 pound will accelerate a 1-slug body 1 foot per second per second.

If, now, it is desired to convert the units of a given dimensional quantity from one of these systems to another, use must be made of internationally adopted standards for the relative magnitudes of the corresponding units. The centimeter, which represents approximately one-billionth of the distance at sea level from the equator to either pole, is defined as one-hundredth as long as the international prototype meter preserved in metal at Sèvres, France. The foot is defined as a unit which is 30.48+ times as long as the centimeter. The gram, which corresponds very closely to the mass of 1 cubic centimeter of water at 4° Centigrade, is defined as one-thousandth of the mass of the international prototype kilogram, also preserved in metal at Sèvres. The pound as a unit of mass is then defined as 453.6− times as great as the gram. But since a body having a mass of 1 pound is considered in America as having a standard weight (i.e., gravitational attraction) of 1 pound, the slug must be defined as well in terms of the standard gravitational attraction $g_s = 32.17+$ feet per second per second. In other words, the American slug is 32.17+ times as great as the British pound and hence 14,593− times as great as the metric gram. Likewise, the kilogram as a force unit is $9.81- \times 10^5$ times as great as the dyne.

Use of these standards in performing the desired conversion of units may take either of two alternative forms of algebraic operation. One involves substituting the equivalent number of units of the required system for the unit of the given system. The other consists in multiplying the quantity in question by the ratio of the unit equivalents. Neither form of operation, of course, will alter either the magnitude or the dimension of the quantity upon which it is performed, since one amounts to the interchange of equal values and the other to multiplication by the ratio of equal values (i.e., by unity).

As the latter procedure has been shown by experience to be less subject to inconsistency—and hence error—in manner of application, a series of unit conversion ratios useful in problems of fluid motion has been compiled in Table VI. Each ratio has been so constituted that the numerical factor is greater than 1, the conversion operation being either multiplication or division in such manner that the units from which it is desired to convert will cancel in numerator and denominator.

TABLE VI

UNIT CONVERSION RATIOS

Length			Volume		
$\frac{30.5\text{ cm}}{\text{ft}}$	$\frac{5280\text{ ft}}{\text{mile}}$	$\frac{1.61\text{ km}}{\text{mile}}$	$\frac{35.3\text{ ft}^3}{\text{m}^3}$	$\frac{7.48\text{ gal}}{\text{ft}^3}$	$\frac{28.3\text{ liters}}{\text{ft}^3}$
Velocity		**Flow rate**	**Mass density**	**Specific weight**	**Force intensity**
$\frac{1.467\text{ fps}}{\text{mph}}$	$\frac{1.69\text{ fps}}{\text{knot}}$	$\frac{449\text{ gpm}}{\text{cfs}}$	$\frac{1.94\text{ slugs/ft}^3}{\text{gr/cc}}$	$\frac{62.4\text{ lb/ft}^3}{\text{gr/cc}}$	$\frac{14.23\text{ psi}}{\text{kg/cm}^2}$
Mass			**Force**		
$\frac{1.46\times 10^4\text{ gr}}{\text{slug}}$	$\frac{32.2\text{ lb}}{\text{slug}}$	$\frac{454\text{ gr}}{\text{lb}}$	$\frac{4.45\times 10^5\text{ dynes}}{\text{lb}}$	$\frac{32.2\text{ poundals}}{\text{lb}}$	$\frac{2.205\text{ lb}}{\text{kg}}$
Energy		**Power**		**Viscosity**	
$\frac{778\text{ ft-lb}}{\text{Btu}}$	$\frac{2.66\times 10^6\text{ ft-lb}}{\text{kw-hr}}$	$\frac{550\text{ ft-lb/sec}}{\text{hp}}$	$\frac{746\text{ watts}}{\text{hp}}$	$\frac{478\text{ poises}}{\text{lb-sec/ft}^2}$	$\frac{929\text{ stokes}}{\text{ft}^2\text{/sec}}$

Consider, for example, the conversion of 17.5 feet to the equivalent number of centimeters. Since Table VI indicates that there are approximately 30.5 centimeters per foot of length, multiplication of 17.5 feet by the ratio $\frac{30.5\text{ cm}}{\text{ft}}$, followed by cancelation of like units in numerator and denominator, will yield the desired result:

$$17.5\text{ ft} = 17.5\text{ ft} \times \frac{30.5\text{ cm}}{\text{ft}} = 17.5 \times 30.5\text{ cm} = 534\text{ cm}$$

As a second example, assume that the stress intensity of 643 dynes per square centimeter must be converted to the equivalent number of pounds per square foot. According to the table, there are essentially

4.45×10^5 dynes per pound and, as before, 30.5 centimeters per foot. Hence, dividing by $\dfrac{4.45 \times 10^5 \text{ dynes}}{\text{lb}}$ and multiplying by $\left(\dfrac{30.5 \text{ cm}}{\text{ft}}\right)^2$,

$$643 \frac{\text{dynes}}{\text{cm}^2} = 643 \frac{\text{dynes}}{\text{cm}^2} \times \frac{\text{lb}}{4.45 \times 10^5 \text{ dynes}} \times \left(\frac{30.5 \text{ cm}}{\text{ft}}\right)^2$$

$$= \frac{643 \times \overline{30.5}^2}{4.45 \times 10^5} \frac{\text{lb}}{\text{ft}^2} = 1.344 \text{ psf}$$

55. PROPERTIES OF COMMON FLUIDS

Density. The measurement of mass usually involves comparison of the gravitational attraction exerted upon the body of matter in question with that exerted upon a reference body of known mass, through use of either a beam or a spring balance. Since, for accuracy, the reference measurement should be made at the same locality, the exact magnitude of the local gravitational attraction is evidently of no consequence. Division of the mass of the body of matter by the volume of the space which it occupies under the required conditions then yields its mass density.

The density of every liquid varies somewhat with temperature and pressure. Since such variation plays only a minor role in practical computations, the values for typical liquids at 60° Fahrenheit and 14.7 pounds per square inch absolute given in Table VII will prove

TABLE VII

Density Characteristics of Common Liquids under Atmospheric Pressure at 60° Fahrenheit

Liquid	Density ρ slug/ft³	Specific Weight γ lb/ft³	Liquid	Density ρ slug/ft³	Specific Weight γ lb/ft³
Alcohol. ethyl	1.53	49.3	Mercury	26.3	847
Benzene	1.71	54.9	Oil		
Brine (20% NaCl)	2.23	71.6	lubricating	1.65–1.70	53–55
Carbon tetra-			crude	1.65–1.80	53–58
chloride	3.09	99.5	fuel	1.80–1.90	58–61
Gasoline	1.28–1.34	41–43	Water		
Glycerine	2.45	78.8	fresh	1.94	62.4
Kerosene	1.51–1.59	49–51	salt	1.99	64.0

sufficiently accurate for most purposes. The same is not true of gases, however, since a gas varies in density with both temperature and pressure in close accordance with the ideal-gas equation

$$\rho = \frac{p}{g_s RT} \tag{226}$$

in which the magnitudes of both p and T must be referred to the respective absolute zeros (459° below zero Fahrenheit in the latter instance). Table VIII therefore lists the densities of common gases at

TABLE VIII

DENSITY CHARACTERISTICS OF COMMON GASES UNDER ATMOSPHERIC PRESSURE AT 60° FAHRENHEIT

Gas	Density ρ slug/ft³	Specific Weight γ lb/ft³	Gas Constant R ft/°F	Adiabatic Constant k
Acetylene	0.00215	0.0693	59.3	1.26
Air	0.00237	0.0763	53.3	1.40
Ammonia	0.00141	0.0455	89.5	1.31
Carbon dioxide	0.00363	0.117	34.9	1.28
Helium	0.000329	0.0106	386.	1.66
Hydrogen	0.000165	0.00531	767.	1.40
Methane (natural gas)	0.00132	0.0424	96.3	1.32
Oxygen	0.00262	0.0844	48.3	1.40
Nitrogen	0.00229	0.0739	55.1	1.40
Sulfur dioxide	0.00537	0.173	23.6	1.26

60° Fahrenheit and 14.7 pounds per square inch absolute, together with values of R, from which the densities at any desired temperatures and pressures may be computed.

Specific weight. The weight per unit volume of any substance must be computed from the density of the substance and the local magnitude of the gravitational acceleration. Variation in the latter quantity is closely indicated by the empirical relationship

$$g = 32.1721 - 0.08211 \cos 2\phi - 0.000003h \tag{233}$$

in which ϕ is latitude and h is elevation in feet above mean sea level. As will be seen from inspection of this equation, the standard gravitational acceleration $g_s = 32.1739$ feet per second per second is very nearly equal to that at sea level and 45° latitude, while the commonly used value $g = 32.2$ feet per second per second does not differ by even 1 per cent from that corresponding to most localities in the United States. For this reason the values for γ given in Tables VII and VIII will be found sufficiently accurate for engineering calculations.

Viscosity. Fluid viscosity has become important in so many fields of practice during recent decades that confusing units of measurement are still encountered. The ease of timing efflux through the outlet of a tank led to the adoption of the efflux period for a given quantity of fluid as a relative viscosity measure in the classification of oils, but the shortness of standard outlet tubes necessarily yields non-linear functions between time and viscosity in the lower range. However, use of a long capillary tube, a rotating cylinder, or a falling sphere, together with the corresponding equation of viscous motion, will permit direct evaluation of the dynamic (or the kinematic) viscosity in basic *L-T-F* or *L-T-M* units. In the metric system the viscosity units themselves have conveniently been given names; that is, 1 dyne-second per square centimeter is customarily called a *poise* and 1 square centimeter per second is called a *stoke*. Similar terms for the viscosity units in the English and American systems have not yet been adopted. In passing it might be noted that the kinematic viscosity of liquids may be evaluated with close approximation from the efflux time for a Saybolt Universal viscosimeter through use of the empirical formula $\nu = 2.37 \times 10^{-6}t - 1.94 \times 10^{-3}/t$, in which ν is measured in square feet per second and t in seconds.

In Figs. 194 and 195 are plotted curves of dynamic viscosity versus temperature for gases and liquids commonly encountered, from which it will be seen that, although a considerable change with temperature invariably takes place, the difference in molecular spacing between liquids and gases results in opposite trends of the two systems of curves. Except under very extreme conditions, pressure is found to have no effect upon either the form or the position of these curves; they may therefore be considered independent of pressure for all practical purposes.

The curves of kinematic viscosity, reproduced in Fig. 196 for both liquids and gases, indicate the same general trends. However, it will be noted that the relative positions of the liquid and the gas curves are radically different from those of Figs. 194 and 195, gases evidently being considerably more viscous per unit density than many common liquids. Attention should also be called to the fact that these curves are specifically for conditions of atmospheric pressure, values for the

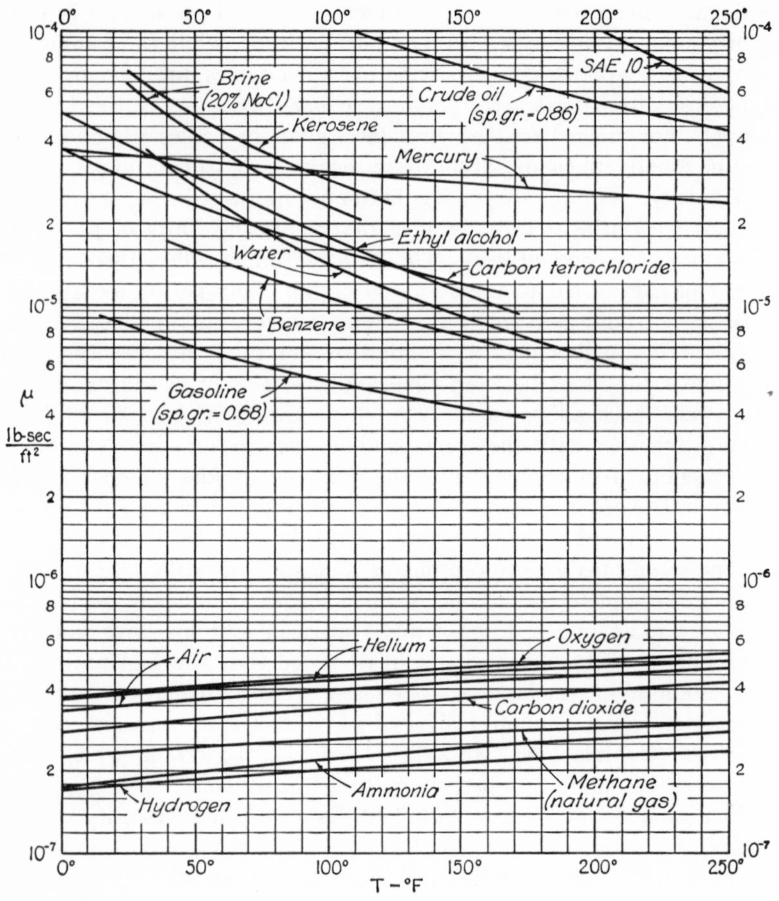

FIG. 194. Dynamic viscosity versus temperature for common gases and liquids.

gases at other pressures being obtainable from Fig. 194 through division by the corresponding density.

Surface tension. The measurement of surface energy may be performed by the bubble method described in Chapter X, although considerably more refined apparatus, not warranting discussion herein, is usually employed in precise work. As will be noted from Table IX, the coefficients of surface tension of common liquids other than mercury do not differ greatly, so that little could be gained experimentally by use of different liquids in the effort to cover a wide Weber-number range. In fact, the simple addition of aerosol to water will yield practically as great a variation in σ as could otherwise be obtained

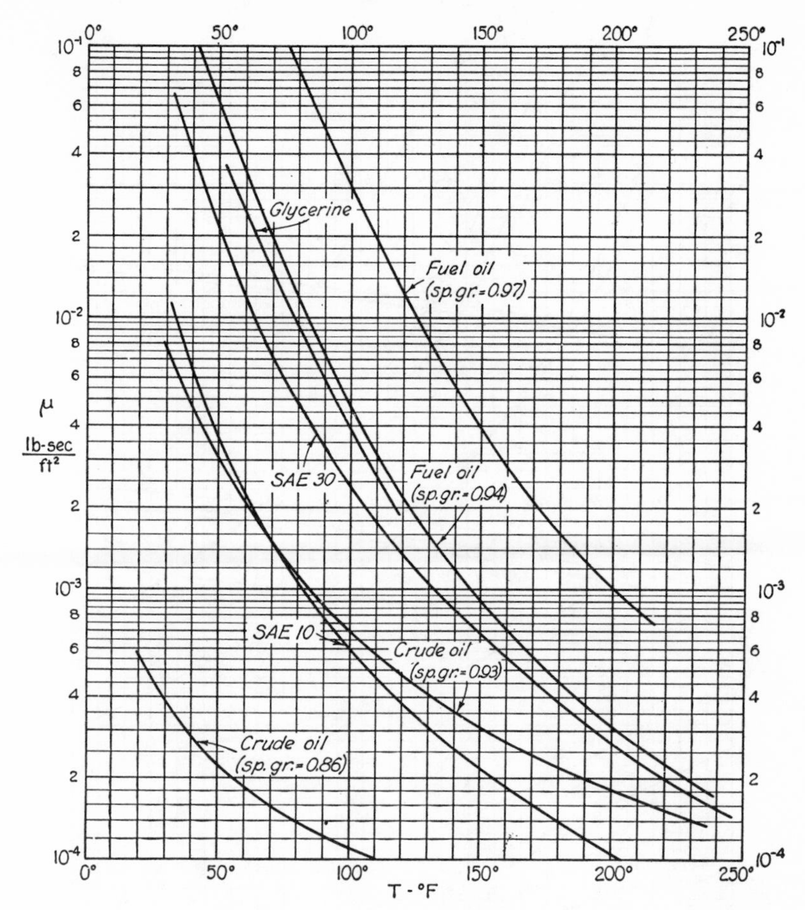

FIG. 195. Dynamic viscosity versus temperature for typical grades of oil.

TABLE IX

SURFACE TENSION OF COMMON LIQUIDS IN CONTACT WITH AIR AT 68° FAHRENHEIT

Liquid	Surface Tension σ lb/ft	Liquid	Surface Tension σ lb/ft
Alcohol, ethyl	0.00153	Mercury, in air	0.0352
Benzene	0.00198	in water	0.0269
Carbon tetrachloride	0.00183	in vacuum	0.0333
Kerosene	0.0016–0.0022	Oil, lubricating	0.0024–0.0026
Water	0.00498	crude	0.0016–0.0026

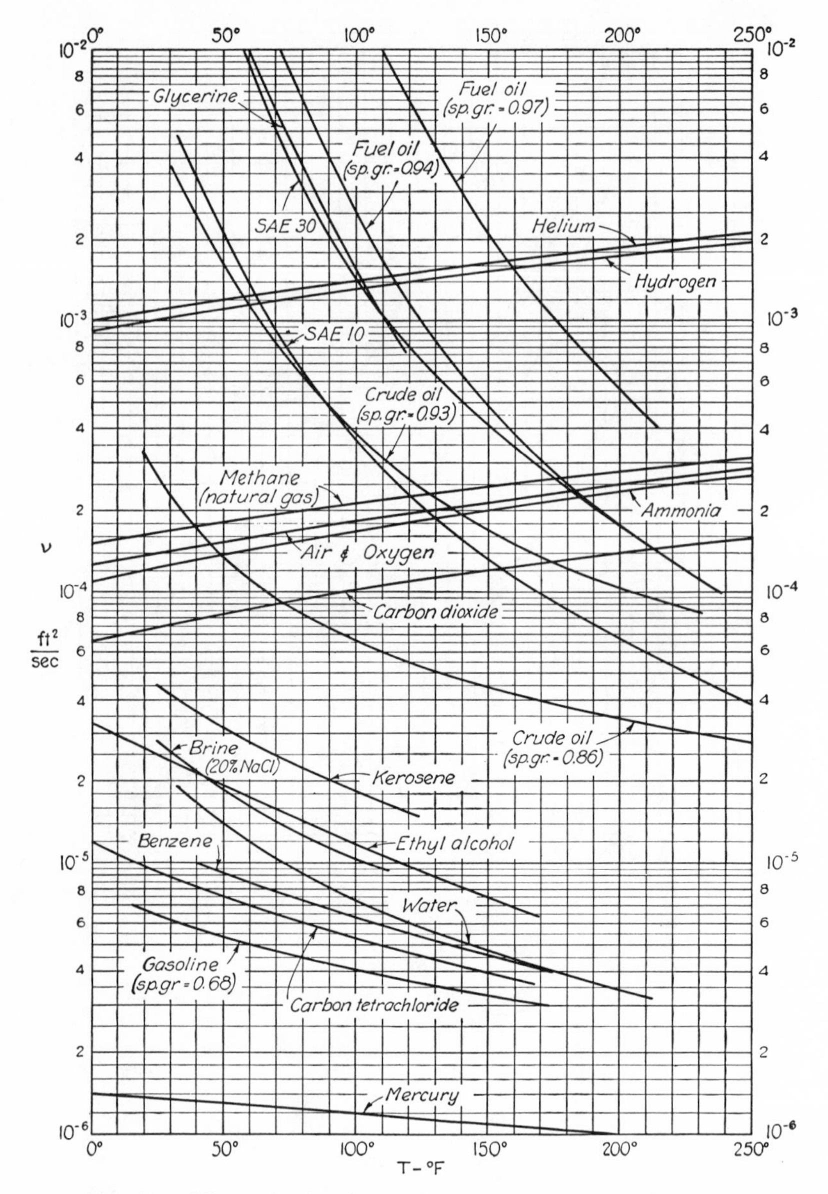

Fig. 196. Kinematic viscosity versus temperature for common fluids.

Elastic modulus. So far as the bulk modulus of elasticity of any gas is concerned, reasonably exact evaluations may be made by means of the ideal-gas law $\rho = p_{abs}/g_s R T_{abs}$ and the definition equation for the modulus. The latter, however, depends upon whether isothermal (constant-temperature) or adiabatic (constant-heat) conditions prevail during the volume change: if the change is isothermal, $E = p_{abs}$; if it is adiabatic, $E = kp_{abs}$ (k being the ratio of the specific heats for constant pressure and constant volume, as listed in Table VIII).

The elastic characteristics of liquids in relation to temperature and pressure are by no means so definitely established. Although it is known that both pressure and temperature influence the magnitude of E, systematic data for even the most common liquids are very incomplete. Approximate values are given for water at atmospheric pressure in Table XI, a rise in E of about 2 per cent occurring for each 1000-pounds-per-square-inch rise in pressure intensity. Ocean water has an elastic modulus which averages about 9 per cent higher than that of fresh water under comparable conditions. If the nominal value of E for fresh water is taken as 300,000 pounds per square inch, corresponding values for other common liquids will be as follows: salt water, 330,000; glycerine, 630,000; mercury, 3,800,000; and oil, 180,000 to 270,000 pounds per square inch.

Properties of air and water. Although scientists and engineers engaged in the various professions dealing with fluid motion are confronted with a steadily growing number of commercially important

TABLE X

MECHANICAL PROPERTIES OF AIR AT ATMOSPHERIC PRESSURE

Temperature T ° Fahrenheit	Density ρ slug/ft^3	Specific Weight γ lb/ft^3	Dynamic Viscosity μ lb-sec/ft^2	Kinematic Viscosity ν ft^2/sec
0	0.00268	0.0862	3.28×10^{-7}	1.26×10^{-4}
20	0.00257	0.0827	3.50	1.36
40	0.00247	0.0794	3.62	1.46
60	0.00237	0.0763	3.74	1.58
80	0.00228	0.0735	3.85×10^{-7}	1.69×10^{-4}
100	0.00220	0.0709	3.96	1.80
120	0.00215	0.0684	4.07	1.89
150	0.00204	0.0651	4.23	2.07
200	0.00187	0.0601	4.49×10^{-7}	2.40×10^{-4}

TABLE XI

MECHANICAL PROPERTIES OF WATER AT ATMOSPHERIC PRESSURE

Temperature T ° Fahrenheit	Density ρ slug/ft^3	Specific Weight γ lb/ft^3	Dynamic Viscosity μ lb-sec/ft^2	Kinematic Viscosity ν ft^2/sec	Surface Tension σ lb/ft	Vapor Pressure p_v psia	Elastic Modulus E psi
32	1.94	62.4	3.75×10^{-5}	1.93×10^{-5}	0.00518	0.08	289,000
40	1.94	62.4	3.24	1.67	0.00514	0.11	296,000
50	1.94	62.4	2.74	1.41	0.00508	0.17	305,000
60	1.94	62.4	2.34	1.21	0.00503	0.26	312,000
70	1.94	62.3	2.04×10^{-5}	1.05×10^{-5}	0.00497	0.36	319,000
80	1.93	62.2	1.80	0.930	0.00492	0.51	325,000
90	1.93	62.1	1.59	0.823	0.00486	0.70	329,000
100	1.93	62.0	1.42	0.736	0.00479	0.96	331,000?
120	1.92	61.7	1.17×10^{-5}	0.610×10^{-5}	0.00466	1.7	333,000?
150	1.90	61.2	0.906	0.476	0.00446	3.7	328,000?
180	1.88	60.6	0.726	0.385	0.00426	7.5	318,000?
212	1.86	59.8	0.594	0.319	0.00403	14.7	303,000?

fluids, the fact remains that the abundance of air and water and their essentiality to civilized life make them by far the most predominant in routine problems of flow. For this reason Tables X and XI list the mechanical properties of air and water, respectively, for the range of temperatures normally encountered in engineering practice; atmospheric pressure intensity is used as the basis of these values. In the tabulation for water, it will be noted, the vapor pressure has also been included, owing to its importance in problems of cavitation.

SELECTED REFERENCES

WEBER, ERNEST. "Physical Units and Standards," Section 3 of Eshbach's *Handbook of Engineering Fundamentals.* Wiley, 1936.

International Critical Tables. McGraw-Hill, 1926–1930.

Smithsonian Physical Tables. 8th edition, Smithsonian Institution, 1933.

INDEX